D0151698

Handbook of Algorithms and Data Structures

In Pascal and C

Second Edition

INTERNATIONAL COMPUTER SCIENCE SERIES

SELECTED TITLES IN THE SERIES

Programming Language Translation: A Practical Approach *P D Terry*

Data Abstraction in Programming Languages *J M Bishop*

The Specification of Computer Programs *W M Turski and T S E Maibaum*

Syntax Analysis and Software Tools *K J Gough*

Functional Programming *A J Field and P G Harrison*

The Theory of Computability: Programs, Machines, Effectiveness and Feasibility
R Sommerhalder and S C van Westrhenen

An Introduction to Functional Programming through Lambda Calculus *G Michaelson*

High-Level Languages and their Compilers *D Watson*

Programming in Ada (3rd Edn) *J G P Barnes*

Elements of Functional Programming *C Reade*

Software Development with Modula-2 *D Budgen*

Program Derivation: The Development of Programs from Specifications *R G Dromey*

Object-Oriented Programming with Simula *B Kirkerud*

Program Design with Modula-2 *S Eisenbach and C Sadler*

Real Time Systems and Their Programming Languages *A Burns and A Wellings*

Fortran 77 Programming (2nd Edn) *T M R Ellis*

Prolog Programming for Artificial Intelligence (2nd Edn) *I Bratko*

Logic for Computer Science *S Reeves and M Clarke*

Computer Architecture *M De Blasi*

The Programming Process *J T Latham, V J Bush and I D Cottam*

Handbook of Algorithms and Data Structures

In Pascal and C

Second Edition

G.H. Gonnet
ETH, Zurich

R. Baeza-Yates
University of Chile, Santiago

ADDISON-WESLEY
PUBLISHING
COMPANY

Wokingham, England • Reading, Massachusetts • Menlo Park, California • New York
Don Mills, Ontario • Amsterdam • Bonn • Sydney • Singapore
Tokyo • Madrid • San Juan • Milan • Paris • Mexico City • Seoul • Taipei

Cover designed by Crayon Design of Henley-on-Thames and
printed by The Riverside Printing Co. (Reading) Ltd.
Printed in Great Britain by Mackays of Chatham plc, Chatham, Kent.

First edition published 1984. Reprinted 1985.
Second edition printed 1991.

British Library Cataloguing in Publication Data
Gonnet, G. H. (Gaston H.)
Handbook of algorithms and data structures : in Pascal and C.-2nd. ed.
1. Programming. Algorithms
I. Title II. Baeza-Yates, R. (Ricardo)
005.1

ISBN 0-201-41607-7

Library of Congress Cataloging in Publication Data
Gonnet, G. H. (Gaston H.)
Handbook of algorithms and data structures : in Pascal and C / G.H. Gonnet, R. Baeza-Yates. - - 2nd ed.
p. cm. - - (International computer science series)
Includes bibliographical references (p.) and index.
ISBN 0-201-41607-7
1. Pascal (Computer program language) 2. (Computer program language) 3. Algorithms. 4. Data structures (Computer science)
I. Baeza-Yates, R. (Ricardo) II. Title. III. Series.
QA76.73.P2G66 1991 90-26318
005. 13′3--dc20 CIP

To my boys: Miguel, Pedro Julio and Ignacio
and my girls: Ariana and Marta

Preface

Preface to the first edition

Computer Science has been, throughout its evolution, more an art than a science. My favourite example which illustrates this point is to compare a major software project (like the writing of a compiler) with any other major project (like the construction of the CN tower in Toronto). It would be absolutely unthinkable to let the tower fall down a few times while its design was being debugged: even worse would be to open it to the public before discovering some other fatal flaw. Yet this mode of operation is being used everyday by almost everybody in software production.

Presently it is very difficult to 'stand on your predecessor's shoulders', most of the time we stand on our predecessor's toes, at best. This handbook was written with the intention of making available to the computer scientist, instructor or programmer the wealth of information which the field has generated in the last 20 years.

Most of the results are extracted from the given references. In some cases the author has completed or generalized some of these results. Accuracy is certainly one of our goals, and consequently the author will cheerfully pay $2.00 for each first report of any type of error appearing in this handbook.

Many people helped me directly or indirectly to complete this project. Firstly I owe my family hundreds of hours of attention. All my students and colleagues had some impact. In particular I would like to thank Maria Carolina Monard, Nivio Ziviani, J. Ian Munro, Per-Åke Larson, Doron Rotem and Derick Wood. Very special thanks go to Frank W. Tompa who is also the coauthor of chapter 2. The source material for this chapter appears in a joint paper in the November 1983 issue of *Communications of the ACM*.

Montevideo
December 1983

G.H. Gonnet

Preface to the second edition

The first edition of this handbook has been very well received by the community, and this has given us the necessary momentum for writing a second edition. In doing so, R. A. Baeza-Yates has joined me as a coauthor. Without his help this version would have never appeared.

This second edition incorporates many new results and a new chapter on text searching. The area of text managing, in particular searching, has risen in importance and matured in recent times. The entire subject of the handbook has matured too; our citations section has more than doubled in size. Table searching algorithms account for a significant part of this growth.

Finally we would like to thank the over one hundred readers who notified us about errors and misprints, they have helped us tremendously in correcting all sorts of blemishes. We are especially grateful for the meticulous, even amazing, work of Lynne Balfe, the proofreader. We will continue cheerfully to pay $4.00 (increased due to inflation) for each first report of an error.

Zürich
December 1990 — G.H. Gonnet

Santiago de Chile
December 1990 — R.A. Baeza-Yates

Contents

Trademark notice

1 Introduction

This handbook is intended to contain most of the information available on algorithms and their data structures; thus it is designed to serve a wide spectrum of users, from the programmer who wants to code efficiently to the student or researcher who needs information quickly.

The main emphasis is placed on algorithms. For these we present their description, code in one or more languages, theoretical results and extensive lists of references.

1.1 Structure of the chapters

The handbook is organized by topics. Chapter 2 offers a formalization of the description of algorithms and data structures; Chapters 3 to 7 discuss searching, sorting, selection, arithmetic and text algorithms respectively. Appendix I describes some probability distributions encountered in data processing; Appendix II contains a collection of asymptotic formulas related to the analysis of algorithms; Appendix III contains the main list of references and Appendix IV contains alternate code for some algorithms.

The chapters describing algorithms are divided into sections and subsections as needed. Each algorithm is described in its own subsection, and all have roughly the same format, though we may make slight deviations or omissions when information is unavailable or trivial. The general format includes:

(1) Definition and explanation of the algorithm and its classification (if applicable) according to the basic operations described in Chapter 2.

(2) Theoretical results on the algorithm's complexity. We are mainly interested in measurements which indicate an algorithm's running time and

its space requirements. Useful quantities to measure for this information include the number of comparisons, data accesses, assignments, or exchanges an algorithm might make. When looking at space requirements, we might consider the number of words, records, or pointers involved in an implementation. Time complexity covers a much broader range of measurements. For example, in our examination of searching algorithms, we might be able to attach meaningful interpretations to most of the combinations of the

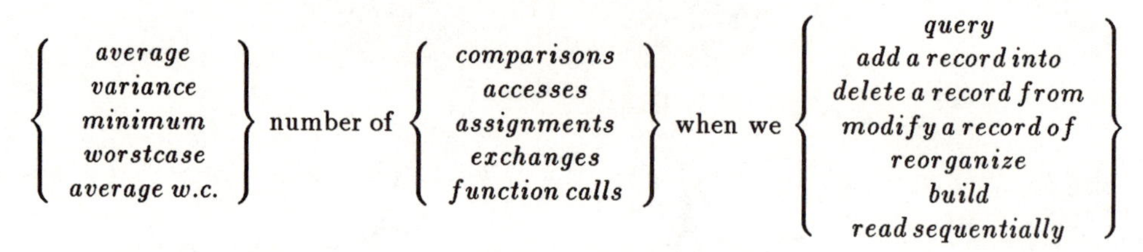

the structure. Other theoretical results may also be presented, such as enumerations, generating functions, or behaviour of the algorithm when the data elements are distributed according to special distributions.

(3) The algorithm. We have selected Pascal and C to describe the algorithms. Algorithms that may be used in practice are described in one or both of these languages. For algorithms which are only of theoretical interest, we do not provide their code. Algorithms which are coded both in Pascal and in C will have one code in the main text and the other in Appendix IV.

(4) Recommendations. Following the algorithm description we give several hints and tips on how to use it. We point out pitfalls to avoid in coding, suggest when to use the algorithm and when not to, say when to expect best and worst performances, and provide a variety of other comments.

(5) Tables. Whenever possible, we present tables which show exact values of complexity measures in selected cases. These are intended to give a feeling for how the algorithm behaves. When precise theoretical results are not available we give simulation results, generally in the form $xxx \pm yy$ where the value yy is chosen so that the resulting interval has a confidence level of 95%. In other words, the actual value of the complexity measure falls out of the given interval only once every 20 simulations.

(6) Differences between internal and external storage. Some algorithms may perform better for internal storage than external, or vice versa. When this is true, we will give recommendations for applications in each case. Since most of our analysis up to this point will implicitly assume that internal memory is used, in this section we will look more closely at the external case (if appropriate). We analyze the algorithm's behaviour

when working with external storage, and discuss any significant practical considerations in using the algorithm externally.

(7) With the description of each algorithm we include a list of relevant references. General references, surveys, or tutorials are collected at the end of chapters or sections. The third appendix contains an alphabetical list of all references with cross-references to the relevant algorithms.

1.2 Naming of variables

The naming of variables throughout this handbook is a compromise between uniformity of notation and accepted terminology in the specific areas.

Except for very few exceptions, explicitly noted, we use:

n for the number of objects or elements or components in a structure;
m for the size of a structure;
b for bucket sizes, or maximum number of elements in a physical block;
d for the digital cardinality or size of the alphabet.

The complexity measures are also named uniformly throughout the handbook. Complexity measures are named X_n^Z and should be read as 'the number of Xs performed or needed while doing Z onto a structure of size n'. Typical values for X are:

A : accesses, probes or node inspections;
C : comparisons or node inspections;
E : external accesses;
h : height of a recursive structure (typically a tree);
I : iterations (or number of function calls);
L : length (of path or longest probe sequence);
M : moves or assignments (usually related to record or key movements);
T : running time;
S : space (bytes or words).

Typical values for Z are:

null (no superscript): successful search (or default operation, when there is only one possibility);
' unsuccessful search;
C : construction (building) of structure;
D : deletion of an element;
E : extraction of an element (mostly for priority queues);
I : insertion of a new element;

M : merging of structures;
Opt : optimal construction or optimal structure (the operation is usually implicit);
MM : minimax, or minimum number of X's in the worst case: this is usually used to give upper and lower bounds on the complexity of a problem.

Note that X_n^I means number of operations done to insert an element into a structure of size n or to insert the $n+1$st element.

Although these measures are random variables (as these depend on the particular structure on which they are measured), we will make exceptions for C_n and C_n' which most of the literature considers to be expected values.

1.3 Probabilities

The probability of a given event is denoted by $Pr\{event\}$. Random variables follow the convention described in the preceding section. The expected value of a random variable X is written $E[X]$ and its variance is $\sigma^2(X)$. In particular, for discrete variables X

$$E[X] = \mu_1' = \sum_i i\, Pr\{X = i\}$$

$$\sigma^2(X) = \sum_i i^2 Pr\{X = i\} - E[X]^2 = E[X^2] - E[X]^2$$

We will always make explicit the probability universe on which expected values are computed. This is ambiguous in some cases, and is a ubiquitous problem with expected values.

To illustrate the problem without trying to confuse the reader, suppose that we fill a hashing table with keys and then we want to know about the average number of accesses to retrieve one of the keys. We have two potential probability universes: the key selected for retrieval (the one inserted first, the one inserted second, ...) and the actual values of the keys, or their probing sequence. We can compute expected values with respect to the first, the second, or both universes. In simpler terms, we can find the expected value of any key for a given file, or the expected value of a given key for any file, or the expected value of any key for any file.

Unless otherwise stated, (1) the distribution of our elements is always random independent uniform $U(0,1)$; (2) the selection of a given element is uniform discrete between all possible elements; (3) expected values which relate to multiple universes are computed with respect to all universes. In terms of the above example, we will compute expected values with respect to randomly selected variables drawn from a uniform $U(0,1)$ distribution.

1.4 Asymptotic notation

Most of the complexity measures in this handbook are asymptotic in the size of the problem. The asymptotic notation we will use is fairly standard and is given below:

$$f(n) = O(g(n))$$

implies that there exists k and n_0 such that $|f(n)| < kg(n)$ for $n > n_0$.

$$f(n) = o(g(n)) \rightarrow \lim_{n \rightarrow \infty} \frac{f(n)}{g(n)} = 0$$

$$f(n) = \Theta(g(n))$$

implies that there exists k_1, k_2, $(k_1 \times k_2 > 0)$ and n_0 such that $k_1 g(n) < f(n) < k_2 g(n)$ for $n > n_0$, or equivalently that $f(n) = O(g(n))$ and $g(n) = O(f(n))$.

$$f(n) = \Omega(g(n)) \rightarrow g(n) = O(f(n))$$

$$f(n) = \omega(g(n)) \rightarrow g(n) = o(f(n))$$

$$f(n) \approx g(n) \rightarrow f(n) - g(n) = o(g(n))$$

We will freely use arithmetic operations with the order notation, for example,

$$f(n) = h(n) + O(g(n))$$

means

$$f(n) - h(n) = O(g(n))$$

Whenever we write $f(n) = O(g(n))$ it is with the understanding that we know of no better asymptotic bound, that is, we know of no $h(n) = o(g(n))$ such that $f(n) = O(h(n))$.

1.5 About the programming languages

We use two languages to code our algorithms: Pascal and C. After writing many algorithms we still find situations for which neither of these languages present a very 'clean' or understandable code. Therefore, whenever possible, we use the language which presents the shortest and most readable code. We intentionally allow our Pascal and C style of coding to resemble each other.

A minimal number of Pascal programs contain **goto** statements. These statements are used in place of the equivalent C statements **return** and **break,** and are correspondingly so commented. Indeed we view their absence from Pascal as a shortcoming of the language. Another irritant in coding some algorithms in Pascal is the lack of order in the evaluation of logical expressions. This is unfortunate since such a feature makes algorithms easier to understand. The typical stumbling block is

```
while (p <> nil) and (key <> p↑.k) do ...
```

Such a statement works in C if we use the sequential and operator (&&), but for Pascal we have to use instead:

```
while p <> nil do begin
      if key = p↑.k then goto 999   {*** break *** } ;
      ...
999:
```

Other minor objections are: the inability to compute addresses of non-heap objects in Pascal (which makes treatment of lists more difficult); the lack of variable length strings in Pascal; the lack of a **with** statement in C; and the lack of **var** parameters in C. (Although this is technically possible to overcome, it obscures the algorithms.)

Our Pascal code conforms, as fully as possible, to the language described in the *Pascal User Manual and Report* by K. Jensen and N. Wirth. The C code conforms to the language described in *The C Programming Language* by B.W. Kernighan and D.M. Ritchie.

1.6 On the code for the algorithms

Except for very few algorithms which are obviously written in pseudo-code, the algorithms in this handbook were run and tested under two different compilers. Actually the same text which is printed is used for compiling, for testing, for running simulations and for obtaining timings. This was done in an attempt to eliminate (or at least drastically reduce!) errors.

Each family of algorithms has a 'tester set' which not only checks for correct behaviour of the algorithm, but also checks proper handling of limiting conditions (will a sorting routine sort a null file? one with one element? one with all equal keys? ...).

In most cases the algorithms are described as a function or a procedure or a small set of functions or procedures. In a few cases, for very simple algorithms, the code is described as in-line code, which could be encapsulated in a procedure or could be inserted into some other piece of code.

Some algorithms, most notably the searching algorithms, are building blocks or components of other algorithms or programs. Some standard actions should not be specified for the algorithm itself, but rather will be specified once that the algorithm is 'composed' with other parts (chapter 2 defines composition in more detail). A typical example of a standard action is an error condition. The algorithms coded for this handbook always use the same names for these standard actions.

Error detection of an unexpected condition during execution. Whenever *Error* is encountered it can be substituted by any block of statements. For example our testers print an appropriate message.

found(record) function call that is executed upon completion of a successful search. Its argument is a record or a pointer to a record which contains the searched key.

notfound(key) function called upon an unsuccessful search. Its argument is the key which was not found.

A special effort has been made to avoid duplication of these standard actions for identical conditions. This makes it easier to substitute blocks of code for them.

1.7 Complexity measures and real timings

For some families of algorithms we include a comparison of real timings. These timings are to be interpreted with caution as they reflect only one sample point in the many dimensions of hardwares, compilers, operating systems, and so on. Yet we have equally powerful reasons to present at least one set of real complexities.

The main reasons for including real timing comparisons are that they take into account:

(1) the actual cost of operations,

(2) hidden costs, such as storage allocation, and indexing.

The main objections, or the factors which may invalidate these real timing tables, are:

(1) the results are compiler dependent: although the same compiler is used for each language, a compiler may favour one construct over others;

(2) the results are hardware dependent;

(3) in some cases, when large amounts of memory are used, the timings may be load dependent.

The timings were done on a Sun 3 running the SunOS 4.1 operating system. Both C and Pascal compilers were run with the optimizer, or object code improver, to obtain the best implementation for the algorithms.

There were no attempts made to compare timings across languages. All the timing results are computed relative to the fastest algorithm. To avoid the incidence of start up-costs, loading, and so on, the tests were run on problems of significant size. Under these circumstances, some $O(n^2)$ algorithms appear to perform very poorly.

2 Basic Concepts

2.1 Data structure description

The formal description of data structure implementations is similar to the formal description of programming languages. In defining a programming language, one typically begins by presenting a syntax for valid programs in the form of a grammar and then sets further validity restrictions (for example, usage rules for symbolic names) which give constraints that are not captured by the grammar. Similarly, a valid data structure implementation will be one that satisfies a syntactic grammar and also obeys certain constraints. For example, for a particular data structure to be a valid weight-balanced binary tree, it must satisfy the grammatical rules for binary trees and it must also satisfy a specific balancing constraint.

2.1.1 Grammar for data objects

A sequence of real numbers can be defined by the BNF production

<S> ::= [real , <S>] | nil

Thus a sequence of reals can have the form nil, [real,nil], [real,[real,nil]], and so on. Similarly, sequences of integers, characters, strings, boolean constants, ... could be defined. However, this would result in a bulky collection of production rules which are all very much alike. One might first try to eliminate this repetitiveness by defining

<S> ::= [<D> , <S>] | nil

where <D> is given as the list of data types

<D> ::= real | int | bool | string | char

However, this pair of productions generates unwanted sequences such as

[real,[int,nil]]

as well as the homogeneous sequences desired.

To overcome this problem, the syntax of a data object class can be defined using a **W-grammar** (also called a two-level or van Wijngaarden grammar). Actually the full capabilities of W-grammars will not be utilized; rather the syntax will be defined using the equivalent of standard BNF productions together with the uniform replacement rule as described below.

A W-grammar generates a language in two steps (levels). In the first step, a collection of generalized rules is used to create more specific production rules. In the second step, the production rules generated in the first step are used to define the actual data structures.

First, the problem of listing repetitive production rules is solved by starting out with generalized rule-forms known as **hyperrules**, rather than the rules themselves. The generalized form of a sequence S is given by the hyperrule

s − D : [D, s − D] ; nil

The set of possible substitutions for **D** are now defined in a **metaproduction**, as distinguished from a conventional BNF-type production. For example, if **D** is given as

D :: **real**; **int**; **bool**; **string**; **char**; ⋯

a sequence of real numbers is defined in two steps as follows. The first step consists of choosing a value to substitute for **D** from the list of possibilities given by the appropriate metaproduction; in this instance, **D → real.** Next invoke the uniform replacement rule to substitute the string **real** for **D** everywhere it appears in the hyperrule that defines **s − D**. This substitution gives

s − real : [real , s − real] ; nil

Thus the joint use of the metaproduction and the hyperrule generates an ordinary BNF-like production defining real sequences. The same two statements can generate a production rule for sequences of any other valid data type (integer, character, ...).

Figures 2.1 and 2.2 contain a W-grammar which will generate many conventional data objects. As further examples of the use of this grammar, consider the generation of a binary tree of real numbers. With **D → real** and **LEAF → nil**, HR[3] generates the production rule

bt − real − nil : [real , bt − real − nil , bt − real − nil] ; nil

Since **bt − real − nil** is one of the legitimate values for **D** according to M[1] let **D → bt − real − nil** from which HR[1] indicates that such a binary tree is a legitimate data structure.

Metaproductions

M[1]	**D** ::	**real; int; bool; string; char; ...;**	# atomic data types
		$\{\mathbf{D}\}_{\mathbf{N}}^{\mathbf{N}}$;	# array
		REC ; (REC) ;	# record
		[D] ;	# reference
		s – D;	# sequence
		gt – D – LEAF;	# general tree
		DICT;	# dictionary structures
		... .	# other structure classes such as graphs, sets, priority queues.
M[2]	**DICT** ::	$\{\mathbf{KEY}\}_{\mathbf{N}}^{\mathbf{N}}$; **s – KEY;**	# sequential search
		bt – KEY – LEAF;	# binary tree
		mt – N – KEY – LEAF;	# multiway tree
		tr – N – KEY.	# digital tree
M[3]	**REC** ::	**D; D, REC.**	# record definition
M[4]	**LEAF** ::	**nil; D.**	
M[5]	**N** ::	**DIGIT; DIGIT N.**	
M[6]	**DIGIT** ::	**0; 1; 2; 3; 4; 5; 6; 7; 8; 9.**	
M[7]	**KEY** ::	**real; int; string; char; (KEY, REC).**	# search key

Figure 2.1: Metaproductions for data objects.

Secondly consider the specification for a hash table to be used with direct chaining. The production

$$\mathbf{s - (string, int)} \;:\; [\,\mathbf{(string, int)}\,,\; \mathbf{s - (string, int)}\,]\;;\; \mathbf{nil}$$

and M[1] yield

$$\mathbf{D} \rightarrow \{\mathbf{s - (string, int)}\}_{0}^{96}$$

Thus HR[1] will yield a production for an array of sequences of string/integer pairs usable, for example, to record NAME/AGE entries using hashing.

Finally consider a production rule for structures to contain B-trees (Section 3.4.2) of strings using HR[4] and the appropriate metaproductions to yield

$$\mathbf{mt - 10 - string - nil} \;:$$

$$[\mathbf{int}, \{\mathbf{string}\}_{1}^{10}, \{\mathbf{mt - 10 - string - nil}\}_{0}^{10}]\;;\; \mathbf{nil}$$

Hyperrules		
HR[1]	**data structure** :	**D.**
HR[2]	**s – D** :	**[D, s – D]; nil.**
HR[3]	**bt – D – LEAF** :	**[D, bt – D – LEAF, bt – D – LEAF]; LEAF.**
HR[4]	**mt – N – D – LEAF** :	**[int,** $\{D\}_1^N$, $\{mt - N - D - LEAF\}_0^N$**]; LEAF.**
HR[5]	**gt – D – LEAF** :	**[D, s – gt – D – LEAF]; LEAF.**
HR[6]	**tr – N – D** :	**[**$\{tr - N - D\}_1^N$**]; [D]; nil.**

Figure 2.2: Hyperrules for data objects.

In this multitree, each node contains 10 keys and has 11 descendants. Certain restrictions on B-trees, however, are not included in this description (that the number of actual keys is to be stored in the **int** field in each node, that this number must be between 5 and 10, that the actual keys will be stored contiguously in the keys-array starting at position 1, ...); these will instead be defined as constraints (see below).

The grammar rules that we are using are inherently ambiguous. This is not inconvenient; as a matter of fact it is even desirable. For example, consider

$$\mathbf{D} \rightarrow \{\mathbf{D}\}_{\mathbf{N}}^{\mathbf{N}} \rightarrow \{\mathbf{real}\}_1^{10} \tag{2.1}$$

and

$$\mathbf{D} \rightarrow \mathbf{DICT} \rightarrow \{\mathbf{KEY}\}_{\mathbf{N}}^{\mathbf{N}} \rightarrow \{\mathbf{real}\}_1^{10} \tag{2.2}$$

Although both derivation trees produce the same object, the second one describes an array used as a sequential implementation of a dictionary structure, while the first may just be a collection of real numbers. In other words, the derivation tree used to produce the data objects contains important semantic information and should not be ignored.

2.1.2 Constraints for data objects

Certain syntactic characteristics of data objects are difficult or cumbersome to define using formal grammars. A semantic rule or constraint may be regarded as a boolean function on data objects ($S : D \rightarrow$ bool) that indicates which are valid and which are not. Objects that are valid instances of a data structure implementation are those in the intersection of the set produced by the W-grammars and those that satisfy the constraints.

Below are some examples of semantic rules which may be imposed on data structures. As phrased, these constraints are placed on data structures that have been legitimately produced by rules given in the previous section.

2.1.2.1 Sequential order

Many data structures are kept in some fixed order (for example, the records in a file are often arranged alphabetically or numerically according to some key). Whatever work is done on such a file should not disrupt this order. This definition normally applies to $\mathbf{s-D}$ and $\{\mathbf{D}\}_{\mathbf{N}}^{\mathbf{N}}$.

2.1.2.2 Uniqueness

Often it is convenient to disallow duplicate values in a structure, for example in representing sets. At other times the property of uniqueness can be used to ensure that records are not referenced several times in a structure (for example, that a linear chain has no cycles or that every node in a tree has only one parent).

2.1.2.3 Hierarchical order

For all nodes, the value stored at any adjacent node is related to the value at the node according to the type of adjacency. This definition normally applies to $\mathbf{bt-D-LEAF}$, $\mathbf{mt-N-D-LEAF}$ and $\mathbf{gt-D-LEAF}$.

Lexicographical trees
A lexicographical tree is a tree that satisfies the following condition for every node s: if s has n keys $(key_1, key_2, ..., key_n)$ stored in it, s must have $n+1$ descendant subtrees $t_0, t_1, \ldots, t_n$. Furthermore, if d_0 is any key in any node of t_0, d_1 any key in any node of t_1, and so on, the inequality $d_0 \leq key_1 \leq d_1 \leq ... \leq key_n \leq d_n$ must hold.

Priority queues
A priority queue can be any kind of recursive structure in which an order relation has been established between each node and its descendants. One example of such an order relation would be to require that $key_p \leq key_d$, where key_p is any key in a parent node, and key_d is any key in any descendant of that node.

2.1.2.4 Hierarchical balance

Height balance
Let s be any node of a tree (binary or multiway). Define $h(s)$ as the height of the subtree rooted in s, that is, the number of nodes in the tallest branch starting at s. One structural quality that may be required is that the height of a tree along any pair of adjacent branches be approximately the same. More formally, the height balance constraint is $| h(s_1) - h(s_2) | \leq \delta$ where s_1 and s_2 are any two subtrees of any node in the tree, and δ is a constant giving

the maximum allowable height difference. In B-trees (see Section 3.4.2) for example, $\delta = 0$, while in AVL-trees $\delta = 1$ (see Section 3.4.1.3).

Weight balance
For any tree, the weight function $w(s)$ is defined as the number of external nodes (leaves) in the subtree rooted at s. A weight balance condition requires that for any two nodes s_1 and s_2, if they are both subtrees of any other node in the tree, $r \le w(s_1)/w(s_2) \le 1/r$ where r is a positive constant less than 1.

2.1.2.5 Optimality

Any condition on a data structure which minimizes a complexity measure (such as the expected number of accesses or the maximum number of comparisons) is an optimality condition. If this minimized measure of complexity is based on a worst-case value, the value is called the **minimax**; when the minimized complexity measure is based on an average value, it is the **minave.**

In summary, the W-grammars are used to define the general shape or pattern of the data objects. Once an object is generated, its validity is checked against the semantic rules or constraints that may apply to it.

References:
[Pooch, U.W. *et al.*, 73], [Aho, A.V. *et al.*, 74], [Rosenberg, A.L., 74], [Rosenberg, A.L., 75], [Wirth, N., 76], [Claybrook, B.G., 77], [Hollander, C.R., 77], [Honig, W.L. *et al.*, 77], [MacVeigh, D.T., 77], [Rosenberg, A.L. *et al.*, 77], [Cremers, A.B. *et al.*, 78], [Gotlieb, C.C. *et al.*, 78], [Rosenberg, A.L., 78], [Bobrow, D.G. *et al.*, 79], [Burton, F.W., 79], [Rosenberg, A.L. *et al.*, 79], [Rosenberg, A.L. *et al.*, 80], [Vuillemin, J., 80], [Rosenberg, A.L., 81], [O'Dunlaing, C. *et al.*, 82], [Gonnet, G.H. *et al.*, 83], [Wirth, N., 86].

2.2 Algorithm descriptions

Having defined the objects used to structure data, it is appropriate to describe the algorithms that access them. Furthermore, because data objects are not static, it is equally important to describe data structure manipulation algorithms.

An algorithm computes a function that operates on data structures. More formally, an algorithm describes a map $S \to R$ or $S \times P \to R$, where S, P, and R are all data structures; S is called the input structure, P contains parameters (for example, to specify a query), and R is the result. The two following examples illustrate these concepts:

(1) Quicksort is an algorithm that takes an array and sorts it. Since there are no parameters,

Quicksort: array → sorted-array

(2) B-tree insertion is an algorithm that inserts a new record P into a B-tree S, giving a new B-tree as a result. In functional notation,

B-tree-insertion: B-tree × new-record → B-tree

Algorithms compute functions over data structures. As always, different algorithms may compute the same functions; $\sin(2x)$ and $2\sin(x)\cos(x)$ are two expressions that compute the same function. Since equivalent algorithms have different computational requirements however, it is not merely the function computed by the algorithm that is of interest, but also the algorithm itself.

In the following section, we describe a few basic operations informally in order to convey their flavour.

References:
[Aho, A.V. *et al.*, 74], [Wirth, N., 76], [Bentley, J.L., 79], [Bentley, J.L., 79], [Saxe, J.B. *et al.*, 79], [Bentley, J.L. *et al.*, 80], [Bentley, J.L. *et al.*, 80], [Remy, J.L., 80], [Mehlhorn, K. *et al.*, 81], [Overmars, M.H. *et al.*, 81], [Overmars, M.H. *et al.*, 81], [Overmars, M.H. *et al.*, 81], [Overmars, M.H. *et al.*, 81], [Overmars, M.H., 81], [Rosenberg, A.L., 81], [Overmars, M.H. *et al.*, 82], [Gonnet, G.H. *et al.*, 83], [Chazelle, B. *et al.*, 86], [Wirth, N., 86], [Tarjan, R.E., 87], [Jacobs, D. *et al.*, 88], [Manber, U., 88], [Rao, V.N.S. *et al.*, 88], [Lan, K.K., 89], [Mehlhorn, K. *et al.*, 90].

2.2.1 Basic (or atomic) operations

A primary class of basic operations manipulate atomic values and are used to focus an algorithm's execution on the appropriate part(s) of a composite data object. The most common of these are as follows:

Selector and constructor
A selector is an operation that allows access to any of the elements corresponding to the right-hand side of a production rule from the corresponding left-hand side object. A constructor is an operation that allows us to assemble an element on the left-hand side of a production given all the corresponding elements on the right. For example, given a $\{\mathbf{string}\}_1^5$ and an integer, we can select the ith element, and given two **bt – real – nil** and a **real** we can construct a new **bt – real – nil**.

Replacement non-scalar × selector × value → non-scalar
A replacement operator removes us from pure functions by introducing the assignment statements. This operator introduces the possibility of cyclic and shared structures. For example, given a **bt-D-LEAF** we can form a threaded

binary tree by replacing the **nil** values in the leaves by (tagged) references back to appropriate nodes in the tree.

Ranking set of scalars × scalar → integer
This operation is defined on a set of scalars $X_1, X_2, ..., X_n$ and uses another scalar X as a parameter. Ranking determines how many of the X_j values are less than or equal to X, thus determining what rank X would have if it were ordered with the other values. More precisely, ranking is finding an integer i such that there is a subset $A \subseteq \{X_1, X_2, ..., X_n\}$ for which $|A| = i$ and $X_j \in A$ if and only if $X_j \leq X$. Ranking is used primarily in directing multiway decisions. For example, in a binary decision, $n = 1$, and i is zero if $X < X_1$, one otherwise.

Hashing value × range → integer
Hashing is an operation which normally makes use of a record key. Rather than using the actual key value however, an algorithm invokes hashing to transform the key into an integer in a prescribed range by means of a hashing function and then uses the generated integer value.

Interpolation numeric-value × parameters → integer
Similarly to hashing, this operation is typically used on record keys. Interpolation computes an integer value based on the input value, the desired range, the values of the smallest and largest of a set of values, and the probability distribution of the values in the set. Interpolation normally gives the statistical mode of the location of a desired record in a random ordered file, that is, the most probable location of the record.

Digitization scalar → sequence of scalars
This operation transforms a scalar into a sequence of scalars. Numbering systems that allow the representation of integers as sequences of digits and strings as sequences of characters provide natural methods of digitization.

Testing for equality value × value → boolean
Rather than relying on multiway decisions to test two values for equality, a distinct operation is included in the basic set. Given two values of the same type (for example, two integers, two characters, two strings), this operation determines whether they are equal. Notice that the use of multiway branching plus equality testing closely matches the behaviour of most processors and programming languages which require two tests for a three-way branch (less than, equal, or greater than).

2.2.2 Building procedures

Building procedures are used to combine basic operations and simple algorithms to produce more complicated ones. In this section, we will define four building procedures: **composition**, **alternation**, **conformation** and **self-organization**.

General references:
[Darlington, J., 78], [Barstow, D.R., 80], [Clark, K.L. *et al.*, 80], [van Leeuwen, J. *et al.*, 80], [Merritt, S.M., 85].

2.2.2.1 Composition

Composition is the main procedure for producing algorithms from atomic operations. Typically, but not exclusively, the composition of $F_1 : S \times P \rightarrow R$ and $F_2 : S \times P \rightarrow R$ can be expressed in a functional notation as $F_2(F_1(S, P_1), P_2)$. A more general and hierarchical description of composition is that the description of F_2 uses F_1 instead of a basic operation.

Although this definition is enough to include all types of composition, there are several common forms of composition that deserve to be identified explicitly.

Divide and conquer
This form uses a composition involving two algorithms for any problems that are greater than a critical size. The first algorithm splits a problem into (usually two) smaller problems. The composed algorithm is then recursively applied to each non-empty component, using recursion termination (see below) when appropriate. Finally the second algorithm is used to assemble the components' results into one result. A typical example of divide and conquer is Quicksort (where the termination alternative may use a linear insertion sort). Diagrammatically:

Divide and conquer

```
solve-problem(A):
    if size(A) <= Critical-Size
        then End-Action
        else begin
            Split-problem;
            solve-problem(A1);
            solve-problem(A2);
            . . .
            Assemble-Results
            end;
```

Special cases of divide and conquer, when applied to trees, are tree traversals.

Iterative application
This operates on an algorithm and a sequence of data structures. The algorithm is iteratively applied using successive elements of the sequence in place of the single element for which it was written. For example, insertion sort iteratively inserts an element into a sorted sequence.

Iterative application

```
solve-problem(S):
    while not empty(S) do begin
        Apply algorithm to next element of sequence S;
        Advance S
        end;
    End-Action
```

Alternatively, if the sequence is in an array:

Iterative application (arrays)

```
solve-problem(A):
    for i:=1 to size(A) do
        Action on A[i];
    End-Action
```

Tail recursion
This method is a composition involving one algorithm that specifies the criterion for splitting a problem into (usually two) components and selecting one of them to be solved recursively. A classical example is binary search.

Tail recursion

```
solve-problem(A):
    if size(A) <= Critical-Size
        then End-Action
        else begin
            Split and select subproblem i;
            solve-problem(A_i)
            end
```

Alternatively, we can unwind the recursion into a while loop:

Tail recursion

```
solve-problem(A):
    while size(A) > Critical-Size do begin
        Split and select subproblem i;
        A := A_i
        end;
    End-Action
```

It should be noted that tail recursion can be viewed as a variant of divide and conquer in which only one of the subproblems is solved recursively. Both divide and conquer and tail recursion split the original problem into subproblems of the same type. This splitting applies naturally to recursive data structures such as binary trees, multiway trees, general trees, digital trees, or arrays.

Inversion
This is the composition of two search algorithms that are then used to search for sets of records based on values of secondary keys. The first algorithm is used to search for the selected attribute (for example, find the 'inverted list' for the attribute 'hair colour' as opposed to 'salary range') and the second algorithm is used to search for the set with the corresponding key value (for instance, 'blonde' as opposed to 'brown'). In general, inversion returns a set of records which may be further processed (for example, using intersection, union, or set difference).

Inverted search

```
inverted-search(S, A, V):
    {*** Search the value V of the attribute A in
        the structure S *** }
    search (search(S, A), V)
```

The structure S on which the inverted search operates has to reflect these two searching steps. For the generation of S, the following metaproductions should be used:

$$\mathbf{S} \rightarrow \mathbf{D} \rightarrow \mathbf{DICT} \rightarrow \cdots(\mathbf{KEY}^{\mathbf{attr}}, \mathbf{D}^{\mathbf{attr}}) \cdots$$

$$\mathbf{D}^{\mathbf{attr}} \rightarrow \mathbf{DICT} \rightarrow \cdots(\mathbf{KEY}^{\mathbf{value}}, \mathbf{D}^{\mathbf{value}}) \cdots$$

$\mathbf{D}^{\mathbf{value}} \rightarrow \mathbf{SET} \rightarrow \cdots$

Digital decomposition

This is applied to a problem of size n by attacking preferred-size pieces (for example, pieces of size equal to a power of two). An algorithm is applied to all these pieces to produce the desired result. One typical example is binary decomposition.

Digital decomposition

```
Solve-problem(A, n)
    {*** n has a digital decomposition n = n_k β_k + ... + n_1 β_1 + n_0 *** }
    Partition the problem into subsets
        A = ∪_{i=0}^{k} ∪_{j=1}^{n_i} A_i^j;
        {*** where size(A_i^j) = β_i *** }
    for i:= 0 to k while not completed do
        simpler-solve(A_i^1, A_i^2, ..., A_i^{n_i});
```

Merge

The merge technique applies an algorithm and a discarding rule to two or more sequences of data structures ordered on a common key. The algorithm is iteratively applied using successive elements of the sequences in place of the single elements for which it was written. The discarding rule controls the iteration process. For example, set union, intersection, merge sort, and the majority of business applications use merging.

Merge

```
Merge(S_1, S_2, ..., S_k):
        while at least one S_i is not empty do
                kmin := minimum value of keys in S_1, ..., S_k;
                for i := 1 to k do
                        if kmin = head(S_i)
                                then t[i] := head(S_i)
                                else t[i] := nil;
                processing-rule( t[ 1], t[ 2],..., t[k]);
        End-Action
```

Randomization

This is used to improve a procedure or to transform a procedure into a probabilistic algorithm. This is appealing when the underlying procedure may fail, may not terminate, or may have a very bad worst case.

Randomization

```
solve-problem (A)
    repeat begin
        randomize(A);
        solve(randomized(A), t(A) units-of-time);
        end until Solve-Succeeds or Too-Many-Iterations;
    if Too-Many-Iterations
        then return(No-Solution-Exists)
        else return(Solution);
```

The conclusion that there is no solution is reached with a certain probability, hopefully very small, of being wrong. Primality testing using Fermat's little result is a typical example of this type of composition.

References:
[Bentley, J.L. *et al.*, 76], [Yao, A.C-C., 77], [Bentley, J.L. *et al.*, 78], [Dwyer, B., 81], [Chazelle, B., 83], [Lesuisse, R., 83], [Walah, T.R., 84], [Snir, M., 86], [Karlsson, R.G. *et al.*, 87], [Veroy, B.S., 88].

2.2.2.2 Alternation

The simplest building operation is alternation. Depending on the result of a test or on the value of a discriminator, one of several alternative algorithms is invoked. For example, based on the value of a command token in a batch updating interpreter, an insertion, modification, or deletion algorithm could be invoked; based on the success of a search in a table, the result could be processed or an error handler called; or based on the size of the input set, an $O(N^2)$ or an $O(N \log N)$ sorting algorithm could be chosen.

There are several forms of alternation that appear in many algorithms; these are elaborated here.

Superimposition
This combines two or more algorithms, allowing them to operate on the same data structure more or less independently. Two algorithms F_1 and F_2 may be superimposed over a structure S if $F_1(S, Q_1)$ and $F_2(S, Q_2)$ can both operate together. A typical example of this situation is a file that can be searched by one attribute using F_1 and by another attribute using F_2. Unlike other forms of alternation, the alternative to be used cannot be determined from the state of the structure itself; rather superimposition implies the capability of using

any alternative on any instance of the structure involved. Diagrammatically:

Superimposition

---

solve–problem(A):
  case 1: *solve–problem*$_1$*(A)*;
  case 2: *solve–problem*$_2$*(A)*;
  ...
  case n: *solve–problem*$_n$*(A)*

Interleaving
This operation is a special case of alternation in which one algorithm does not need to wait for other algorithms to terminate before starting its execution. For example one algorithm might add records to a file while a second algorithm makes deletions; interleaving the two would give an algorithm that performs additions and deletions in a single pass through the file.

Recursion termination
This is an alternation that separates the majority of the structure manipulations from the end actions. For example, checking for end of file on input, for reaching a leaf in a search tree, or for reduction to a trivial sublist in a binary search are applications of recursion termination. It is important to realize that this form of alternation is as applicable to iterative processes as recursive ones. Several examples of recursion termination were presented in the previous section on composition (see, for example, divide and conquer).

2.2.2.3 Conformation

If an algorithm builds or changes a data structure, it is sometimes necessary to perform more work to ensure that semantic rules and constraints on the data structure are not violated. For example, when nodes are inserted into or deleted from a tree, the tree's height balance may be altered. As a result it may become necessary to perform some action to restore balance in the new tree. The process of combining an algorithm with a 'clean-up' operation on the data structure is called **conformation** (sometimes **organization** or **reorganization**). In effect, conformation is a composition of two algorithms: the original modification algorithm and the constraint satisfaction algorithm. Because this form of composition has an acknowledged meaning to the algorithm's users, it is convenient to list it as a separate class of building operation rather than as a variant of composition. Other examples of conformation include reordering elements in a modified list to restore lexicographic order, percolating newly inserted elements to their appropriate locations in a priority queue, and removing all dangling (formerly incident) edges from a graph after a vertex is deleted.

2.2.2.4 Self-organization

This is a supplementary heuristic activity that an algorithm may often perform in the course of querying a structure. Not only does the algorithm do its primary work, but it also reaccommodates the data structure in a way designed to improve the performance of future queries. For example, a search algorithm may relocate the desired element once it is found so that future searches through the file will locate the record more quickly. Similarly, a page management system may mark pages as they are accessed, in order that 'least recently used' pages may be identified for subsequent replacement.

Once again, this building procedure may be viewed as a special case of composition (or of interleaving); however, its intent is not to build a functionally different algorithm, but rather to augment an algorithm to include improved performance characteristics.

2.2.3 Interchangeability

The framework described so far clearly satisfies two of its goals: it offers sufficient detail to allow effective encoding in any programming language, and it provides a uniformity of description to simplify teaching. It remains to be shown that the approach can be used to discover similarities among implementations as well as to design modifications that result in useful new algorithms.

The primary vehicle for satisfying these goals is the application of interchangeability. Having decomposed algorithms into basic operations used in simple combinations, one is quickly led to the idea of replacing any component of an algorithm by something similar.

The simplest form of interchangeability is captured in the static objects' definition. The hyperrules emphasize similarities among the data structure implementations by indicating the universe of uniform substitutions that can be applied. For example, in any structure using a sequence of reals, the hyperrule for **s – D** together with that for **D** indicates that the sequence of reals can be replaced by a sequence of integers, a sequence of binary trees, and so on. Algorithms that deal with such modified structures need, at most, superficial changes for manipulating the new sequences, although more extensive modifications may be necessary in parts that deal directly with the components of the sequence.

The next level of interchangeability results from the observation that some data structure implementations can be used to simulate the behaviour of others. For example, wherever a bounded sequence is used in an algorithm, it may be replaced by an array, relying on the sequentiality of the integers to access the array's components in order. Sequences of unbounded length may be replaced by sequences of arrays, a technique that may be applied to adapt an algorithm designed for a one-level store to operate in a two-level memory environment wherein each block will hold one array. This notion of interchangeability is the one usually promoted by researchers using abstract data types; their claim is that the algorithms should have been originally specified in terms of abstract sequences. We feel that the approach presented here does not contradict those claims, but rather that many algorithms already exist for specific representations, and that an operational approach to specifying algorithms, together with the notion of interchangeability, is more likely to appeal to data structure practitioners. In cases where data abstraction has been applied, this form of interchangeability can be captured in a meta-production, as was done for **DICT** in Figure 2.1.

One of the most common examples of this type of interchange is the implementation of linked lists and trees using arrays. For example, an $\mathbf{s} - \mathbf{D}$ is implemented as an $\{\mathbf{D}, \mathbf{int}\}_1^\mathbf{N}$ and a $\mathbf{bt} - \mathbf{D} - \mathbf{nil}$ as an $\{\mathbf{D}, \mathbf{int}, \mathbf{int}\}_1^\mathbf{N}$. In both cases the integers play the same role as the pointers: they select a record of the set. The only difference is syntactic, for example

```
p↑.next   ->  next[p]
p↑.right  ->  right[p].
```

Typically the value 0 is reserved to simulate a null pointer.

The most advanced form of interchangeability has not been captured by previous approaches. There are classes of operations that have similar intent yet behave very differently. As a result, replacing some operations by others in the same class may produce startling new algorithms with desirable properties. Some of these equivalence classes are listed below.

Basic algorithms	{hashing; interpolation; direct addressing }
	{collision resolution methods }
	{binary partition; Fibonaccian partition; median partition; mode partition }
Semantic rules	{height balance; weight balance }
	{lexicographical order; priority queues }
	{ordered hashing; Brent's hashing; binary tree hashing }
	{minimax; minave }

3 Searching Algorithms

3.1 Sequential search

3.1.1 Basic sequential search

This very basic algorithm is also known as the **linear search** or **brute force search**. It searches for a given element in an array or list by looking through the records sequentially until it finds the element or reaches the end of the structure. Let n denote the size of the array or list on which we search. Let A_n be a random variable representing the number of comparisons made between keys during a successful search and let A_n' be a random variable for the number of comparisons in an unsuccessful search. We have

$$Pr\{A_n = i\} = \frac{1}{n} \qquad (1 \le i \le n)$$

$$E[A_n] = \frac{n+1}{2}$$

$$\sigma^2(A_n) = \frac{n^2-1}{12}$$

$$A_n' = n$$

Below we give code descriptions of the sequential search algorithm in several different situations. The first algorithm (two versions) searches an array $r[i]$ for the first occurrence of a record with the required key; this is known as **primary key search**. The second algorithm also searches through an array, but does not stop until it has found every occurrence of the desired key; this is known as **secondary key search**. The third algorithm inserts a new key into the array without checking if the key already exists (this must be done for primary keys). The last two algorithms deal with the search for primary and secondary keys in linked lists.

Sequential search in arrays (non-repeated keys)

```
function search(key : typekey; var r : dataarray) : integer;
var i : integer;

begin
i := 1;
while  (i<n) and (key <> r[i].k) do i := i+1;
if r[i].k=key  then search := i    {*** found(r[i]) ***}
             else      search := -1;  {*** notfound(key) ***}
end;
```

For a faster inner loop, if we are allowed to modify location $n + 1$, then:

Sequential search in arrays (non-repeated keys)

```
function search(key : typekey; var r : dataarray) : integer;
var i : integer;

begin
r[n+1].k := key;
i := 1;
while key <> r[i].k do i := i+1;
if i <= n then search := i     {*** found(r[i]) ***}
        else search := -1;  {*** notfound(key) ***}
end;
```

Sequential search in arrays (secondary keys)

```
for i:=1 to n do
    if key = r[i].k then found(r[i]);
```

Insertion of a new key in arrays (secondary keys)

```
procedure insert(key : typekey; var r : dataarray);

begin
if n>=m then Error {*** Table is full ***}
else begin
     n := n+1;
     r[n].k := key
     end
end;
```

Sequential search in lists (non-repeated keys)

```
datarecord *search(key, list)
typekey key;  datarecord *list;

{ datarecord *p;
for (p=list; p != NULL && key != p ->k; p = p ->next);
return(p);
}
```

Sequential search in lists (secondary keys)

```
p := list;
while p <> nil do
     begin
     if key = p↑.k then found(p↑);
     p := p↑.next
     end;
```

The sequential search is the simplest search algorithm. Although it is not very efficient in terms of the average number of comparisons needed to find a record, we can justify its use when:

(1) our files only contain a few records (say, $n \leq 20$);

(2) the search will be performed only infrequently;

(3) we are looking for secondary keys and a large number of hits ($O(n)$) is expected;

(4) testing extremely complicated conditions.

The sequential search can also look for a given range of keys instead of one unique key, at no significant extra cost. Another advantage of this search algorithm is that it imposes no restrictions on the order in which records are stored in the list or array.

The efficiency of the sequential search improves somewhat when we use it to examine external storage. Suppose each physical I/O operation retrieves b records; we say that b is the **blocking factor** of the file, and we refer to each block of b records as a **bucket**. Assume that there are a total of n records in the external file we wish to search and let $k = \lfloor n/b \rfloor$. If we use E_n as a random variable representing the number of external accesses needed to find a given record, we have

$$E[E_n] = k + 1 - \frac{kb(k+1)}{2n} \approx \frac{k+1}{2}$$

$$\sigma^2(E_n) = \frac{bk(k+1)}{n}\left[\frac{2k+1}{6} - \frac{kb(k+1)}{4n}\right] \approx \frac{k^2}{12}$$

References:
[Knuth, D.E., 73], [Berman, G. *et al.*, 74], [Knuth, D.E., 74], [Clark, D.W., 76], [Wise, D.S., 76], [Reingold, E.M. *et al.*, 77], [Gotlieb, C.C. *et al.*, 78], [Hansen, W.J., 78], [Flajolet, P. *et al.*, 79], [Flajolet, P. *et al.*, 79], [Kronsjo, L., 79], [Flajolet, P. *et al.*, 80], [Willard, D.E., 82], [Sedgewick, R., 88].

3.1.2 Self-organizing sequential search: move-to-front method

This algorithm is basically the sequential search, enhanced with a simple heuristic method for improving the order of the list or array. Whenever a record is found, that record is moved to the front of the table and the other records are slid back to make room for it. (Note that we only need to move the elements which were ahead of the given record in the table; those elements further on in the table need not be touched.) The rationale behind this procedure is that if some records are accessed more often than others, moving those records to the front of the table will decrease the time for future searches. It is, in fact, very common for records in a table to have unequal probabilities of being accessed; thus, the move-to-front technique may often reduce the average access time needed for a successful search.

We will assume that there exists a probability distribution in which $Pr\{accessing\ key\ K_i\} = p_i$. Further we will assume that the keys are numbered in such a way that $p_1 \geq p_2 \geq \ldots \geq p_n > 0$. With this model we have

$$E[A_n] = C_n = \frac{1}{2} + \sum_{i,j} \frac{p_i p_j}{p_i + p_j}$$

$$\sigma^2(A_n) = (2 - C_n)(C_n - 1) + 4 \sum_{i<j<k} \frac{p_i p_j p_k}{p_i + p_j + p_k} \left(\frac{1}{p_i + p_j} + \frac{1}{p_i + p_k} + \frac{1}{p_j + p_k} \right)$$

$$A_n' = n$$

$$C_n \leq \frac{\pi}{2} C_n^{Opt} = \frac{\pi}{2} \sum i p_i = \frac{\pi}{2} \mu_1'$$

where $\mu_1' = C_n^{Opt} = \sum i p_i$ is the first moment of the distribution.

If we let $T(z) = \sum_{i=1}^{n} z^{p_i}$ then

$$C_n = \int_0^1 z[T'(z)]^2 \, dz$$

Let $C_n(t)$ be the average number of additional accesses required to find a record, given that t accesses have already been made. Starting at $t = 0$ with a randomly ordered table we have

$$| C_n(t) - C_n | = O(n^2/t)$$

Below we give a code description of the move-to-front algorithm as it can be implemented to search linked lists. This technique is less suited to working with arrays.

Self-organizing (move-to-front) sequential search (lists)

```
function search(key : typekey; var head : list) : list;
label 999;
var p, q : list;

begin

if head = nil then search := nil
else if key = head↑.k then search := head
else begin
     {*** Find record ***}
     p := head;
```

```
        while p↑.next <> nil do
            if p↑.next↑.k = key then begin
                {*** Move to front of list ***}
                q := head;
                head := p↑.next;
                p↑.next := p↑.next↑.next;
                head↑.next := q;
                search := head;
                goto 999 {*** Break ***}
                end
            else p := p↑.next;
        search := nil
        end;
    999:
    end;
```

Insertion of a new key on a linked list

```
    function insert(key : typekey; head : list) : list;
    var p : list;

    begin
    n := n+1;
    new(p);
    p↑.k := key;
    p↑.next := head;
    insert := p;
    end;
```

There are more sophisticated heuristic methods of improving the order of a list than the move-to-front technique; however, this algorithm can be recommended as particularly appropriate when we have reasons to suspect that the accessing probabilities for individual records will change with time.

Moreover, the move-to-front approach will quickly improve the organization of a list when the accessing probability distribution is very skewed.

If we can guarantee that the search will be successful we can obtain an efficient array implementation by sliding elements back while doing the search.

When searching a linked list, the move-to-front heuristic is preferable to the transpose heuristic (see Section 3.1.3).

Below we give some efficiency measures for this algorithm when the accessing probabilities follow a variety of distributions.

Zipf's law (harmonic): $p_i = (iH_n)^{-1}$

$$C_n = \frac{1}{2} + \frac{(2n+1)H_{2n} - 2(n+1)H_n}{H_n} = \frac{2\ln(2)n}{H_n} - \frac{1}{2} + o(1)$$

Lotka's law: $p_i = (i^2 H_n^{(2)})^{-1}$

$$C_n = \frac{3}{\pi} \ln n - 0.00206339\ldots + O(\frac{\ln n}{n})$$

Exponential distribution: $p_i = (1-a)a^{i-1}$

$$C_n = -\frac{2\ln 2}{\ln a} - \frac{1}{2} - \frac{\ln a}{24} - \frac{\ln^3 a}{2880} + O(\ln^5 a)$$

Wedge distribution: $p_i = \frac{2(n+1-i)}{n(n+1)}$

$$\begin{aligned} C_n &= \left(\frac{4n+2}{3} - \frac{1}{8n(n+1)}\right) H_n - \frac{(2n+1)(8n^2+14n-3)}{12n(n+1)} H_{2n} \\ &\quad + \frac{4n+4}{3} - \frac{13}{12(n+1)} \\ &= \frac{4(1-\ln 2)}{3} n - H_n + \frac{5(1-\ln 2)}{3} + \frac{H_n}{n} + O(n^{-1}) \end{aligned}$$

Generalized Zipf's: $p_i \propto i^{-\lambda}$

$$C_n \le \frac{1}{\lambda}\left(\psi\left(\frac{\lambda+1}{2\lambda}\right) - \psi\left(\frac{1}{2\lambda}\right)\right)\mu_1' \le \frac{\pi}{2}\mu_1'$$

where μ_1' is the optimal cost (see Section 3.1.4). The above formula is maximized for $\lambda = 2$, and this is the worst-case possible probability distribution.

Table 3.1 gives the relative efficiency of move-to-front compared to the optimal arrangement of keys, when the list elements have accessing probabilities which follow several different 'folklore' distributions.

References:
[McCabe, J., 65], [Knuth, D.E., 73], [Hendricks, W.J., 76], [Rivest, R.L., 76], [Bitner, J.R., 79], [Gonnet, G.H. *et al.*, 79], [Gonnet, G.H. *et al.*, 81], [Tenenbaum, A.M. *et al.*, 82], [Bentley, J.L. *et al.*, 85], [Hester, J.H. *et al.*, 85], [Hester, J.H. *et al.*, 87], [Chung, F.R.K. *et al.*, 88], [Makinen, E., 88].

3.1.3 Self-organizing sequential search: transpose method

This is another algorithm based on the basic sequential search and enhanced by a simple heuristic method of improving the order of the list or array. In this model, whenever a search succeeds in finding a record, that record is

Table 3.1: Relative efficiency of move-to-front.

	C_n/C_n^{Opt}			
n	*Zipf's law*	80%–20% *rule*	*Bradford's law* ($b = 3$)	*Lotka's law*
5	1.1921	1.1614	1.1458	1.2125
10	1.2580	1.2259	1.1697	1.2765
50	1.3451	1.3163	1.1894	1.3707
100	1.3623	1.3319	1.1919	1.3963
500	1.3799	1.3451	1.1939	1.4370
1000	1.3827	1.3468	1.1942	1.4493
10000	1.3858	1.3483	1.1944	1.4778

transposed with the record that immediately precedes it in the table (provided of course that the record being sought was not already in the first position). As with the move-to-front (see Section 3.1.2) technique, the object of this rearrangement process is to improve the average access time for future searches by moving the most frequently accessed records closer to the beginning of the table. We have

$$E[A_n] \;=\; C_n \;=\; Prob(I_n)\sum_{\pi}\left(\left(\prod_{i=1}^{n} p_i^{i-\pi(i)}\right)\sum_{j=1}^{n} p_j\pi(j)\right)$$

where π denotes any permutation of the integers 1,2,...,n. $\pi(j)$ is the location of the number j in the permutation π, and $Prob(I_n)$ is given by

$$Prob(I_n) \;=\; \left(\sum_{\pi}\prod_{i=1}^{n} p_i^{i-\pi(i)}\right)^{-1}$$

This expected value of the number of the accesses to find an element can be written in terms of permanents by

$$C_n \;=\; \frac{\sum_{k=1}^{n} perm(P_k)}{perm(P)}$$

where P is a matrix with elements $p_{i,j} = p_i^{-j}$ and P_k is a matrix like P except that the kth row is $p_{k,j} = jp_k^{1-j}$. We can put a bound on this expected value by

$$C_n \;\le\; \frac{2n}{n+1}C_n^{Opt} \;<\; 2\mu_1'$$

where μ_1' is the optimal search time (see Section 3.1.4).

In general the transpose method gives better results than the move-to-front (MTF) technique for stable probabilities. In fact, for all record accessing probability distributions, we have

$$C_n^{transpose} \leq C_n^{MTF}$$

When we look at the case of the unsuccessful search, however, both methods have the identical result

$$A_n' = n$$

Below we give a code description of the transpose algorithm as it can be applied to arrays. The transpose method can also be implemented efficiently for lists, using an obvious adaptation of the array algorithm.

Self-organizing (transpose) sequential search (arrays)

```
function search(key : typekey; var r : dataarray) : integer;
var i : integer;
    tempr : datarecord;

begin
i := 1;
while (i<n) and (r[i].k <> key) do i := i+1;

if key = r[i].k then
    begin
    if i>1 then
        begin
        {*** Transpose with predecessor ***}
        tempr := r[i];
        r[i] := r[i-1];
        r[i-1] := tempr;
        i := i-1
        end;
    search := i    {*** found(r[i]) ***}
    end
else search := -1;  {*** notfound(key) ***}
end;
```

It is possible to develop a better self-organizing scheme by allocating extra storage for counters which record how often individual elements are accessed; however, it is conjectured that the transpose algorithm is the optimal heuristic organization scheme when allocating such extra storage is undesirable.

It should be noted that the transpose algorithm may take quite some time to rearrange a randomly ordered table into close to optimal order. In fact, it may take $\Omega(n^2)$ accesses to come within a factor of $1+\epsilon$ of the final steady state.

Because of this slow adaptation ability, the transpose algorithm is not recommended for applications where accessing probabilities may change with time.

For sequential searching of an array, the transpose heuristic is preferable over the move-to-front heuristic.

Table 3.2 gives simulation results of the relative efficiency of the transpose method compared to the optimal arrangement of keys, when the list elements have accessing probabilities which follow several different 'folklore' distributions. It appears that for all smooth distributions, the ratio between transpose and the optimal converges to 1 as $n \to \infty$.

Table 3.2: Simulation results on the relative efficiency of transpose.

	C_n/C_n^{Opt}			
n	*Zipf's law*	80%–20% *rule*	*Bradford's law* ($b=3$)	*Lotka's law*
5	1.109897	1.071890	1.097718	1.110386
10	1.08490±0.00003	1.06788±0.00004	1.07073 ±0.00002	1.10041±0.00005
50	1.03213±0.00004	1.03001±0.00006	1.01949 ±0.00004	1.01790±0.00007
100	1.01949±0.00004	1.01790±0.00007	1.011039±0.000009	1.0645±0.0003
500	1.00546±0.00003	1.00458±0.00005	1.002411±0.000004	1.0503±0.0011
1000	1.00311±0.00004	1.00252±0.00005	1.001231±0.000003	1.0444±0.0021

References:

[Hendricks, W.J., 76], [Rivest, R.L., 76], [Tenenbaum, A.M., 78], [Bitner, J.R., 79], [Gonnet, G.H. *et al.*, 79], [Gonnet, G.H. *et al.*, 81], [Bentley, J.L. *et al.*, 85], [Hester, J.H. *et al.*, 85], [Hester, J.H. *et al.*, 87], [Makinen, E., 88].

3.1.4 Optimal sequential search

When we know the accessing probabilities for a set of records in advance, and we also know that these probabilities will not change with time, we can minimize the average number of accesses in a sequential search by arranging the records in order of decreasing accessing probability (so that the most often required record is first in the table, and so on). With this preferred ordering of the records, the efficiency measures for the sequential search are

$$E[A_n] = \mu_1' = \sum_{i=1}^{n} i p_i$$

$$\sigma^2(A_n) = \sum_{i=1}^{n} i^2 p_i - (\mu_1')^2$$

$$A'_n = n$$

Naturally, these improved efficiencies can only be achieved when the accessing probabilities are known in advance and do not change with time. In practice, this is often not the case. Further, this ordering requires the overhead of sorting all the keys initially according to access probability. Once the sorting is done, however, the records do not need reorganization during the actual search procedure.

3.1.5 Jump search

Jump search is a technique for searching a sequential ordered file. This technique is applicable whenever it is possible to jump or skip a number of records at a cost that is substantially less than the sequential read of those records.

Let a be the cost of a jump and b the cost of a sequential search. If the jumps have to be of a fixed size, then the optimum jump size is $\sqrt{na/b}$, and consequently

$$C_n = \sqrt{nab} + O(1)$$

Doing uniform size jumps is not the best strategy. It is better to have larger jumps at the beginning and smaller jumps at the end, so that the average and worst-case searching times are minimized.

For the optimal strategy the ith jump should be of $\sqrt{2an/b} - ai/b$ records; then:

$$C_n = \sqrt{\frac{8abn}{9}} + O(1)$$

Jump search algorithm

```
read_first_record;
while key > r.k do Jump_records;
while key < r.k do read_previous_record;
if key=r.k then    found(r)
        else notfound(key);
```

This method can be extended to an arbitrary number of levels; at each level the size of the jump is different. Ultimately when we use $\log_2 n$ levels this algorithm coincides with binary search (see Section 3.2.1).

There are two situations in which this algorithm becomes an appealing alternative:

(1) Tape reading where we can command to skip physical records almost without cost to the computer.

(2) Sequential files with compressed and/or encoded information when the cost of decompressing and/or decoding is very high.

When binary search is possible there is no reason to use jump searching.

References:
[Six, H., 73], [Shneiderman, B., 78], [Janko, W., 81], [Leipala, T., 81], [Guntzer, U. *et al.*, 87].

General references:
[Shneiderman, B., 73], [Lodi, E. *et al.*, 76], [Shneiderman, B. *et al.*, 76], [Wirth, N., 76], [Nevalainen, O. *et al.*, 77], [Allen, B. *et al.*, 78], [Claybrook, B.G. *et al.*, 78], [McKellar, A.C. *et al.*, 78], [Standish, T.A., 80], [Mehlhorn, K., 84], [Manolopoulos, Y.P. *et al.*, 86], [Wirth, N., 86], [Papadakis, T. *et al.*, 90], [Pugh, W., 90].

3.2 Sorted array search

The following algorithms are designed to search for a record in an array whose keys are arranged in order. Without loss of generality we will assume an increasing order.

We will discuss only the searching algorithms. The insertion of new elements or direct construction of a sorted array of size m is the same for all algorithms. These searching algorithms are not efficient when the table undergoes a lot of insertions and deletions. Both updating operations cost $O(n)$ work each. It is then implicit that these tables are rather static.

Insertion into an ordered array

```
procedure insert(key : typekey; var r : dataarray);
label 999;
var i : integer;

begin
i := n;
if n>=m then Error {*** Table full ***}
else begin
    n := n+1;
    while i>0 do
        if r[i].k > key then begin
```

```
            r[i+1] := r[i];
            i := i-1
            end
        else goto 999; {*** break ***}

    {*** Insert new record ***}
    999:
    r[i+1].k := key
    end
end;
```

The above algorithm will not detect the insertion of duplicates, that is, elements already present in the table. If we have all the elements available at the same time, it is advantageous to sort them in order, as opposed to inserting them one by one.

General references:
[Peterson, W.W., 57], [Price, C.E., 71], [Overholt, K.J., 73], [Horowitz, E. *et al.*, 76], [Guibas, L.J. *et al.*, 77], [Flajolet, P. *et al.*, 79], [Flajolet, P. *et al.*, 80], [Mehlhorn, K., 84], [Linial, N. *et al.*, 85], [Manolopoulos, Y.P. *et al.*, 86], [Yuba, T. *et al.*, 87], [Pugh, W., 90].

3.2.1 Binary search

This algorithm searches a sorted array using the **tail recursion** technique. At each step of the search, a comparison is made with the middle element of the array. Then the algorithm decides which half of the array should contain the required key, and discards the other half. The process is repeated, halving the number of records to be searched at each step until only one key is left. At this point one comparison is needed to decide whether the searched key is present in the file. If the array contains n elements and $k = \lfloor \log_2 n \rfloor$ then we have:

$$\lfloor \log_2 (n+1) \rfloor \;\leq\; A_n \;=\; A_n' \;\leq\; k+1$$

$$C_n \;=\; C_n' \;=\; k+2-\frac{2^{k+1}}{n+1} \;\approx\; \log_2 n$$

$$\sigma^2(A_n') \;\leq\; \frac{1}{12}$$

If we use three-way comparisons and stop the search on equality, the number of comparisons for the successful search changes to:

$$1 \;\leq\; A_n \;\leq\; k+1$$

$$E[A_n] \;=\; C_n \;=\; k+1-\frac{2^{k+1}-k-2}{n} \;\approx\; \log_2(n)-1+\frac{k+2}{n}$$

$$\sigma^2(A_n) \;=\; \frac{3\times 2^{k+1}-(k+2)^2-2}{n}-\left(\frac{2^{k+1}-k-2}{n}\right)^2$$

$$\approx\; 2.125 \pm .125+o(1)$$

$$C_n \;=\; \left(1+\frac{1}{n}\right)C_n'-1$$

(The random variables A_n and A_n' are as defined in Section 3.1; C_n and C_n' are the expected values of A_n and A_n' respectively.)

Binary search algorithm

```
function search(key : typekey; var r : dataarray) : integer;
var high, j, low : integer;

begin
low := 0;
high := n;
while high-low > 1 do begin
    j := (high+low) div 2;
    if key <= r[j].k then high := j
                     else low  := j
    end;
if r[high].k = key then search := high  {*** found(r[high]) ***}
                   else search := -1;   {*** notfound(key) ***}
end;
```

There are more efficient search algorithms than the binary search but such methods must perform a number of special calculations: for example, the interpolation search (see Section 3.2.2) calculates a special interpolation function, while hashing algorithms (see Section 3.3) must compute one or more hashing functions. The binary search is an optimal search algorithm when we restrict our operations only to comparisons between keys.

Binary search is a very stable algorithm: the range of search times stays very close to the average search time, and the variance of the search times is $O(1)$. Another advantage of the binary search is that it is well suited to searching for keys in a given range as well as searching for one unique key.

One drawback of the binary search is that it requires a sorted array. Thus additions, deletions, and modifications to the records in the table can be expensive, requiring work on the scale of $O(n)$.

Table 3.3 gives figures showing the performance of the three-way comparison binary search for various array sizes.

Table 3.3: Exact results for binary search.

n	C_n	$\sigma^2(A_n)$	C_n'
5	2.2000	0.5600	2.6667
10	2.9000	0.8900	3.5455
50	4.8600	1.5204	5.7451
100	5.8000	1.7400	6.7327
500	7.9960	1.8600	8.9780
1000	8.9870	1.9228	9.9770
5000	11.3644	2.2004	12.3619
10000	12.3631	2.2131	13.3618

References:
[Arora, S.R. *et al.*, 69], [Flores, I. *et al.*, 71], [Jones, P.R., 72], [Knuth, D.E., 73], [Overholt, K.J., 73], [Aho, A.V. *et al.*, 74], [Berman, G. *et al.*, 74], [Bentley, J.L. *et al.*, 76], [Reingold, E.M. *et al.*, 77], [Gotlieb, C.C. *et al.*, 78], [Flajolet, P. *et al.*, 79], [Kronsjo, L., 79], [Leipala, T., 79], [Yao, A.C-C., 81], [Erkioe, H. *et al.*, 83], [Lesuisse, R., 83], [Santoro, N. *et al.*, 85], [Arazi, B., 86], [Baase, S., 88], [Brassard, G. *et al.*, 88], [Sedgewick, R., 88], [Manber, U., 89].

3.2.2 Interpolation search

This is also known as the **estimated entry search**. It is one of the most natural ways to search an ordered table which contains numerical keys. Like the binary search (see Section 3.2.1), it uses the 'tail recursion' approach, but in a more sophisticated way. At each step of the search, the algorithm makes a guess (or interpolation) of where the desired record is apt to be in the array, basing its guess on the value of the key being sought and the values of the first and last keys in the table. As with the binary search, we compare the desired key with the key in the calculated probe position; if there is no match, we discard the part of the file we know does not contain the desired key and probe the rest of the file using the same procedure recursively.

Let us suppose we have normalized the keys in our table to be real numbers in the closed interval [0,1] and let $\alpha \in [0,1]$ be the key we are looking for. For any integer $k \leq n$, the probability of needing more than k probes to find α is given by

$$Pr\{A_n > k\} \approx \prod_{i=1}^{k}(1 - \frac{1}{2}\xi^{2^{-i}})$$

where $\xi = \frac{2}{\pi n \alpha(1-\alpha)}$.

$$\begin{aligned} E[A_n] &= \log_2 \log_2 n + O(1) \\ &\approx \log_2 \log_2 (n+3) \\ \sigma^2(A_n) &= O(\log_2 \log_2 n) \end{aligned}$$

$$\begin{aligned} E[A'_n] &= \log_2 \log_2 n + O(1) \\ &\approx \log_2 \log_2 n + 0.58 \end{aligned}$$

When implementing the interpolation search, we must make use of an interpolating formula. This is a function $\phi(\alpha, n)$ which takes as input the desired key $\alpha(\alpha \in [0, 1])$ and the array of length n, and which yields an array index between 1 and n, essentially a guess at where the desired array element is. Two of the simplest linear interpolation formulas are $\phi(\alpha, n) = \lceil n\alpha \rceil$ and $\phi(\alpha, n) = \lfloor n\alpha + 1 \rfloor$. Below we give a description of the interpolation search.

Interpolation search algorithm

```
function search(key : typekey; var r : dataarray) : integer;
var high, j, low : integer;

begin
low := 1;
high := n;
while (r[high].k >= key) and (key > r[low].k) do
    begin
    j := trunc((key-r[low].k) / (r[high].k-r[low].k) *
                (high-low)) + low;
    if      key > r[j].k then low  := j+1
    else if key < r[j].k then high := j-1
                         else low  := j
    end;
if r[low].k = key then search := low {*** found(r[low]) ***}
              else search := -1; {*** notfound(key) ***}
end;
```

The interpolation search is asymptotically optimal among all algorithms which search arrays of numerical keys. However, it is very sensitive to a non-uniform [0,1] distribution of the keys. Simulations show that the interpolation search can lose its $O(\log \log n)$ behaviour under some non-uniform key distributions.

While it is relatively simple in theory to adjust the algorithm to work suitably even when keys are not distributed uniformly, difficulties can arise in practice. First of all, it is necessary to know how the keys are distributed and this information may not be available. Furthermore, unless the keys follow a very simple probability distribution, the calculations required to adjust the algorithm for non-uniformities can become quite complex and hence impractical.

Interpolation search will not work if key values are repeated.

Table 3.4 gives figures for the efficiency measures of the interpolation search for various array sizes. The most important cost in the algorithm is the computation of the interpolation formula. For this reason, we will count the number of times the body of the while loop is executed (A_n). The amount L_n is the average of the worst-case A_n for every file.

Table 3.4: Simulation results for interpolation search.

n	$E[A_n]$	L_n	$E[A_n']$
5	0.915600±0.000039	1.45301±0.00014	1.28029±0.00009
10	1.25143±0.00010	2.18449±0.00024	1.50459±0.00015
50	1.91624±0.00029	3.83115±0.00083	2.02709±0.00032
100	2.15273±0.00040	4.5588±0.0013	2.23968±0.00042
500	2.60678±0.00075	6.1737±0.0029	2.67133±0.00073
1000	2.7711±0.0010	6.8265±0.0040	2.83241±0.00094
5000	3.0962±0.0018	8.2185±0.0084	3.1551±0.0017
10000	3.2173±0.0023	8.749 ±0.012	3.2760±0.0022
50000	3.4638±0.0043	9.937 ±0.025	3.5221±0.0043

From the above results we can see that the value for $E[A_n]$ is close to the value of $\log_2 \log_2 n$; in particular under the arbitrary assumption that

$$E[A_n] = \alpha \log_2 \log_2 n + \beta$$

for $n \geq 500$, then

$$\alpha = 1.0756 \pm 0.0037 \quad \beta = -0.797 \pm 0.012$$

References:

[Kruijer, H.S.M., 74], [Waters, S.J., 75], [Whitt, J.D. *et al.*, 75], [Yao, A.C-C. *et al.*, 76], [Gonnet, G.H., 77], [Perl, Y. *et al.*, 77], [Gotlieb, C.C. *et al.*, 78], [Perl, Y. *et al.*, 78], [Franklin, W.R., 79], [van der Nat, M., 79], [Burton, F.W. *et al.*, 80], [Gonnet, G.H. *et al.*, 80], [Ehrlich, G., 81], [Lewis, G.N. *et al.*, 81], [Burkhard, W.A., 83], [Mehlhorn, K. *et al.*, 85], [Santoro, N. *et al.*, 85], [Manolopoulos, Y.P. *et al.*, 87], [Carlsson, S. *et al.*, 88], [Manber, U., 89].

3.2.3 Interpolation–sequential search

This algorithm is a combination of the interpolation (see Section 3.2.2) and sequential search methods (see Section 3.1). An initial interpolation probe is made into the table, just as in the interpolation algorithm; if the given element is not found in the probed position, the algorithm then proceeds to search through the table sequentially, forwards or backwards depending on which direction is appropriate. Let A_n and A_n' be random variables representing the number of array accesses for successful and unsuccessful searches respectively. We have

$$E[A_n] = 1 + \frac{2}{n}\sum_{k=1}^{n-1} \frac{\Gamma(n)}{\Gamma(k)\Gamma(n-k)}(k/n)^k(1-k/n)^{n-k}$$
$$= 1 + \left(\frac{n\pi}{32}\right)^{1/2}(1 - \frac{7}{12n}) + O(n^{-1})$$

$$E[A_n'] = \frac{2}{n+1} + 2\sum_{k=1}^{n-1}\left(\frac{k}{n}I_{k/n}(k+1, n-k) - \frac{k+1}{n+1}I_{k/n}(k+2, n-k)\right)$$
$$= \left(\frac{n\pi}{32}\right)^{1/2} + O(1)$$

As with the standard interpolation search (see Section 3.2.2), this method requires an interpolation formula ϕ such as $\phi(\alpha, n) = \lceil n\alpha \rceil$ or $\phi(\alpha, n) = \lfloor n\alpha + 1 \rfloor$; for the code below we use the latter.

Interpolation–sequential search

```
function search(key : typekey; var r : dataarray) : integer;
var j : integer;

begin
if n > 1 then
    begin
    {*** initial probe location ***}
    j := trunc((key−r[1].k) / (r[n].k−r[1].k) * (n−1)) + 1;
    if key < r[j].k then
            while (j>1) and (key<r[j].k) do j := j−1
        else while (j<n) and (key>r[j].k) do j := j+1
    end
else j := 1;
```

```
if r[j].k = key then search := j    {*** found(r[j]) ***}
             else search := -1;  {*** notfound(key) ***}
end;
```

Asymptotically, this algorithm behaves significantly worse than the pure interpolation search. Note however, that for $n < 500$ it is still more efficient than binary search.

When we use this search technique with external storage, we have a significant improvement over the internal case. Suppose we have storage buckets of size b (that is, each physical I/O operation reads in a block of b records); then the number of external accesses the algorithm must make to find a record is given by

$$E[E_n] = 1 + \frac{1}{b}\left(\frac{n\pi}{32}\right)^{1/2} + O(n^{-1/2})$$

In addition to this reduction the accessed buckets are contiguous and hence the seek time may be reduced.

Table 3.5 lists the expected number of accesses required for both successful and unsuccessful searches for various table sizes.

Table 3.5: Exact results for interpolation–sequential search.

n	$E[A_n]$	$E[A'_n]$
5	1.5939	1.9613
10	1.9207	2.3776
50	3.1873	3.7084
100	4.1138	
500	7.9978	
1000	10.9024	
5000	23.1531	
10000	32.3310	

References:
[Gonnet, G.H. *et al.*, 77].

3.3 Hashing

Hashing or scatter storage algorithms are distinguished by the use of a **hashing function**. This is a function which takes a key as input and yields an integer in a prescribed range (for example, $[0, m-1]$) as a result. The function is designed so that the integer values it produces are uniformly distributed throughout the range. These integer values are then used as indices for an array of size m called the **hashing table**. Records are both inserted into and retrieved from the table by using the hashing function to calculate the required indices from the record keys.

When the hashing function yields the same index value for two different keys, we have a **collision**. Keys which collide are usually called **synonyms**. A complete hashing algorithm consists of a hashing function and a method for handling the problem of collisions. Such a method is called a **collision resolution scheme**.

There are two distinct classes of collision resolution schemes. The first class is called **open-addressing**. Schemes in this class resolve collisions by computing new indices based on the value of the key; in other words, they 'rehash' into the table. In the second class of resolution schemes, all elements which 'hash' to the same table location are linked together in a chain.

To insert a key using open-addressing we follow a sequence of probes in the table. This sequence of probe positions is called a **path**. In open-addressing a key will be inserted in the first empty location of its path. There are at most $m!$ different paths through a hashing table and most open-addressing methods use far less paths than $m!$ Several keys may share a common path or portions of a path. The portion of a path which is fully occupied with keys will be called a **chain**.

The undesirable effect of having chains longer than expected is called **clustering**. There are two possible definitions of clustering.

(1) Let $p = \Theta(m^k)$ be the maximum number of different paths. We say that a collision resolution scheme has $k+1$ clustering if it allows p different **circular paths**. A circular path is the set of all paths that are obtained from circular permutations of a given path. In other words, all the paths in a circular path share the same order of table probing except for their starting position.

(2) If the path depends exclusively on the first k initial probes we say that we have k-clustering.

It is generally agreed that linear probing suffers from primary clustering, quadratic and double hashing from secondary clustering, and uniform and random probing from no clustering.

Assume our hashing table of size m has n records stored in it. The quantity $\alpha = n/m$ is called the **load factor** of the table. We will let A_n be a random variable which represents the number of times a given algorithm must access the hashing table to locate any of the n elements stored there. It is expected that some records will be found on the first try, while for others we may have to either rehash several times or follow a chain of other records before we locate the record we want. We will use L_n to denote the length of the longest probe sequence needed to find any of the n records stored in the table. Thus our random variable A_n will have the range

$$1 \le A_n \le L_n$$

Its actual value will depend on which of the n records we are looking for.

In the same way, we will let A'_n be a random variable which represents the number of accesses required to insert an $n+1$th element into a table already containing n records. We have

$$1 \le A'_n \le n+1$$

The search for a record in the hashing table starts at an initial probe location calculated by the hashing function, and from there follows some prescribed sequence of accesses determined by the algorithm. If we find an empty location in the table while following this path, we may conclude that the desired record is not in the file. Thus it is important that an open-addressing scheme be able to tell the difference between an empty table position (one that has not yet been allocated) and a table position which has had its record deleted. The probe sequence may very well continue past a deleted position, but an empty position marks the end of any search. When we are inserting a record into the hashing table rather than searching for one, we use the first empty or deleted location we find.

Let

$$C_n = E[A_n]$$

and

$$C'_n = E[A'_n].$$

C_n denotes the expected number of accesses needed to locate any individual record in the hashing table while C'_n denotes the expected number of accesses needed to insert a record. Thus

$$C_n = \frac{1}{n}\sum_{i=0}^{n-1} E[A'_i] = \frac{1}{n}\sum_{i=0}^{n-1} C'_i$$

Below we give code for several hash table algorithms. In all cases we will search in an array of records of size m, named r, with the definition in Pascal being

Search array definition

```
type datarecord = record ... k : typekey; ... end;
    dataarray = array [0..m-1] of datarecord;

var n : integer;    {*** Number of keys in hash table ***}
```

```
procedure insert(new : typekey; var r : dataarray);
function search(key : typekey; var r : dataarray) : -1 .. m-1;

{*** auxiliary functions ***}
function deleted(r[i] : datarecord) : boolean;
function empty(r[i] : datarecord) : boolean;
function hashfunction(key : typekey) : 0 .. m-1;
function increment(key : typekey) : 1 .. m-1;
```

and in C being

Search array definition

```
typedef struct { ... typekey k; ... } datarecord, dataarray[ ];
typedef int boolean;

int  n;    /*** Number of keys in hash table ***/

    void insert(new, r)  typekey new;  dataarray r;
    int  search(key, r)  typekey key;  dataarray r;

    /*** auxiliary functions ***/
    boolean deleted(ri)  datarecord *ri;
    boolean empty(ri)  datarecord *ri;
    int hashfunction(key)  typekey key;
    int increment(key)  typekey key;
```

The key value being searched is stored in the variable *key*. There exist functions (or default values) that indicate whether an entry is empty ($empty(r[i])$) or indicate whether a value has been deleted ($deleted(r[i])$). The hashing functions yield values between 0 and $m-1$. The increment functions, used for several double-hashing algorithms, yield values between 1 and $m-1$.

General references:
[Peterson, W.W., 57], [Schay, G. *et al.*, 63], [Batson, A., 65], [Chapin, N., 69], [Chapin, N., 69], [Bloom, B.H., 70], [Coffman, E.G. *et al.*, 70], [Collmeyer, A.J. *et al.*, 70], [Knott, G.D., 71], [Nijssen, G.M., 71], [Nijssen, G.M., 71], [Price, C.E., 71], [Williams, J.G., 71], [Webb, D.A., 72], [Bays, C., 73], [Knuth, D.E., 73], [Aho, A.V. *et al.*, 74], [Bayer, R., 74], [Montgomery, A.Y., 74], [Rothnie, J.B. *et al.*, 74], [Bobrow, D.G., 75], [Deutscher, R.F. *et al.*, 75], [Ghosh, S.P. *et al.*, 75], [Maurer, W.D. *et al.*, 75], [Goto, E. *et al.*, 76], [Guibas, L.J., 76], [Horowitz, E. *et al.*, 76], [Sassa, M. *et al.*, 76], [Severance, D.G. *et al.*,

76], [Clapson, P., 77], [Reingold, E.M. *et al.*, 77], [Rosenberg, A.L. *et al.*, 77], [Gotlieb, C.C. *et al.*, 78], [Guibas, L.J., 78], [Halatsis, C. *et al.*, 78], [Kollias, J.G., 78], [Kronsjo, L., 79], [Mendelson, H. *et al.*, 79], [Pippenger, N., 79], [Romani, F. *et al.*, 79], [Scheurmann, P., 79], [Larson, P., 80], [Lipton, R.J. *et al.*, 80], [Standish, T.A., 80], [Tai, K.C. *et al.*, 80], [Bolour, A., 81], [Litwin, W., 81], [Tsi, K.T. *et al.*, 81], [Aho, A.V. *et al.*, 83], [Nishihara, S. *et al.*, 83], [Reingold, E.M. *et al.*, 83], [Larson, P., 84], [Mehlhorn, K., 84], [Torn, A.A., 84], [Devroye, L., 85], [Szymanski, T.G., 85], [Badley, J., 86], [Jacobs, M.C.T. *et al.*, 86], [van Wyk, C.J. *et al.*, 86], [Felician, L., 87], [Ramakrishna, M.V., 87], [Ramakrishna, M.V. *et al.*, 88], [Ramakrishna, M.V., 88], [Christodoulakis, S. *et al.*, 89], [Manber, U., 89], [Broder, A.Z. *et al.*, 90], [Cormen, T.H. *et al.*, 90], [Gil, J. *et al.*, 90].

3.3.1 Practical hashing functions

For all the hashing algorithms we assume that we have a hashing function which is 'good', in the sense that it distributes the values uniformly over the table size range m. In probabilistic terms for random keys k_1 and k_2 this is expressed as

$$Pr\{h(k_1) = h(k_2)\} \leq \frac{1}{m}$$

A universal class of hashing functions is a class with the property that given any input, the average performance of all the functions is good. The formal definition is equivalent to the above if we consider h as a function chosen at random from the class. For example, $h(k) = (ak + b) \bmod m$ with integers $a \neq 0$ and b is a universal class of hash functions.

Keys which are integers or can be represented as integers, are best hashed by computing their residue with respect to m. If this is done, m should be chosen to be a prime number.

Keys which are strings or sequences of words (including those which are of variable length) are best treated by considering them as a number base b. Let the string s be composed of k characters $s_1 s_2 \ldots s_k$. Then

$$h(s) = \left(\sum_{i=0}^{k-1} B^i s_{k-i} \right) \bmod m$$

To obtain a more efficient version of this function we can compute

$$h(s) = \left(\left(\sum_{i=0}^{k-1} B^i s_{k-i} \right) \bmod 2^w \right) \bmod m$$

where w is the number of bits in a computer word, and the $\bmod 2^w$ operation is done by the hardware. For this function the value $B = 131$ is recommended, as B^i has a maximum cycle $\bmod 2^k$ for $8 \leq k \leq 64$.

Hashing function for strings

```
int hashfunction(s)
char *s;

{ int i;
  for(i=0; *s; s++) i = 131*i + *s;
  return(i % m);
}
```

References:
[Maurer, W.D., 68], [Bjork, H., 71], [Lum, V.Y. *et al.*, 71], [Forbes, K., 72], [Lum, V.Y. *et al.*, 72], [Ullman, J.D., 72], [Gurski, A., 73], [Knuth, D.E., 73], [Lum, V.Y., 73], [Knott, G.D., 75], [Sorenson, P.G. *et al.*, 78], [Bolour, A., 79], [Carter, J.L. *et al.*, 79], [Devillers, R. *et al.*, 79], [Wegman, M.N. *et al.*, 79], [Papadimitriou, C.H. *et al.*, 80], [Sarwate, D.V., 80], [Mehlhorn, K., 82], [Ajtai, M. *et al.*, 84], [Wirth, N., 86], [Brassard, G. *et al.*, 88], [Fiat, A. *et al.*, 88], [Ramakrishna, M.V., 88], [Sedgewick, R., 88], [Fiat, A. *et al.*, 89], [Naor, M. *et al.*, 89], [Schmidt, J.P. *et al.*, 89], [Siegel, A., 89], [Mansour, Y. *et al.*, 90], [Pearson, P.K., 90], [Schmidt, J.P. *et al.*, 90].

3.3.2 Uniform probing hashing

Uniform probing hashing is an open-addressing scheme which resolves collisions by probing the table according to a permutation of the integers $[1,m]$. The permutation used depends only on the key of the record in question. Thus for each key, the order in which the table is probed is a random permutation of all table locations. This method will equally likely use any of the $m!$ possible paths.

Uniform probing is a theoretical hashing model which has the advantage of being relatively simple to analyze. The following list summarizes some of the pertinent facts about this scheme:

$$Pr\{A_n' > k\} \;=\; \frac{n^{\underline{k}}}{m^{\underline{k}}}$$

where $n^{\underline{k}}$ denotes the descending factorial, that is, $n^{\underline{k}} = n(n-1)\cdots(n-k+1)$.

$$E[A_n] \;=\; C_n \;=\; \frac{m+1}{n}(H_{m+1} - H_{m-n+1}) \;\approx\; -\alpha^{-1}\ln(1-\alpha)$$

$$\sigma^2(A_n) \;=\; \frac{2(m+1)}{m-n+2} - C_n(C_n+1)$$

$$\approx\; \frac{2}{1-\alpha} + \alpha^{-1}\ln(1-\alpha) - \alpha^{-2}\ln^2(1-\alpha)$$

$$C_m = \frac{m+1}{m}(H_{m+1}-1) = \ln m + \gamma - 1 + o(1)$$

$$\sigma^2(A_m) = m+1-C_m(C_m+1) = m - \ln^2 m + (1-2\gamma)\ln m + O(1)$$

$$C_n^{worst\ file} = \frac{n+1}{2}$$

$$E[A_n'] = C_n' = \frac{m+1}{m-n+1} \approx \frac{1}{1-\alpha}$$

$$\sigma^2(A_n') = \frac{(m+1)n(m-n)}{(m-n+1)^2(m-n+2)} \approx \frac{\alpha}{(1-\alpha)^2}$$

$$C_m' = m$$

$$C_n'^{\ worst\ file} = C_n' = \frac{m+1}{m-n+1}$$

$$L_n = \max_{0 \le i < n} A_i'$$

$$1 \le L_n \le n$$

$$E[L_n] = -\log_\alpha m + \log_\alpha(-\log_\alpha m) + O(1)$$

$$E[L_m] = 0.631587... \times m + O(1)$$

$$E[keys\ requiring\ i\ accesses] = \frac{n^{\underline{i}}}{m^{\underline{i}}}\left(\frac{(m-n)i+m+1}{i(i+1)}\right)$$

Table 3.6 gives figures for some of the quantities we have been discussing in the cases $m = 100$ and $m = \infty$.

Table 3.6: Exact results for uniform probing hashing.

	$m = 100$			$m = \infty$		
α	C_n	$\sigma^2(A_n)$	C_n'	C_n	$\sigma^2(A_n)$	C_n'
50%	1.3705	0.6358	1.9804	1.3863	0.6919	2.0
80%	1.9593	3.3837	4.8095	2.0118	3.9409	5.0
90%	2.4435	8.4190	9.1818	2.5584	10.8960	10.0
95%	2.9208	17.4053	16.8333	3.1534	26.9027	20.0
99%	3.7720	44.7151	50.0	4.6517	173.7101	100.0

It does not seem practical to implement a clustering-free hashing function.

Double hashing (see Section 3.3.5) behaves very similarly to uniform probing. For all practical purposes they are indistinguishable.

References:
[Furukawa, K., 73], [Knuth, D.E., 73], [Ajtai, M. *et al.*, 78], [Gonnet, G.H., 80], [Gonnet, G.H., 81], [Greene, D.H. *et al.*, 82], [Larson, P., 83], [Yao, A.C-C., 85], [Ramakrishna, M.V., 88], [Schmidt, J.P. *et al.*, 90].

3.3.3 Random probing hashing

This is an open-addressing hashing scheme in which collisions are resolved by additional probes into the table. The sequence of these probes is considered to be random and depends only on the value of the key. The difference between this scheme and uniform probing is that here some positions may be repeated in the probe sequence, whereas in uniform probing no position is examined more than once. Random probing is another theoretical model which is relatively simple to analyze.

The pertinent formulas for this scheme are given by:

$$Pr\{A_n' > k\} = \alpha^k$$

$$E[A_n] = C_n = \frac{m}{n}(H_m - H_{m-n}) = -\alpha^{-1}\ln(1-\alpha) + O\left(\frac{1}{m-n}\right)$$

$$\begin{aligned}\sigma^2(A_n) &= \frac{2m^2}{n}\left(H_m^{(2)} - H_{m-n}^{(2)}\right) - C_n(C_n+1) \\ &= \frac{2}{1-\alpha} + \alpha^{-1}\ln(1-\alpha) - \alpha^{-2}\ln^2(1-\alpha) + O\left(\frac{1}{m-n}\right)\end{aligned}$$

$$C_m = H_m = \ln m + \gamma + O(m^{-1})$$

$$\begin{aligned}\sigma^2(A_m) &= 2m\, H_m^{(2)} - H_m - H_m^2 \\ &= \frac{\pi^3}{3} - \ln^2 m - (1+2\gamma)\ln m + O(1)\end{aligned}$$

$$C_n^{worst\ file} = \infty$$

$$1 \le A_n' \le \infty$$

$$E[A_n'] = C_n' = \frac{1}{1-\alpha}$$

$$\sigma^2(A_n') = \frac{\alpha}{(1-\alpha)^2}$$

All collision resolution schemes that do not take into account the future probe sequences of the colliding records have the same expected successful search time under random probing.

Table 3.7: Exact results for random probing hashing.

	$m = 100$			$m = \infty$		
α	C_n	$\sigma^2(A_n)$	C_n'	C_n	$\sigma^2(A_n)$	C_n'
50%	1.3763	0.6698	2.0	1.3863	0.6919	2.0
80%	1.9870	3.7698	5.0	2.0118	3.9409	5.0
90%	2.5093	10.1308	10.0	2.5584	10.8960	10.0
95%	3.0569	23.6770	20.0	3.1534	26.9027	20.0
99%	4.2297	106.1598	100.0	4.6517	173.7101	100.0

Table 3.7 gives figures for some of the basic complexity measures in the case of $m = 100$ and $m = \infty$.

Notice that the asymptotic results ($m \to \infty; \alpha$ fixed) coincide with uniform probing, while for finite values of m, uniform probing gives better results.

Random probing could be implemented using pseudo-random probe locations; it does not seem, however, to be a good alternative to the double hashing algorithm described in Section 3.3.5.

References:
[Morris, R., 68], [Furukawa, K., 73], [Larson, P., 82], [Celis, P. *et al.*, 85], [Celis, P., 85], [Celis, P., 86], [Poblete, P.V. *et al.*, 89], [Ramakrishna, M.V., 89].

3.3.4 Linear probing hashing

Linear probing is an open-addressing hashing algorithm that resolves collisions by probing to the next table location modulo m. In other words, it probes sequentially through the table starting at the initial hash index, possibly running until it reaches the end of the table, rolling to the beginning of the table if necessary, and continuing the probe sequence from there. This method resolves collisions using only one circular path. For this model:

$$\begin{aligned} E[A_n] = C_n &= \frac{1}{2}\left(1 + \sum_{k \geq 0} \frac{(n-1)^{\underline{k}}}{m^k}\right) \\ &= \frac{1}{2}\left(1 + \frac{1}{1-\alpha}\right) - \frac{1}{2(1-\alpha)^3 m} + O(m^{-2}) \end{aligned}$$

$$\sigma^2(A_n) = \frac{1}{6} + \frac{m}{n}\left(\sum_{k \geq 0} \frac{k^2+k+3}{6} \frac{n^{\underline{k}}}{m^k} - \frac{1}{2}\right) - (C_n)^2$$

$$= \frac{\alpha(\alpha^2 - 3\alpha + 6)}{12(1-\alpha)^3} - \frac{3\alpha + 1}{2(1-\alpha)^5 m} + O(m^{-2})$$

$$C_n^{(worst\ file)} = \frac{n+1}{2}$$

$$C_m = \sqrt{\pi m/8} + \frac{1}{3} + O(m^{-1/2})$$

$$E[A_n'] = C_n' = \frac{1}{2}\left(1 + \sum_{k \geq 0} \frac{(k+1)n^{\underline{k}}}{m^k}\right)$$

$$= \frac{1}{2}\left(1 + \frac{1}{(1-\alpha)^2}\right) - \frac{3\alpha}{2(1-\alpha)^4 m} + O(m^{-2})$$

$$\sigma^2(A_n') = \frac{1}{6} + \sum_{k \geq 0} \frac{(k+1)(k^2+3k+5)}{12} \frac{n^{\underline{k}}}{m^k} - (C_n')^2$$

$$= \frac{3(1-\alpha)^{-4}}{4} - \frac{2(1-\alpha)^{-3}}{3} - \frac{1}{12} - \frac{\alpha(8\alpha+9)}{2(1-\alpha)^6 m} + O(m^{-2})$$

$$L_n = O(\log n) \qquad (\alpha < 1)$$

$$C_n'^{(worst\ file)} = 1 + \frac{n(n+1)}{2m}$$

We denote the hashing table as an array r, with each element $r[i]$ having a key k.

Linear probing hashing: search

```
function search(key : typekey; var r : dataarray) : integer;
var i, last : integer;

begin
i := hashfunction(key) ;
last := (i+n-1) mod m;
while (i<>last) and (not empty(r[i])) and (r[i].k<>key) do
    i := (i+1) mod m;
if r[i].k=key then search := i    {*** found(r[i]) ***}
        else search := -1; {*** notfound(key) ***}
end;
```

Linear probing hashing: insertion

```
procedure insert(key : typekey; var r : dataarray);
var i, last : integer;

begin
i := hashfunction(key) ;
last := (i+m-1) mod m;
while (i<>last) and (not empty(r[i]))
    and (not deleted(r[i])) and (r[i].k<>key) do
        i := (i+1) mod m;
if empty(r[i]) or deleted(r[i]) then
        begin
        {*** insert here ***}
        r[i].k := key;
        n := n+1
        end
else Error {*** table full, or key already in table ***};
end;
```

Linear probing hashing uses one of the simplest collision resolution techniques available, requiring a single evaluation of the hashing function. It suffers, however, from a piling-up phenomenon called **primary clustering**. The longer a contiguous sequence of keys grows, the more likely it is that collisions with this sequence will occur when new keys are added to the table. Thus the longer sequences grow faster than the shorter ones. Furthermore, there is a greater probability that longer chains will coalesce with other chains, causing even more clustering. This problem makes the linear probing scheme undesirable with a high load factor α.

It should be noted that the number of accesses in a successful or unsuccessful search has a very large variance. Thus it is possible that there will be a sizable difference in the number of accesses needed to find different elements.

It should also be noted that given any set of keys, the order in which the keys are inserted has no effect on the total number of accesses needed to install the set.

An obvious variation on the linear probing scheme is to move backward through the table instead of forward, when resolving collisions. Linear probing can also be used with an increment $q > 1$ such that q is co-prime with m. More generally, we could move through a unique permutation of the table entries, which would be the same for all the table; only the starting point of the permutation would depend on the key in question. Clearly, all these variations would exhibit exactly the same behaviour as the standard linear probing model.

As noted previously, deletions from the table must be marked as such for the algorithm to work correctly. The presence of deleted records in the table is called **contamination**, a condition which clearly interferes with the efficiency of an unsuccessful search. When new keys are inserted after deletions, the successful search is also deteriorated.

Up until now, we have been considering the shortcomings of linear probing when it is used to access internal storage. With external storage, the performance of the scheme improves significantly, even for fairly small storage buckets. Let b be the blocking factor, that is, the number of records per storage bucket. We find that the number of external accesses (E_n) is

$$E_n = 1 + \frac{A_n - 1}{b}$$

while the number of accesses required to insert an $n+1$th record is

$$E_n' = 1 + \frac{A_n' - 1}{b}$$

Furthermore, for external storage, we may change the form of the algorithm so that we scan each bucket completely before examining the next bucket. This improves the efficiency somewhat over the simplest form of the linear probing algorithm.

Table 3.8 gives figures for the efficiency of the linear probing scheme with $m = 100$, and $m = \infty$.

Table 3.8: Exact results for linear probing hashing.

	$m = 100$			$m = \infty$		
α	C_n	$\sigma^2(A_n)$	C_n'	C_n	$\sigma^2(A_n)$	C_n'
50%	1.4635	1.2638	2.3952	1.5	1.5833	2.5
80%	2.5984	14.5877	9.1046	3.0	35.3333	13.0
90%	3.7471	45.0215	19.6987	5.5	308.25	50.5
95%	4.8140	87.1993	32.1068	10.5	2566.58	200.5
99%	6.1616	156.583	50.5	50.5	330833.0	5000.5

References:
[Schay, G. *et al.*, 62], [Buchholz, W., 63], [Tainiter, M., 63], [Konheim, A.G. *et al.*, 66], [Morris, R., 68], [Kral, J., 71], [Knuth, D.E., 73], [van der Pool, J.A., 73], [Bandyopadhyay, S.K., 77], [Blake, I.F. *et al.*, 77], [Lyon, G.E., 78], [Devillers, R. *et al.*, 79], [Larson, P., 79], [Mendelson, H. *et al.*, 80], [Quittner, P. *et al.*, 81], [Samson, W.B., 81], [Larson, P., 82], [Mendelson, H., 83], [Pflug, G.C. *et al.*, 87], [Pittel, B., 87], [Poblete, P.V., 87], [Aldous, D., 88], [Knott, G.D., 88], [Sedgewick, R., 88], [Schmidt, J.P. *et al.*, 90].

3.3.5 Double hashing

Double hashing is an open-addressing hashing algorithm which resolves collisions by means of a second hashing function. This second function is used to calculate an increment less than m which is added on to the index to make successive probes into the table. Each different increment gives a different path, hence this method uses $m-1$ circular paths. We have

$$E[A_n] = C_n = -\alpha^{-1}\ln(1-\alpha) + o(1) \qquad (\alpha < 0.319...)$$

$$E[A_n'] = C_n' = (1-\alpha)^{-1} + o(1) \qquad (\alpha < 0.319...)$$

$$\lim_{n\to\infty} Pr\{L_n = O(\log n)\} = 1$$

Actually, double hashing is not identical to uniform probing (see Section 3.3.2). For example, if $m = 13$ then

$$C_{13}^{Doub.\,hash.} - C_{13}^{Unif.\,prob.} = 0.0009763...$$

$$E[L_{13}^{Doub.\,hash.}] - E[L_{13}^{Unif.\,prob.}] = 0.001371...$$

Below we give descriptions of search and insertion algorithms which implement the double hashing scheme. Both algorithms require the table size m to be a prime number; otherwise there is the possibility that the probe sequence, for some keys, will not cover the entire table.

Double hashing: search

```
function search(key : typekey; var r : dataarray) : integer;
var i, inc, last : integer;

begin
i := hashfunction(key) ;
inc := increment(key) ;
last := (i+(n-1)*inc) mod m;
while (i<>last) and (not empty(r[i])) and (r[i].k<>key) do
     i := (i+inc) mod m;
if r[i].k=key then search := i    {*** found(r[i]) ***}
           else search := -1;  {*** notfound(key) ***}
end;
```

Double hashing: insertion

```
procedure insert(key : typekey; var r : dataarray);
var i, inc, last : integer;

begin
i := hashfunction(key) ;
inc := increment(key);
last := (i+(m-1)*inc) mod m;
while (i<>last) and (not empty(r[i]))
    and (not deleted(r[i])) and (r[i].k<>key) do
        i := (i+inc) mod m;
if empty(r[i]) or deleted(r[i]) then
        begin
        {*** insert here ***}
        r[i].k := key;
        n := n+1
        end
else Error {*** table full, or key already in table ***};
end;
```

Double hashing is a practical and efficient hashing algorithm. Since the increment we use to step through the table depends on the key we are searching for, double hashing does not suffer from primary clustering. This also implies that changing the order of insertion of a set of keys may change the average number of accesses required to do the inserting. Thus several reorganization schemes have been developed to reorder insertion of keys in ways which make double hashing more efficient.

If the initial position and the increment are not independent, the resulting search path cannot be considered random. For example if the initial position and the increment have the same parity, the $i+inc$, $i+3*inc$, $i+5*inc$, etc. will all be even. This is called the parity problem in hashing. This problem is solved by insisting that $hashfunction(k)$ and $increment(k)$ behave like independent random variables.

As with linear probing (see Section 3.3.4), deletion of records leads to contamination and decreases the efficiency of the unsuccessful search. When new keys are inserted after deletions, the successful search is also deteriorated. The unsuccessful search can be improved by keeping in a counter the length of the longest probe sequence (*llps*) in the file. Thus the search algorithm is the same as before, except that the variable *last* is computed as

```
last := (i+(llps-1)*inc) mod m;
```

Whenever we insert a new key we may need to update this counter.

Extensive simulations show that it is practically impossible to establish statistically whether double hashing behaves differently from uniform probing (see Section 3.3.2). For example we would need a sample of 3.4×10^7 files of size 13 to show statistically with 95% confidence that double hashing is different from uniform probing. Table 3.9 list some sample results.

Table 3.9: Simulation results for double hashing.

	$m = 101$			
n	C_n	$\sigma^2(A_n)$	L_n	C'_n
51	1.37679±0.00009	0.6557±0.0003	4.5823±0.0012	2.00159±0.00012
81	1.96907±0.00021	3.4867±0.0020	11.049±0.004	4.87225±0.00088
91	2.45611±0.00036	8.6689±0.0062	18.159±0.009	9.2966±0.0028
96	2.93478±0.00058	17.849±0.016	27.115±0.017	17.0148±0.0073
100	3.7856±0.0013	50.292±0.069	48.759±0.045	51.0
n	$m = 4999$			
2500	1.38617±0.00010	0.6914±0.0003	9.340±0.010	1.99997±0.00012
3999	2.01054±0.00022	3.9285±0.0025	25.612±0.041	4.9952±0.0010
4499	2.55599±0.00039	10.845±0.009	48.78±0.10	9.9806±0.0039
4749	3.14830±0.00073	26.650±0.036	88.59±0.25	19.941±0.015
4949	4.6249±0.0032	166.73±0.75	318.8±2.2	97.93±0.31

References:
[Bell, J.R. *et al.*, 70], [Bookstein, A., 72], [Luccio, F., 72], [Knuth, D.E., 73], [Guibas, L.J., 76], [Guibas, L.J. *et al.*, 78], [Samson, W.B., 81], [Yao, A.C-C., 85], [Lueker, G.S. *et al.*, 88], [Sedgewick, R., 88], [Schmidt, J.P. *et al.*, 90].

3.3.6 Quadratic hashing

Quadratic hashing is an open-addressing algorithm that resolves collisions by probing the table in increasing increments modulo the table size, that is, $h(k)$, $h(k)+1$, $h(k)+4$, $h(k)+9$, ... If the increments are considered to be a random permutation of the integers $1, ..., m$, we obtain the following results

$$E[A_n] = C_n = \frac{n+1}{n}(H_{m+1} - H_{m-n+1}) + 1 - \frac{n}{2(m+1)} + O(n^{-1})$$
$$\approx 1 - \ln(1-\alpha) - \frac{\alpha}{2}$$

$$E[A'_n] = C'_n = \frac{m+1}{m-n+1} - \frac{n}{m+1} + H_{m+1} - H_{m-n+1} + O(m^{-1})$$
$$\approx (1-\alpha)^{-1} - \alpha - \ln(1-\alpha)$$

Quadratic hashing: search

```
function search(key : typekey; var r : dataarray) : integer;
var i, inc : integer;

begin
i := hashfunction(key) ;
inc := 0;
while (inc<m) and (not empty(r[i])) and (r[i].k<>key) do
     begin
     i := (i+inc+1) mod m;
     inc := inc + 2
     end;
if r[i].k=key then search := i    {*** found(r[i]) ***}
          else search := -1;  {*** notfound(key) ***}
end;
```

Quadratic hashing: insertion

```
procedure insert(key : typekey; var r : dataarray);
var i, inc : integer;

begin
i := hashfunction(key);
inc := 0;
while (inc<m) and (not empty(r[i])) and
     (not deleted(r[i])) and (r[i].k<>key) do begin
          i := (i+inc+1) mod m;
          inc := inc + 2;
          end;
if empty(r[i]) or deleted(r[i]) then
          begin
          {*** insert here ***}
          r[i].k := key;
          n := n+1
          end
else Error {*** table full, or key already in table ***};
end;
```

Quadratic hashing requires a single hashing function evaluation per search. It suffers, however, from a slight piling-up phenomenon called **secondary**

clustering.

This algorithm may fail to insert a key after the table is half full. This is due to the fact that the ith probe coincides with the $m - i$th probe. This can be solved by the use of the probe sequence $h(k)$, $h(k) + 1$, $h(k) - 1$, $h(k) + 4$, $h(k) - 4$, ... whenever m is a prime of the form $4k + 1$.

Table 3.10 show some simulation results for quadratic hashing. F_n indicates the average number of times that the algorithm failed during insertion. These simulation results are not in close agreement with the proposed formulas for secondary clustering.

Table 3.10: Simulation results for quadratic hashing.

	$m = 101$			
n	C_n	L_n	C'_n	F_n
51	1.41410±0.00011	4.9875±0.0013	2.11837±0.00008	$< 10^{-6}$
81	2.06278±0.00025	11.5711±0.0043	5.12986±0.00031	$< 10^{-6}$
91	2.56693±0.00040	18.5212±0.0090	9.52385±0.00062	$< 10^{-5}$
96	3.03603±0.00061	26.569±0.015	16.9118±0.0012	< 0.00026
100	3.69406±0.00098	37.217±0.020	38.871287	0.5709±0.0019
	$m = 4999$			
2499	1.42869±0.00012	10.380±0.011	2.13732±0.00010	< 0.000027
3999	2.15350±0.00032	28.165±0.043	5.6080±0.0009	< 0.000055
4499	2.77974±0.00064	51.98±0.11	11.2084±0.0038	< 0.000089
4749	3.4385±0.0012	91.85±0.27	21.6824±0.0094	< 0.00014
4949	4.9699±0.0040	317.3±2.2	99.261±0.027	< 0.00048

References:
[Maurer, W.D., 68], [Bell, J.R., 70], [Day, A.C., 70], [Radke, C.E., 70], [Hopgood, F.R.A. *et al.*, 72], [Knuth, D.E., 73], [Ackerman, A.F., 74], [Ecker, A., 74], [Nishihara, S. *et al.*, 74], [Batagelj, V., 75], [Burkhard, W.A., 75], [Santoro, N., 76], [Wirth, N., 76], [Samson, W.B. *et al.*, 78], [Wirth, N., 86], [Wogulis, J., 89].

3.3.7 Ordered and split-sequence hashing

It is easy to verify that the average access time for uniform probing (see Section 3.3.2), double hashing (see Section 3.3.5), and quadratic hashing (see Section 3.3.6) depends not only on the keys, but also on the order in which these keys are inserted. Although the absolute order of insertion is difficult to alter, the algorithms described in this and the following sections will simulate altering the order of insertion. That is, if convenient, keys already in the table are moved to make room for newly inserted keys.

In this section we present two techniques that assume we can define an order relation on the keys in the table.

Ordered hashing is a composition of a hashing step, followed by double hashing collision resolution. Furthermore, ordered hashing reorders keys to simulate the effect of having inserted all the keys in increasing order. To achieve this effect, during insertion, smaller value keys will cause relocation of larger value keys found in their paths.

For the analysis of ordered hashing we assume, as for uniform probing (see Section 3.3.2), that the hashing function produces probing sequences without clustering. Let x be the probability that a randomly selected key in the file is less than the searched key. Then

$$\alpha = n/m$$

$$Pr\{A_n'(x) > k\} = \frac{n^{\underline{k}}}{m^{\underline{k}}} x^k$$

$$E[A_n'(x)] = \sum_{k\geq 0} \frac{n^{\underline{k}}}{m^{\underline{k}}} x^k = \frac{1}{1-\alpha x} - \frac{\alpha(1-\alpha)x^2}{(1-\alpha x)^3 m} + O(n^{-2})$$

$$C_n' = E[A_n'] = \sum_{k=0}^{n} \frac{n^{\underline{k}}}{m^{\underline{k}}(k+1)}$$

$$= -\alpha^{-1}\ln(1-\alpha) - \frac{1}{n}\left(\frac{3\alpha-2}{2(1-\alpha)} - \frac{1-\alpha}{\alpha}\ln(1-\alpha)\right) + O(n^{-2})$$

$$C_m' = E[A_m'] = H_{m+1}$$

The values for A_n and C_n are the same as those for double hashing (see Section 3.3.5).

Ordered hashing: search

```
function search(key : typekey; var r : dataarray) : integer;
var i, inc, last : integer;

begin
i := hashfunction(key) ;
inc := increment(key) ;
last := (i+(n-1)*inc) mod m;
while (i<>last) and (not empty(r[i])) and (r[i].k<key) do
    i := (i+inc) mod m;
```

```
    if r[i].k=key then search := i    {*** found(r[i]) ***}
            else search := -1; {*** notfound(key) ***}
    end;
```

Ordered hashing: insertion

```
procedure insert(key : typekey; var r : dataarray);
var i : integer;
    temp : typekey;

begin
if n>=m then   Error     {*** table is full ***}
else begin
    i := hashfunction(key) ;
    while (not empty(r[i])) and (not deleted(r[i]))
        and (r[i].k<>key) do begin
        if r[i].k > key then begin
            {*** Exchange key and continue ***}
            temp := key;   key := r[i].k; r[i].k := temp
            end;
        i := (i+increment(key)) mod m
        end;
    if empty(r[i]) or deleted(r[i]) then begin
        {*** do insertion ***}
        r[i].k := key;
        n := n+1
        end
    else Error     {*** key already in table ***}
    end
end;
```

This variation of double hashing (see Section 3.3.5) reduces the complexity of the unsuccessful search to roughly that of the successful search at a small cost during insertion.

Table 3.11 shows simulation results for ordered hashing. We present the values for C_n' since the values for C_n and L_n are expected to be the same as those for double hashing.

Split-sequence hashing chooses one of two possible collision resolution sequences depending on the value of the key located at the initial probe position. When we search for a key k, we first compare k with the key k' stored in position $h(k)$. If $k = k'$ or $h(k)$ is empty, the search ends. Otherwise we

Table 3.11: Simulation results on unsuccessful searches for ordered hashing.

	C'_n		
n	$m = 101$	n	$m = 4999$
51	1.38888±0.00008	2500	1.38639±0.00007
81	2.00449±0.00022	3999	2.01137±0.00022
91	2.53016±0.00039	4499	2.55787±0.00041
96	3.07959±0.00063	4749	3.15161±0.00071
100	4.2530±0.0014	4949	4.6415±0.0021

follow one of two possible probe sequences depending on $k < k'$ or $k > k'$.

For example, split linear probing uses an increment q_1 if $k < k'$, or q_2 if $k > k'$, where q_1 and q_2 are both co-prime with m. Similarly, we can define split quadratic hashing, split double hashing, and so on.

Simulations show that split linear probing hashing can improve the average search time of linear probing by more than 50% for values of α near 1, for random keys.

References:
[Amble, O. *et al.*, 74], [Lodi, E. *et al.*, 85].

3.3.8 Reorganization schemes

3.3.8.1 Brent's algorithm

Brent's reorganization scheme is based on double hashing (see Section 3.3.5). This scheme will place a new key by moving forward at most one other key. The placement is done such that the total number of accesses (new key and old keys moved forward) is minimized. This is achieved by searching for the first empty location in the probing path of the new key or the paths of any of the keys in the path of the new key.

Considering uniform probing, and $\alpha = n/m$ (the load factor), then

$$C_n \approx 1 + \frac{\alpha}{2} + \frac{\alpha^3}{4} + \frac{\alpha^4}{15} - \frac{\alpha^5}{18} + \frac{2\alpha^6}{15} + \frac{9\alpha^7}{80} - \frac{293\alpha^8}{5670} - \frac{319\alpha^9}{5600} + \cdots$$

$$C_m \approx 2.4941\ldots$$

Table 3.12 shows some values for C_α.

It has been conjectured and verified by simulation that

$$L_m = O(\sqrt{m})$$

Table 3.12: Exact values for C_α.

α	C_α
0.50	1.2865
0.80	1.5994
0.90	1.8023
0.95	1.9724
0.99	2.2421
1.00	2.4941

The values for the unsuccessful search are identical to those for double hashing (see Section 3.3.5).

Brent's reorganization hashing: insertion

```
procedure insert(key : typekey; var r : dataarray);
label 999;
var i, ii, inc, init, j, jj : integer;

begin
init := hashfunction(key);
inc  := increment(key);
for i:=0 to n do
    for j:=i downto 0 do begin
        jj := (init + inc*j) mod m;
        ii := (jj + increment(r[jj].k) * (i-j)) mod m;
        if empty(r[ii]) or deleted(r[ii]) then begin
            {*** move record forward ***}
            r[ii] := r[jj];
            {*** insert new in r[jj] ***}
            r[jj].k := key;
            n := n+1;
            goto 999  {*** return ***}
            end
        end;
Error {*** table full ***};
999:
end;
```

The above algorithm will not detect the insertion of duplicates, that is, elements already present in the table.

The searching algorithm is identical to double hashing (see Section 3.3.5).

This method improves the successful search at the cost of additional work during the insertion of new keys. This reorganization scheme allows us to completely fill a hashing table, while still keeping the average number of accesses bounded by 2.5. The length of the longest probe sequence, which is the actual worst case for a random file, is also significantly reduced.

For a stable table where its elements will be searched several times after insertion, this reorganization will prove very efficient.

Table 3.13 summarizes simulation results for Brent's reorganization scheme. The columns headed by I_n count the number of elements accessed to insert a new key in the table. I_n gives an accurate idea of the cost of the reorganization. Note that the variance on the number of accesses is also greatly reduced. The simulation results are in excellent agreement with the predicted theoretical results.

Table 3.13: Simulation results for Brent's hashing.

$m = 101$				
n	C_n	$\sigma^2(A_n)$	L_n	I_n
51	1.27590±.00005	0.28021±.00007	2.9782±.0004	1.48412±.00012
81	1.57687±.00009	0.76473±.00020	4.8400±.0010	2.49529±.00035
91	1.76674±.00011	1.25604±.00038	6.2819±.0015	3.50016±.00063
96	1.91961±.00014	1.82723±.00062	7.7398±.0021	4.6333±.0010
100	2.13671±.00018	3.1374±.0014	10.7624±.0040	7.1536±.0023
101	2.24103±.00022	4.1982±.0024	13.0843±.0060	9.1732±.0038
$m = 4999$				
2499	1.28628±.00005	0.29164±.00007	4.5115±.0030	1.49637±.00012
3999	1.60044±.00009	0.80739±.00021	7.7687±.0064	2.55468±.00036
4499	1.80448±.00012	1.35682±.00041	10.587±.010	3.64497±.00067
4749	1.97535±.00014	2.03962±.00071	13.876±.015	4.9424±.0011
4949	2.24480±.00021	3.9949±.0021	24.240±.037	8.4245±.0032
4999	2.47060±.00030	10.195±.018	85.72±.29	18.468±.027

References:

[Brent, R.P., 73], [Feldman, J.A. *et al.*, 73], [Knuth, D.E., 73], [Tharp, A.L., 79].

3.3.8.2 Binary tree hashing

Binary tree hashing is based on double hashing (see Section 3.3.5). This scheme will insert a new key in the table by moving forward, if necessary, other keys. The placement is done such that the total number of accesses (new key and old keys moved forward) is minimized. This is achieved by searching for empty locations in the probing path of the new key or the paths of the keys in its path or the paths of any keys in the path of the path, and so on. The name 'binary tree' comes from the fact that the algorithm probes locations following a binary tree pattern.

Considering uniform probing, and $\alpha = n/m$ (the load factor), then

$$C_n \approx 1 + \frac{\alpha}{2} + \frac{\alpha^3}{4} + \frac{\alpha^4}{15} - \frac{\alpha^5}{18} + \frac{2\alpha^6}{105} + \frac{83\alpha^7}{720} + \frac{613\alpha^8}{5760} - \frac{69\alpha^9}{1120} + \cdots$$

$$C_m \approx 2.13414...$$

If M_n is the number of keys that are moved forward for an insertion, then

$$M_n \approx \frac{\alpha^2}{3} - \frac{\alpha^3}{4} + \frac{2\alpha^4}{15} + \frac{\alpha^5}{9} + \frac{8\alpha^6}{105} - \frac{101\alpha^7}{720} - \frac{506\alpha^8}{2835} - \cdots$$

$$M_m \approx 0.38521...$$

Table 3.14 shows exact values for these complexity measures.

Table 3.14: Exact values for comparisons and moves.

α	C_α	M_α
0.50	1.28517	0.06379
0.80	1.57886	0.17255
0.90	1.75084	0.24042
0.95	1.88038	0.29200
0.99	2.04938	0.35819
1.00	2.13414	0.38521

It is conjectured, and supported by simulation, that

$$L_m = \log_2 m + O(1)$$

Binary tree reorganization hashing: insertion

```
procedure insert(key : typekey; var r : dataarray);
var i, inc, init, j : integer;

    function SearchMove (init, inc, level : integer) : integer;
    {*** Find the first hole (empty location) at the given depth
         in the binary tree spanned by a key ***}
    label    999;
    var i, incl, j, k : integer;
    begin
```

```
        i := (init + inc*level) mod m;
        if empty(r[i]) or deleted(r[i]) then SearchMove := i
        else begin
            for j:=level-1 downto 0 do begin
                i := (init + inc*j) mod m;
                inc1 := increment(r[i].k);
                k := SearchMove((i+inc1) mod m, inc1, level-j-1);
                if k>-1 then begin
                    {*** A hole was found, move forward ***}
                    r[k] := r[i];
                    SearchMove := i;
                    goto 999 {*** return ***}
                    end
                end;
            {*** Could not find hole ***}
            SearchMove := -1;
            end;
        999:
        end;

    begin
    init := hashfunction(key);
    inc  := increment(key);
    i := 0;      j := -1;
    while (i<=n) and (j<0) and (n<m) do begin
        j := SearchMove(init, inc, i);
        i := i+1
        end;
    if j>-1 then begin
        {*** A hole was found, insert key ***}
        r[j].k := key;
        n := n+1
        end
    else Error    {*** table is full ***};
    end;
```

Binary tree reorganization hashing: movement of entries

```
    function SearchMove (init, inc, level : integer) : integer;
    {*** Find the first hole (empty location) at the given depth
        in the binary tree spanned by a key ***}
    label   999;
```

```
var i, incl, j, k : integer;
begin
i := (init + inc*level) mod m;
if empty(r[i]) or deleted(r[i]) then SearchMove := i
else begin
    for j:=level-1 downto 0 do begin
        i := (init + inc*j) mod m;
        incl := increment(r[i].k);
        k := SearchMove((i+incl) mod m, incl, level-j-1);
        if k>-1 then begin
            {*** A hole was found, move forward ***}
            r[k] := r[i];
            SearchMove := i;
            goto 999 {*** return ***}
            end
        end;
    {*** Could not find hole ***}
    SearchMove := -1;
    end;
999:
end;
```

The above algorithm will not detect the insertion of duplicates, that is, elements already present in the table.

This reorganization scheme significantly reduces the number of accesses for a successful search at the cost of some additional effort during the insertion of new keys. This algorithm is very suitable for building static tables, which will be searched often.

Table 3.15 summarizes simulation results for the binary tree hashing reorganization scheme. The column headed by I_n counts the average number of elements accessed to insert a new key in the table. I_n gives an accurate idea of the cost of the reorganization. Note that the expected length of the longest probe sequence (L_n) is very short. On the other hand, the cost of inserting new elements is particularly high for full or nearly full tables. The simulation results are in excellent agreement with the predicted theoretical results.

References:
[Gonnet, G.H. *et al.*, 77], [Mallach, E.G., 77], [Rivest, R.L., 78], [Gonnet, G.H. *et al.*, 79], [Lyon, G.E., 79], [Madison, J.A.T., 80].

3.3.8.3 Last-come-first-served hashing

In open-addressing hashing, a new element is usually inserted in the first empty location found in its probe sequence (or first-come-first-served). The

Table 3.15: Simulation results for binary tree hashing.

$m = 101$				
n	C_n	L_n	I_n	M_n
51	1.27475±.00005	2.9310±.0004	1.48633±.00011	0.061774±.000023
81	1.55882±.00008	4.3938±.0007	2.56876±.00038	0.165760±.000039
91	1.72359±.00010	5.2899±.0010	3.83135±.00085	0.228119±.000049
96	1.84624±.00011	6.0181±.0013	5.6329±.0019	0.273611±.000058
100	1.99963±.00017	7.0822±.0022	12.837±.014	0.327670±.000082
101	2.06167±.00023	7.6791±.0034	31.54±.29	0.34760±.00011
$m = 4999$				
2499	1.28485±.00005	4.3213±.0026	1.49835±.00012	0.063668±.000024
3999	1.57955±.00008	6.6825±.0051	2.62862±.00040	0.171101±.000041
4499	1.75396±.00010	8.1678±.0071	3.98929±.00092	0.236601±.000052
4749	1.88698±.00013	9.4163±.0094	6.0202±.0021	0.285576±.000063
4949	2.06221±.00019	11.403±.016	15.729±.017	0.347749±.000093
4999	2.14844±.00067	13.344±.069	495±49	0.37645±.00032

last-come-first-served (LCFS) technique exchanges the new element with the first element in its probe sequence, if there is a collision. The displaced key is then considered the new element, and the insertion continues in the same fashion. Therefore, an element stored in a table location is always the *last* one to have probed there.

Assuming random probing, we have

$$E[A_n] = C_n = \frac{m}{n}(H_m - H_{m-n})$$

$$\begin{aligned}\sigma^2(A_n) &= \frac{(m-1-C_n(m-n+1))C_n}{m+1} + \frac{m^2}{n(m+1)}(H_m^{(2)} - H_{m-n}^{(2)}) \\ &\approx -\frac{\ln(1-\alpha)}{\alpha} - \frac{1-\alpha}{\alpha^2}\ln(1-\alpha) + O(1/m)\end{aligned}$$

$$\sigma^2(A_m) = \ln m + \gamma + \frac{\pi^2}{6} + O\left(\frac{\ln^2 m}{m}\right)$$

$$E[L_n] \le 1 + \Gamma^{-1}(\alpha m)\left(1 + \frac{\ln\ln(1/(1-\alpha))}{\ln\Gamma^{-1}(\alpha m)} + O\left(\frac{1}{\ln^2\Gamma^{-1}(\alpha m)}\right)\right)$$

where $\alpha = n/m$. In comparison with random probing, the successful search time is the same, but the variance is logarithmic instead of linear.

We can take advantage of this small variance by doing a **centred search**. That is, instead of searching the probe sequence $h_1, h_2, ...$, we search the probe sequence in decreasing probability order, according to the probability of finding the key in the ith location of the sequence. For LCFS hashing, the probability distribution is a positive Poisson distribution with parameter $\lambda = -\ln(1-\alpha)$. Instead of following the optimal order, it is simpler to use a **mode-centred search**. In this case the mode is $d = \max(1, \lfloor\lambda\rfloor)$. Thus, we search the probe sequence in the order $d,\ d+1,\ d-1,\ d+2,\ d-2, ...,\ 2d-1,\ 1,\ 2d,\ 2d+1, ...$. For $\alpha < 1 - e^2 \approx 0.86466$, mode-centred search is equivalent to the standard search. For $\alpha \geq 1 - e^2$, we have

$$C_n < 1 - \frac{2(1-\alpha)}{\alpha}\lfloor -\ln(1-\alpha)\rfloor + \frac{2(10e^{-3}+1)}{\sqrt{3}\alpha}\sqrt{-\ln(1-\alpha)}$$

This bound is not tight, but shows that C_n for mode-centred search is roughly the square root of that of the standard algorithm when the table becomes full.

A generalization of LCFS hashing is to probe s times before displacing a stored key. In this case, the optimal s is $s = \lfloor -\alpha^{-1}\ln(1-\alpha)\rfloor$. However, the variance only decreases by a constant smaller than 1.473, for any α.

A disadvantage of LCFS, is that the number of data movements is larger than in random probing.

References:
[Cunto, W. *et al.*, 88], [Poblete, P.V. *et al.*, 89].

3.3.8.4 Robin Hood hashing

Robin Hood hashing is another technique used to reduce the variance of the expected successful search time. During an insertion, when two keys collide, the key that has probed the most number of locations stays in that position, and the other continues probing. The name of this method reflects the fact that the key with the longer probe sequence (the poor) wins over the other key (the rich). Under random probing, we have the following results

$$E[A_n] = C_n = \frac{m}{n}(H_m - H_{m-n})$$

$$\sigma^2(A_n) < \sigma^2(A_m) \leq 1.883$$

$$E[L_n] < 3C_n + \lceil\ln(m-2)\rceil \quad \text{for } n \leq m$$

As for LCFS, we can replace the standard search by a centred search. For the optimal order we have $C_n \leq 2.57$. Using a **mean-centred search** we have $C_n \leq 2.84$.

A disadvantage of Robin Hood hashing is that during an insertion we have to compute the length of the probe sequence for one of the keys. This can be done by traversing the probe sequence of that key until the current location is found. For double hashing, this can also be obtained by performing a division over a finite field.

References:
[Celis, P. *et al.*, 85], [Celis, P., 85], [Celis, P., 86].

3.3.8.5 Self-adjusting hashing

This family of algorithms applies the ideas of self-organizing lists (Sections 3.1.2 and 3.1.3) to hashing. In any form of hashing with chaining, we can directly apply the move-to-front or transpose methods to every chain during a successful search (see Sections 3.3.10 to 3.3.12). This is a form of composition, as described in Section 2.2.2.

In the case of open-addressing techniques, during a successful search, we can exchange elements with the same hashing value in the probe sequence of the search key (using either the move-to-front or the transpose method). In the particular case of linear probing (Section 3.3.4), the condition of having the same hashing value is not needed because the probe sequence for all keys follows the same pattern. Although the move-to-front technique may work well for some cases, it is better to use the transpose technique. For the latter case, simulations show that the average search time improves for the Zipf and 80%–20% probability distributions, using either linear probing (Section 3.3.4) or random probing (Section 3.3.3).

This technique can also be combined with split-sequence hashing (Section 3.3.7). However, the improvements are modest compared with the complexity of the code.

References:
[Pagli, L., 85], [Wogulis, J., 89].

3.3.9 Optimal hashing

Optimal hashing based on double hashing (see Section 3.3.5) or uniform probing (see Section 3.3.2) is the logical conclusion of the previous reorganization algorithms. Two complexity measures can be minimized: the average number of probes (C_n^{Opt}), or the number of accesses in the longest probe sequence (L_n^{Opt}).

The insertion algorithm is translated into an assignment problem (assignment of keys into table locations) and the cost of each assignment of K_i to location j is the number of probes necessary to locate the key K_i into location j.

For the minimax arrangement for random probing (see Section 3.3.3) we have

$$\ln m + \gamma + \frac{1}{2} + o(1) \leq E[L_m]$$

$$\lim_{m \to \infty} Pr\{L_m \leq 4 \ln m\} = 1$$

For the minimax arrangement for uniform probing (see Section 3.3.2), we have the lower bound

$$\sum_{i=0}^{m-1}(-1)^{m-i-1}\binom{m}{i}\sum_{k\geq 0}\left(\frac{i^{\underline{k}}}{m^{\underline{k}}}\right)^m \leq E[L_m]$$

$$\lceil -\alpha^{-1}\ln(1-\alpha)\rceil \leq E[L_n]$$

$$E[L_n] \leq 1+\Gamma^{-1}(n)\left(1+\frac{\ln\ln(1/(1-\alpha))}{\ln\Gamma^{-1}(n)}+O\left(\frac{1}{\ln^2\Gamma^{-1}(n)}\right)\right)$$

For the minimum-average arrangement for random probing (see Section 3.3.3) and for uniform probing (see Section 3.3.2) we have:

$$1.688382... \leq C_m = O(1)$$

These optimal algorithms are mostly of theoretical interest. The algorithms to produce these optimal arrangements may require $O(m)$ additional space during the insertion of new elements.

Tables 3.16 and 3.17 show some simulation results on optimal arrangements.

Table 3.16: Simulation results for optimal hashing (minave).

n	m	α	C_n^{Opt}	L_n
798	997	80%	1.4890±0.0041	4.40±0.11
897	997	90%	1.6104±0.0043	5.147±0.089
947	997	95%	1.6892±0.0059	5.68±0.12
987	997	99%	1.7851±0.0058	6.77±0.13
19	19	100%	1.729±0.011	4.385±0.071
41	41	100%	1.783±0.011	5.29±0.11
101	101	100%	1.798±0.011	6.30±0.18
499	499	100%	1.824±0.011	7.92±0.36
997	997	100%	1.8279±0.0064	8.98±0.38

References:
[Gonnet, G.H. *et al.*, 77], [Gonnet, G.H., 77], [Lyon, G.E., 78], [Gonnet, G.H. *et al.*, 79], [Gonnet, G.H., 81], [Krichersky, R.E., 84], [Yao, A.C-C., 85], [Poblete, P.V. *et al.*, 89].

3.3.10 Direct chaining hashing

This method makes use of both hashing functions and sequential lists (chains) in the following way. The hashing function first computes an index into the

Table 3.17: Simulation results for optimal hashing (minimax).

n	m	α	C_n	L_n^{Opt}
399	499	80%	1.4938±0.0067	3.000±0.030
449	499	90%	1.6483±0.0079	3.050±0.043
474	499	95%	1.6995±0.0070	3.990±0.020
494	499	99%	1.7882±0.0077	5.120±0.089
19	19	100%	1.749±0.011	3.929±0.062
41	41	100%	1.796±0.010	4.665±0.088
101	101	100%	1.807±0.010	5.53±0.14
499	499	100%	1.8300±0.0081	7.38±0.29

hashing table using the record key. This table location does not hold an actual record, but a pointer to a linked list of all records which hash to that location. This is a composition of hashing with linked lists. The data structure used by this algorithm is described by

$$\{\mathbf{s} - \mathbf{D}\}_0^{\mathbf{N}}$$

where $\mathbf{s} - \mathbf{D}$ represents a linked list of data elements $\mathbf{D}$. Let P_n and P_n' be random variables which represent the number of pointers (chain links) inspected for the successful and unsuccessful searches respectively. Thus

$$P_n = A_n, \quad P_n' = A_n' + 1 .$$

The pertinent facts about this algorithm are listed below:

$$Pr\{chain\ with\ length\ i\} = \binom{n}{i} \frac{(m-1)^{n-i}}{m^n}$$

$$E[A_n] = C_n = 1 + \frac{n-1}{2m} \approx 1 + \frac{\alpha}{2}$$

$$\sigma^2(A_n) = \frac{(n-1)(n-5)}{12m^2} + \frac{n-1}{2m} \approx \frac{\alpha^2}{12} + \frac{\alpha}{2}$$

$$E[A_n'] = C_n' = \frac{n}{m}$$

$$\sigma^2(A_n') = \frac{n(m-1)}{m^2} \approx \alpha$$

$$E[L_n] = \Gamma^{-1}(m)\left(1 + \frac{\ln \alpha}{\ln \Gamma^{-1}(m)} + O\left(\frac{1}{\ln^2 \Gamma^{-1}(m)}\right)\right)$$

$$E[L_m] = \Gamma^{-1}(m) - \frac{3}{2} + \frac{\gamma - 1}{\ln \Gamma^{-1}(m)} + O\left(\frac{1}{\ln m}\right) + Q(\ln\ln m)$$

where $Q(x)$ is a periodic function of x and very small in magnitude.

Let S_r and S_p be the size of a record and the size of a pointer, then the expected storage used, $E[S_n]$, is

$$E[S_n] \; = \; (m+n)S_p \; + \; nS_r$$

Whenever

$$\frac{S_p}{S_r + S_p} < (1 - 1/m)^n \approx e^{-\alpha},$$

this algorithm uses less storage than separate chaining hashing (see Section 3.3.11).

Descriptions of the search and insert algorithms are given below. For this algorithm, we will not use r, the array of records, but *ptrs* an array of heads of linked lists. The nodes of the linked list are the ones which contain the keys.

Direct chaining hashing: search

```
datarecord *search(key, ptrs)
typekey key;  datarecord *ptrs[ ];

{ int i, last;
  datarecord *p;
p = ptrs[hashfunction(key)];
while (p!=NULL && key!=p->k)   p = p->next;
return(p);
}
```

Direct chaining hashing: insertion

```
void insert(key, ptrs)
typekey key;  datarecord *ptrs[ ];

{ extern int n;
  int i;
i = hashfunction(key);
ptrs[i] = NewNode(key, ptrs[i]);
n++;
}
```

The above algorithm will not detect the insertion of duplicates, that is, elements already present in the table.

The direct chaining method has several advantages over open-addressing schemes. It is very efficient in terms of the average number of accesses for both successful and unsuccessful searches, and in both cases the variance of the number of accesses is small. L_n grows very slowly with respect to n.

Unlike the case with open-addressing schemes, contamination of the table because of deletions does not occur; to delete a record all that is required is an adjustment in the pointers of the linked list involved.

Another important advantage of direct chaining is that the load factor α can be greater than 1; that is, we can have $n > m$. This makes the algorithm a good choice for dealing with files which may grow beyond expectations.

There are two slight drawbacks to the direct chaining method. The first is that it requires additional storage for the $(m+n)$ pointers used in linking the lists of records. The second is that the method requires some kind of memory management capability to handle allocation and deallocation of list elements.

This method is very well suited for external search. In this case we will likely keep the array of pointers in main memory. Let E_n^b be the expected number of buckets accessed when direct chaining hashing is used in external storage with bucket size b. Then

$$\alpha = n/m$$

$$\begin{aligned} E_n^b &= \frac{n-1+m(b+1)}{2bm} + \frac{m(b^2-1)}{12bn} \\ &\quad + \frac{m}{bn}\sum_{j=1}^{b-1} \frac{\omega_j}{(1-\omega_j)^2}\left(\frac{m+\omega_j-1}{m}\right)^n \\ &= \frac{\alpha+b+1}{2b} + \frac{b^2-1}{12\alpha b} + O(k^{-\alpha}) \qquad (k>1) \end{aligned}$$

where $\omega_j = e^{\frac{2\pi i j}{b}}$ is a root of unity.

$$E_n^2 = \frac{n-1}{4m} + \frac{3}{4} + \frac{m}{8n}(1-(1-2/m)^n)$$

References:

[Morris, R., 68], [Tai, K.C. *et al.*, 80], [Gonnet, G.H., 81], [Knott, G.D., 84], [Vitter, J.S. *et al.*, 85], [Graham, R.L. *et al.*, 88], [Knott, G.D. *et al.*, 89].

3.3.11 Separate chaining hashing

This method uses a hashing function to probe into an array of keys and pointers. Collisions are resolved by a linked list starting at the table entry.

The data structure used is described by $\{\mathbf{1} - \mathbf{D}\}_0^{\mathbf{N}}$ where

$$\mathbf{l-D} \;:\; (\mathbf{D},[\mathbf{l-D}]);\; (\mathbf{D},\mathbf{nil})\,.$$

Let A_n and A_n' denote the number of accesses to records $\mathbf{l-D}$. The pertinent facts about this algorithm are

$$E[A_n'] \;=\; C_n' \;=\; \frac{n}{m} + (1-1/m)^n \;\approx\; \alpha + e^{-\alpha}$$

$$\sigma^2(A_n') \;=\; \frac{n(m-1)}{m^2} + \frac{m-2n}{m}(1-1/m)^n - (1-1/m)^{2n}$$
$$\approx\; \alpha + (1-2\alpha)e^{-\alpha} - e^{-2\alpha}$$

The values for A_n, L_n and L_m coincide with those for direct chaining hashing (see Section 3.3.10).

Let S_r and S_p be the size of a record and the size of a pointer, then the expected storage used, $E[S_n]$, is

$$E[S_n] \;=\; (n + m(1-1/m)^n)(S_r + S_p) \;\approx\; m(\alpha + e^{-\alpha})(S_r + S_p)$$

Whenever

$$\frac{S_p}{S_p + S_r} > (1-1/m)^n \approx e^{-\alpha}$$

this algorithm uses less storage than direct chaining hashing (see Section 3.3.10).

Descriptions of the search and insert algorithms are given below.

Separate chaining hashing: search

```
datarecord *search(key, r)
typekey key;  dataarray r;

{ datarecord *p;
p = &r[hashfunction(key)];
while (p!=NULL && key!=p ->k)   p = p ->next;
return(p);
}
```

Separate chaining hashing: insertion

```
void insert(key, r)
typekey key;  dataarray r;
```

```
{ extern int n;
  int i;
i = hashfunction(key);
if (empty(r[i])) /*** insert in main array ***/
        r[i].k = key;
    else /*** insert in new node ***/
        r[i].next = NewNode(key, r[i].next);
n++;
}
```

The above algorithm will not detect the insertion of duplicates, that is, elements already present in the table.

This method has several advantages over open-addressing schemes. It is very efficient in terms of the average number of accesses for both successful and unsuccessful searches, and in both cases the variance of the number of accesses is small. The length of the longest probe sequence, that is to say, the actual worst-case, grows very slowly with respect to n.

Unlike open-addressing schemes, contamination of the table because of deletions does not occur.

The load factor can go beyond 1 which makes the algorithm a good choice for tables that may grow unexpectedly.

This method requires some extra storage to allocate space for pointers. It also requires a storage allocation scheme to allocate and return space for records.

As mentioned in Section 3.3.8.5, it is possible to use self-organizing techniques on every chain. For separate chaining, using the transpose technique, we have

$$E[A_n] = C_n \approx \left(1 + \frac{\alpha}{2}\right) / \ln \alpha$$

where $\alpha = n/m > 1$.

Similarly, the split-sequence technique mentioned in Section 3.3.7 can be applied to separate chaining. That is, when we search for a key k, we first compare it with the key k' stored in location $h(k)$. If $k = k'$ or $h(k)$ is empty, the search terminates. Otherwise, we follow one of two lists, depending on whether $k > k'$ or $k < k'$. For this we have

$$\begin{aligned} E[A_n] = C_n &= \frac{1}{3}\left(\frac{n-1}{m} + 4 - \frac{m}{n}\left(1 - \left(1 - \frac{1}{m}\right)^n\right)\right) \\ &\approx \frac{\alpha^2 + 4\alpha - 1 + e^{-\alpha}}{3\alpha} \end{aligned}$$

$$\begin{aligned} E[A'_n] = C'_n &= \frac{1}{2}\left(\left(1 - \frac{1}{m}\right)^n + \frac{n}{m} + 1\right) \\ &\approx \frac{1}{2}\left(\alpha + e^{-\alpha} + 1\right) \end{aligned}$$

References:
[Johnson, L.R., 61], [Morris, R., 68], [Olson, C.A., 69], [Bookstein, A., 72], [van der Pool, J.A., 72], [Bays, C., 73], [Gwatking, J.C., 73], [Knuth, D.E., 73], [van der Pool, J.A., 73], [Behymer, J.A. *et al.*, 74], [Devillers, R. *et al.*, 79], [Quittner, P. *et al.*, 81], [Larson, P., 82], [Norton, R.M. *et al.*, 85], [Ramakrishna, M.V., 88], [Sedgewick, R., 88].

3.3.12 Coalesced hashing

Coalesced hashing is a hashing algorithm which resolves collisions by chaining. The chain pointers point to elements of the hashing array itself. The data structure used by this algorithm is described by $\{\mathbf{D}, \mathbf{int}\}_0^{\mathbf{N}}$ where the **int** is taken to be a 'pointer' to the next element in the chain (an index into the array). The name 'coalesced hashing' comes from the fact that colliding records are stored in the main table, and keys hashing to other locations may share part of a chain.

The complexity measures for this algorithm are:

$$E[A_n] = C_n = 1 + \frac{m}{8n}\left((1+2/m)^n - 1 - \frac{2n}{m}\right) + \frac{n-1}{4m}$$
$$= 1 + \frac{1}{8\alpha}(e^{2\alpha} - 1 - 2\alpha) + \frac{\alpha}{4} + O(m^{-1})$$

$$\sigma^2(A_n) = -\frac{7}{36} + \frac{5\alpha}{24} - \frac{\alpha^2}{16} - \frac{37}{432\alpha} - \frac{1}{64\alpha^2}$$
$$+ \left(-\frac{1}{16} - \frac{1}{16\alpha} + \frac{1}{32\alpha^2}\right)e^{2\alpha} + \frac{4e^{3\alpha}}{27\alpha} - \frac{e^{4\alpha}}{64\alpha^2} + O(m^{-1})$$

$$E[A'_n] = C'_n = 1 + \frac{1}{4}\left((1+2/m)^n - 1 - \frac{2n}{m}\right)$$
$$= 1 + \frac{1}{4}\left(e^{2\alpha} - 1 - 2\alpha\right) + O(m^{-1})$$

$$\sigma^2(A'_n) = \frac{35}{144} - \frac{\alpha}{12} - \frac{\alpha^2}{4} + \frac{2\alpha - 5}{8}e^{2\alpha} + \frac{4}{9}e^{3\alpha} - \frac{e^{4\alpha}}{16} + O(m^{-1})$$

Descriptions of the search and insert algorithms are given below. The insertion algorithm uses the variable *nextfree* to avoid a repetitive search of the table for empty locations. This variable should be initialized to $m - 1$ before starting to fill the table.

Coalesced hashing: search

```
int search(key, r)
typekey key; dataarray r;

{ int i;
i = hashfunction(key);
while (i!=(-1) && !empty(r[i]) && r[i].k!=key)  i = r[i].next;
if (i==(-1) || empty(r[i]))  return(-1);
     else                    return(i);
}
```

Coalesced hashing: insertion

```
void insert(key, r)
typekey key; dataarray r;

{ extern int n, nextfree;
  int i;

i = hashfunction(key);
if (empty(r[i])) {
     r[i].k = key;
     r[i].next = (-1);
     n++;
     }
else { /*** Find end of chain ***/
     while (r[i].next!=(-1) && r[i].k!=key)  i = r[i].next;
     if (r[i].k==key)  Error /*** key already in table ***/;
     else {
          /*** Find next free location ***/
          while (!empty(r[nextfree]) && nextfree>=0)  nextfree--;
          if (nextfree<0)  Error /*** Table is full ***/;
          else {
               r[i].next = nextfree;
               r[nextfree].k = key;
               r[nextfree].next = (-1);
               n++;
               }
          }
     }
}
```

Coalesced hashing is an efficient internal hashing algorithm at the cost of one integer (pointer to an array) per entry. Average and variance values for the successful and unsuccessful cases are very low.

This algorithm has some of the advantages of chaining methods without the need for dynamic storage allocation.

Owing to the use of the variable *nextfree* the first collisions will fill the top of the table more quickly than the rest. This observation leads to a variation of the algorithm called **coalesced hashing with cellar**. In this variation we leave the top part of the table (called the 'cellar') to be filled only by collisions. The hashing function is now restricted to generate addresses in the lower part ('address region') of the table. The algorithms to perform searching and insertion with cellars are identical to the above; the only difference lies in the *hashfunction* which now generates integers over a restricted range.

Let us call m' the total size of the table (m 'address' entries and $m' - m$ 'cellar' entries). Let

$$\alpha = n/m$$

$$\beta = m/m'$$

and λ be defined as the positive solution of $e^{-\lambda} + \lambda = 1/\beta$. Then the complexity measures become:

$$E[A_n] = C_n = 1 + \frac{\alpha}{2} \qquad \text{if } \alpha \le \lambda$$

$$= 1 + \frac{1}{8\alpha}\left(e^{2(\alpha-\lambda)} - 1 - 2(\alpha - \lambda)\right)(3 - 2/\beta + 2\lambda)$$

$$+ \frac{\alpha + 2\lambda - \lambda^2/\alpha}{4} + O\left(\frac{\log m'}{\sqrt{m'}}\right)$$

otherwise;

$$E[A_n'] = C_n' = \alpha + e^{-\alpha} \qquad \text{if } \alpha \le \lambda$$

$$= \frac{1}{\beta} + \frac{1}{4}\left(e^{2(\alpha-\lambda)} - 1\right)(3 - 2/\beta + 2\lambda)$$

$$- \frac{\alpha - \lambda}{2} + O\left(\frac{\log m'}{\sqrt{m'}}\right)$$

otherwise.

For every value of α we could select an optimal value for β which minimizes either the successful or unsuccessful case. The value $\beta = 0.853...$ minimizes the successful case for a full table while $\beta = 0.782...$ does similarly for the unsuccessful case. The value $\beta = 0.86$ appears to be a good compromise for both cases and a wide range of values for α.

References:
[Williams, F.A., 59], [Bays, C., 73], [Knuth, D.E., 73], [Banerjee, J. *et al.*, 75], [Vitter, J.S., 80], [Vitter, J.S., 80], [Vitter, J.S., 81], [Greene, D.H. *et al.*, 82], [Vitter, J.S., 82], [Vitter, J.S., 82], [Chen, W-C. *et al.*, 83], [Vitter, J.S., 83], [Chen, W-C. *et al.*, 84], [Knott, G.D., 84], [Chen, W-C. *et al.*, 86], [Murthy, D. *et al.*, 88].

3.3.13 Extendible hashing

Extendible hashing is a scheme which allows the hashing table to grow and guarantees a maximum of two external accesses per record retrieved. This scheme is best understood in terms of external storage searching. The structure is composed of a **directory** and **leaf-pages**.

$$\textbf{directory} : (\mathbf{N}, \{[\mathbf{leafpage}]\}_{\mathbf{0}}^{\mathbf{2^N-1}}).$$

$$\textbf{leafpage} : (\mathbf{int}, \{\mathbf{KEY}\}_{\mathbf{1}}^{\mathbf{b}}).$$

where the directory consists of a set of pointers to leaf-pages and the leaf-pages are buckets of size b with an additional depth indicator. Both directory and leaf depth indicators show how many bits of the hashing address are being considered; that is, at depth d the hashing function returns a value in $0, ..., 2^d - 1$. The depth of the leaf-pages is always less than or equal to the depth of the directory. Several directory entries may point to the same leaf-page.

Basically this algorithm uses a composition of a hashing step with sequential search in buckets. Every key resides in the bucket pointed by the directory entry indexed by the hashing value of the key. Collisions (overflow in the buckets) are handled by duplicating the directory.

Let $D_b(n)$ be the expected depth of the directory, m_d be the number of entries in the directory, and m_b be the number of leaf-pages (buckets). We will assume that the number of keys is random, Poisson distributed with expected value n, then

$$d(n) = (1 + 1/b) \log_2 n$$

$$D_b(n) = \sum_{k \geq 0} \left\{ 1 - \left(\sum_{j=1}^{b} e^{-n2^{-k}} \frac{(n2^{-k})^j}{j!} \right)^{2^k} \right\}$$

$$= d(n) + \frac{\gamma - \ln((b+1)!)}{b \ln 2} + Q_1(d(n)) + o(1)$$

$$E[m_d] = 1 + \sum_{k \geq 0} 2^k \left\{ 1 - \left(\sum_{j=1}^{b} e^{-n2^{-k}} \frac{(n2^{-k})^j}{j!} \right)^{2^k} \right\}$$

$$= \left(\frac{\Gamma(1-1/b)}{\ln 2((b+1)!)^{1/b}} + Q_2(d(n))\right) n^{1+1/b}(1+o(1))$$
$$\approx \frac{3.92}{b} n^{1+1/b}$$

and

$$E[m_b] = 1 + \sum_{k\geq 0} 2^k \sum_{j>b} e^{-n2^{-k}} \frac{(n2^{-k})^j}{j!}$$
$$= \frac{n}{b}\left(\frac{1}{\ln 2} + Q_3(\log_2 n)\right) + O(n^{1/2}\log n)$$

The functions $Q_i(x)$ are complicated periodic functions with period 1 and average value 0 (that is, $\int_0^1 Q_i(x)\,dx = 0$).

Extendible hashing search

```
i := hashfunction(key) mod md;
read–directory–entry(i) into npage;
read–leaf–page(npage) into r;
i := 1;
while (i<b) and (r[i].k <> key) do i := i+1;
if r[i].k = key then     found(r[i])
    else     notfound(key);
```

The insertions are straightforward if the corresponding bucket is not full. When an insertion causes a bucket to overflow, this bucket will be split into two and its depth increased by one. All the records in the bucket are then rehashed with the new depth. Some of the pointers in the directory pointing to the splitting bucket may have to be redirected to the new bucket. If the depth of the bucket exceeds the depth of the directory, the directory is duplicated and its depth increased by one. Duplicating the directory implies a simple copy of its contents; no buckets are split. Certainly most buckets will be pointed to by two or more directory entries after the directory duplicates.

Assuming the existence of a fixed hash function $h_I(K)$, which returns an integer in a sufficiently large interval, the hashing function for level d can be implemented as

$$h_I(K) \bmod 2^d$$

This method allows graceful growth and shrinkage of an external hashing table. Assuming that the directory cannot be kept in internal storage, this method guarantees access to any record in two external accesses. This makes it a very good choice for organizing external files.

In case the directory can be kept in main memory, we can access records with a single external access, which is optimal.

The directory is $O(b^{-1}n^{1+1/b})$ in size. This means that for very large n or for relatively small bucket sizes, the directory may become too large. It is not likely that such a directory can be stored in main memory.

Insertions may be direct or may require the splitting of a leaf-page or may even require the duplication of the directory. This gives a bad worst-case complexity for insertion of new records.

Deletions can be done easily by marking or even by 'folding' split buckets. Shrinking of the directory, on the other hand, is very expensive and may require $O(n)$ overhead to every deletion is some cases.

Table 3.18 gives numerical values for several measures in extendible hashing with Poisson distributed keys, for two different bucket sizes.

Table 3.18: Exact values for extendible hashing.

	$b = 10$			$b = 50$		
n	$D_b(n)$	$E[m_d]$	$E[m_b]$	$D_b(n)$	$E[m_d]$	$E[m_b]$
100	4.60040	25.8177	14.4954	1.71109	3.42221	2.92498
1000	8.45970	374.563	144.022	5.02284	32.7309	31.0519
10000	12.1860	4860.14	1438.01	8.99995	511.988	265.644
100000	16.0418	68281.7	14492.6	12.0072	4125.43	2860.62

References:

[Fagin, R. *et al.*, 79], [Yao, A.C-C., 80], [Regnier, M., 81], [Scholl, M., 81], [Tamminen, M., 81], [Flajolet, P. *et al.*, 82], [Lloyd, J.W. *et al.*, 82], [Tamminen, M., 82], [Burkhard, W.A., 83], [Flajolet, P., 83], [Lomet, D.B., 83], [Lomet, D.B., 83], [Bechtald, U. *et al.*, 84], [Mullen, J., 84], [Kawagoe, K., 85], [Mullin, J.K., 85], [Ouksel, M., 85], [Tamminen, M., 85], [Veklerov, E., 85], [Enbody, R.J. *et al.*, 88], [Salzberg, B., 88], [Sedgewick, R., 88], [Weems, B.P., 88], [Henrich, A. *et al.*, 89].

3.3.14 Linear hashing

Linear hashing is a scheme which allows the hashing table to grow or shrink as records are inserted or deleted. This growth or shrinkage is continuous, one entry at a time, as opposed to extendible hashing (see Section 3.3.13) where the directory may duplicate due to a single insertion.

This scheme is best understood in terms of external representations. An external **bucket** is a physical record, convenient for performing input/output operations which may contain up to b records.

$$\textbf{linear - hash - file} : \{\textbf{bucket}\}_0^{m-1}.$$

$$\textbf{bucket} : (\{\textbf{KEY}\}_1^b, \textbf{overflow}).$$

$$\textbf{overflow} : [\{\textbf{KEY}\}_1^{b_o}, \textbf{overflow}] ; \textbf{nil}.$$

A bucket may receive more than b records, in which case the excess records are placed in a list of **overflow** buckets.

A file undergoes a **full expansion** when it grows from m_0 to $2m_0$. The process of growing from $2m_0$ to $4m_0$, and so on, is an exact replica of the first full expansion.

This algorithm requires a **control function** $m = g(z)$, which regulates the growth of the array based on the number of records per storage used. We will use a control function that guarantees a constant storage utilization. For the storage utilization we will also consider the overflow buckets, that is, we want to guarantee that

$$\frac{n}{b(m + m_{ov})} = \alpha \text{ or } m = \frac{n}{b\alpha} - m_{ov}$$

where m_{ov} is the number of overflow buckets.

The following formulas indicate the limiting behaviour of linear hashing, that is, the limit when $n, m \to \infty$ while n/m remains constant.

$$C_n = \frac{1}{z_0}\int_{z_0}^{2z_0} (g(z)s(z/2) + (1 - g(z))s(z))\, dz$$

$$C_n' = \frac{1}{z_0}\int_{z_0}^{2z_0} (g(z)u(z/2) + (1 - g(z))u(z))\, dz$$

$$E[m_{ov}] = \frac{1}{6bz_0^2}\int_{z_0}^{2z_0} (2g(z)t(z/2) + (1 - g(z))t(z))\, dz$$

where

$$s(z) = 1 + \frac{1}{zb}\sum_{k\geq 0}(k+1)\sum_{j=1}^{b_o}\left(\frac{kb_o}{2} + j\right) P(b + kb_o + j, z)$$

is the expected number of accesses for a successful search in a single chain with relative load z and similarly

$$u(z) = 1 + \sum_{k\geq 0}(k+1)\sum_{j=1}^{b_o} P(b + kb_o + j, z)$$

for the unsuccessful search and

$$t(z) = b_o\sum_{k\geq 0}(k+1)\sum_{j=1}^{b_o} P(b + kb_o + j, z) = b_o(u(z) - 1)$$

is the expected space taken by the overflow records of a single chain. $P(i,z)$ is a Poisson probability distribution with parameter zb:

$$P(i,z) = \frac{e^{-zb}(zb)^i}{i!}$$

and finally $g(z)$ is the control function resulting from the policy of constant total storage utilization:

$$g(z) = \frac{zb/\alpha - t(z) - b}{2t(z/2) - t(z) + b} \qquad z_0 = g^{-1}(0)$$

The hashing function for this algorithm depends on m and on m_0 as well as on the key. Each time that m increases by one, the hashing function changes, but this change is minimal in the following sense:

$$h_{m+1}(K) = \begin{cases} h_m(K) & \text{iff } h_m(K) \neq m - m_0 \\ m \text{ or } m - m_0 & \text{otherwise} \end{cases}$$

A common implementation of this function, assuming that we have available a basic hash function $h_I(K)$ which transforms the key into an integer in a sufficiently large interval, is:

Hashing function for linear hashing

```
i := hI(key);
if (i mod m0) < m-m0 then    hashfunction := i mod (2*m0);
        else    hashfunction := i mod m0;
```

The algorithm, as described, suffers from a discontinuity. At the beginning of a full expansion most chains are of the same expected length, while at the end of an expansion, the last chains to be expanded will have an expected length about twice the average. This problem may be remedied by splitting the full expansion into partial expansions. For example, we can expand first from m_0 to $3m_0/2$ using the entries from 0 to $m_0/2 - 1$ and from $m_0/2$ to $m_0 - 1$ and secondly from $3m_0/2$ to $2m_0$ using the entries from 0 to $(m_0/2) - 1$, $m_0/2$ to $m_0 - 1$ and m_0 to $(3m_0/2) - 1$. By doing partial expansions the discontinuities are much less pronounced.

Linear hashing is a descendant of **virtual hashing**. In virtual hashing the main difference is that the file is duplicated in size in a single step, when appropriate, rather than entry by entry.

Dynamic hashing is a term used to describe these type of algorithms which will grow/shrink the file while keeping roughly the same access cost. Dynamic hashing is also the name of another predecessor of linear hashing, an algorithm using the following data structure:

$$\textbf{directory} : \{\textbf{bucketbinarytrie}\}_0^{m-1}$$

$$\textbf{bucketbinarytrie} : [\{\textbf{key}\}_1^b]; [\{\textbf{bucketbinarytrie}\}_0^1]$$

where hashing is done at the directory level, and overflow in buckets produce a new internal node in the binary trie (see Section 3.4.4) with the corresponding bucket split.

These methods are supposed to be excellent methods for storing large tables which require quick access in external storage.

References:
[Larson, P., 78], [Litwin, W., 78], [Litwin, W., 79], [Larson, P., 80], [Litwin, W., 80], [Mullin, J.K., 81], [Scholl, M., 81], [Larson, P., 82], [Larson, P., 82], [Lloyd, J.W. *et al.*, 82], [Ramamohanarao, K. *et al.*, 82], [Ouksel, M. *et al.*, 83], [Kjellberg, P. *et al.*, 84], [Mullen, J., 84], [Ramamohanarao, K. *et al.*, 84], [Kawagoe, K., 85], [Larson, P., 85], [Larson, P., 85], [Ramamohanarao, K. *et al.*, 85], [Tamminen, M., 85], [Veklerov, E., 85], [Litwin, W. *et al.*, 86], [Robinson, J.T., 86], [Litwin, W. *et al.*, 87], [Enbody, R.J. *et al.*, 88], [Larson, P., 88], [Larson, P., 88], [Lomet, D.B., 88], [Ouksel, M. *et al.*, 88], [Salzberg, B., 88], [Baeza-Yates, R.A., 89].

3.3.15 External hashing using minimal internal storage

These algorithms are designed to work for external files. Under this assumption, all internal computations are viewed as insignificant when compared to an external access to the file. The goal is to minimize the number of external accesses, at the cost of maintaining some additional 'indexing' information in main memory.

The algorithms described in this section act as 'filters' on the external accesses of most other hashing algorithms (uniform probing, random probing, double hashing, ...). In other words, the searching is conducted as in the basic hashing algorithms, except that instead of accessing the external table directly, we first 'consult' the internal information. When an access to external storage is allowed, it is either guaranteed to succeed, or has a very high probability of succeeding.

These algorithms will use the **signature** of a key. A signature function is a hashing function that returns a sequence of bits. It can be viewed as returning a uniformly distributed real number in [0,1) and the sequence of bits is given by its binary representation.

The minimization of resources can be approached in two distinct ways:

(1) guarantee exactly one external access (optimal) for each key while minimizing the additional internal storage required; or

(2) given a fixed amount of internal storage, minimize the number of external accesses.

Let us call **k-prefix** the first k bits of the signature of a key. To solve the first problem we will construct the following table. For each table location we will code the following information: (1) the location is empty or (2) the location is occupied and the key stored in this location has a prefix of length k. The prefix stored is the shortest possible required to distinguish the stored key from all other keys that probe this location. Note that on average, only $C_n - 1$ other keys probe an occupied location. This algorithm requires building a table of variable length prefixes, hence we will call it **variable-length signatures**.

Let $m_b(n)$ be the average number of internal bits required by these algorithms; if the main hashing algorithm is uniform probing (see Section 3.3.2) or random probing (see Section 3.3.3) we have the following lower and upper bounds (the upper bounds represent the behaviour of the above algorithm):

$$\alpha = n/m$$

$$m_b(n) \geq \frac{n}{\ln 2}\left(\alpha + (1-\alpha)\ln(1-\alpha) - \int_0^\alpha \frac{x \ln x}{1-x}\,dx\right) + O(1)$$

$$m_b(n) \leq \log_2(-\ln(1-\alpha)) + O(1)$$

$$\frac{m\pi^2}{6\ln 2} - \frac{3\log_2 m}{2} + O(1) \leq m_b(m) \leq \log_2 \log_2 m + O(1)$$

A better lower bound is obtained for **memoryless** hashing algorithms. Let us call an algorithm memoryless if it does not store any information gained from previous probes, except the implicit fact that they failed. All the hashing algorithms in this section are memoryless.

$$\log_2 \beta + \frac{1}{\ln 2}\left(\frac{1}{2\beta} + \frac{5}{12\beta^2} + O(\beta^{-3})\right) + O\left(\frac{1}{m-n}\right) \leq m_b(n)$$

where $\beta = -\ln(1-\alpha)$ and

$$\log_2 H_m + O(\frac{1}{\ln m}) \leq m_b(m)$$

For the second problem we now restrict ourselves to using a fixed, small number, d, of bits per location. The goal is now to reduce the number of external accesses. If we store in each location the **d-prefix** of the stored key, we reduce the unnecessary accesses by a factor of 2^d. For this algorithm

$$C_n^d = 1 - 2^{-d}(\ln(1-\alpha)/\alpha + 1) + O\left(\frac{1}{m-n}\right)$$

This algorithm can be extended for external buckets containing b records each. For this extension, we cannot keep a signature of all the records in the bucket, but instead we keep a **separator**. A separator is a prefix long enough to distinguish between the signatures of the records which are stored in the bucket (lower prefixes) and those of the records which overflowed to other buckets (larger prefixes). This algorithm may displace records with high prefixes as records with smaller prefixes are inserted.

Finally, by selecting a fixed length separator and by moving forward records that would force a larger separator, an optimal and very economical algorithm is obtained. In this last case, there is a limit on the load of the external file. In other words, an insertion may fail although there is still room in the table (this happens when all the buckets are full or their separators are fully utilized).

Although these algorithms require internal tables, the actual sizes for real situations are affordable by almost any system. The reduction in the number of external accesses is very attractive. These methods are more economical in internal storage than extendible hashing (see Section 3.3.13) with an internally stored directory.

References:
[Larson, P. *et al.*, 84], [Gonnet, G.H. *et al.*, 88], [Larson, P., 88].

3.3.16 Perfect hashing

A perfect hashing function is a function which yields no collisions. Hence a single table probe is required, and all keys require exactly the same accessing time. Normally, the hashing function has to be tailored to the keys and consequently this algorithm is practical only for static sets of keys (such as keywords for a compiler). A **minimal perfect hashing function** is a perfect hashing function in which the number of keys is equal to the number of table entries ($n = m$).

For an arbitrary set of n keys, single-probe perfect hashing functions require $B_{n,m}$ bits of information (in the form of subtables or selected parameters of the function)

$$B_{n,m} = \log_2 \left(\frac{m^n}{m^{\underline{n}}} \right) = n \log_2 e \ + \ (m-n)\log_2(1-n/m) \ + \ O(1)$$

$$B_{m,m} = m \log_2 e \ + \ O(\log m)$$

Table 3.19 shows different functions that have been proposed for perfect hashing, where k is the key, assumed to be an integer and $a, b, c, ...$ are parameters to be chosen appropriately.

To construct a minimal perfect hashing function efficiently we will use an auxiliary integer array (A) of size m_2 which will store parameters of the hashing function.

The hashing function is $(A[k \bmod m_2] \ k) \bmod m$ where $m_2 \approx m$ and $\gcd(m, m_2) = 1$. This function uses a particular multiplicative factor for each

Table 3.19: Perfect hashing functions.

Hash function	*Comments*	*Reference*
$\lfloor (ak+b)/c \rfloor$	Depends too much on key distribution	[Sprugnoli, 77]
$\lfloor ((ak+b) \bmod c)/d \rfloor$	Good and practical for less than 100 keys	[Sprugnoli, 77]
$\lvert k\rvert + a[k_1] + a[k_{last}]$	Exponential time to compute the a table, may not work for some set of keys	[Cichelli, 80]
$\lfloor (a/(bk+c) \rfloor \bmod d$	Exponential time to compute a, and a may be of $O(n)$ size	[Jaeschke, 81]
$(ak \bmod b) \bmod m$	$O(n^3 \log n)$ to build, $m \approx 6n$	[Fredman *et al.*, 84]
$(k \bmod 2a + 100m + 1) \bmod m$	Uses an extra header table	[Cormack *et al.*, 85]
$(h_0(k) + g(h_1(k)) + g(h_2(k))) \bmod m$	Polynomial time for minimal function	[Sager, 85]
$(A[k \bmod a]\ k) \bmod m$	$O(m^2)$ building time, uses extra array of size a	Section 3.3.16

cluster of keys (all keys having the same $k \bmod m_2$ value form a cluster). The algorithm will use at most m_2 clusters (dimension of the array A).

Perfect hashing search

```
int search(key, r, A)
int key;  dataarray r;  int *A;
```

```
{ int i;
  extern int m, m2;
  i = hashfunction(A[key%m2], key);
  if(r[i].k == key) return(i);
  else              return(-1);
}
```

The insertion algorithm has to insert all keys at once. The building is done by inserting the largest clusters first and the smaller later. The insertion algorithm returns true or false depending on whether it could build the table for the given keys, and the integers m and m_2. If it could not, another m_2 can be tried. The probability of failure is $O(1/m)$.

Perfect hashing insertion

```
int insert(input, n, r, A)
dataarray input, r; int n, *A;

{ extern int m, m2;
  int d, i, ia, ib, iup, j;
  datarecord tempr;

  if(m < n) return(0);
  for(i=0; i<m2; i++) A[i] = 0;
  for(i=0; i<n; i++) A[input[i].k % m2]++;
  /* Shellsort input array based on collision counts */
  for (d=n; d>1;) {
     if (d<5)  d = 1;
        else   d = (5*d-1)/11;
     for (i=n-1-d; i>=0; i--) {
          tempr = input[i];
          ia = tempr.k % m2;
          for (j=i+d; j<n && (A[ia] < A[ib=input[j].k % m2] ||
                               A[ia] == A[ib] && ia > ib); j+=d)
               input[j-d] = input[j];
          input[j-d] = tempr;
          }
     }

  for(i=0; i<n; i=iup) {
     ia = input[i].k % m2;
     iup = i + A[ia];
```

```
        for(A[ia]=ib=1; ib < 9*m; A[ia] += ib++) {
            for(j=i; j<iup && empty(r[hashfunction(A[ia],input[j].k)]);
                j++) r[hashfunction(A[ia],input[j].k)] = input[j];
            if(j >= iup) break;
            for(j--; j >= i; j--)
                r[hashfunction(A[ia],input[j].k)].k = NOKEY;
            }
        if(ib >= 9*m)
            /* Cannot build optimal hashing table with m and m2 */
            return(0);
        }
    return(1);
}
```

References:
[Sprugnoli, R., 77], [Anderson, M.R. *et al.*, 79], [Tarjan, R.E. *et al.*, 79], [Cichelli, R.J., 80], [Jaeschke, G. *et al.*, 80], [Jaeschke, G., 81], [Yao, A.C-C., 81], [Mehlhorn, K., 82], [Bell, R.C. *et al.*, 83], [Du, M.W. *et al.*, 83], [Mairson, H.G., 83], [Chang, C.C., 84], [Fredman, M.L. *et al.*, 84], [Fredman, M.L. *et al.*, 84], [Yang, W.P. *et al.*, 84], [Cercone, N. *et al.*, 85], [Cormack, G.V. *et al.*, 85], [Larson, P. *et al.*, 85], [Sager, T.J., 85], [Yang, W.P. *et al.*, 85], [Aho, A.V. *et al.*, 86], [Berman, F. *et al.*, 86], [Chang, C.C. *et al.*, 86], [Dietzfelbinger, M. *et al.*, 88], [Gori, M. *et al.*, 89], [Ramakrishna, M.V. *et al.*, 89], [Schmidt, J.P. *et al.*, 89], [Brain, M.D. *et al.*, 90], [Pearson, P.K., 90], [Winters, V.G., 90].

3.3.17 Summary

Table 3.20 shows the relative total times for inserting 10007 random keys and performing 50035 searches (five times each key). We also include other searching algorithms, to compare them with hashing.

Table 3.20: Relative total times for searching algorithms.

Algorithm	*C*	*Pascal*
Sequential search in arrays		149
Sequential search in arrays (with sentinel)		90
Self-organizing (transpose)	182	153
Binary search	32	26
Interpolation search		26
Interpolation-sequential search		26
Linear probing hashing	2.4	1.4
Double hashing	2.3	1.4
Quadratic hashing		1
Ordered hashing		1.4
Brent's hashing	2.3	1.4
Binary tree hashing		1.5
Direct chaining hashing	1.2	
Separate chaining hashing	1	
Coalesced hashing	1.3	
Perfect hashing	47	

3.4 Recursive structures search

3.4.1 Binary tree search

The binary tree search is an algorithm for searching a lexicographically ordered binary tree. Without loss of generality we may assume that the left descendant nodes of any node contain keys whose values are less than or equal to the root, and that the right descendant nodes contain keys whose values are greater than the root.

Let A_n be the number of accesses (or node inspections) made in the course of a successful search for a given key in a binary tree of size n, and let A_n' be the number of accesses made in the course of an unsuccessful search of size n.

The symbol $h(n)$ denotes the **height** of a tree of size n, that is, the number of nodes in the longest path through the tree. With this definition, a null tree has height 0, a single node tree has height 1. The **depth** of a node in a tree is the distance from the root to that node; thus the depth of the root is 0.

Several variations on the basic binary tree structure arise with the introduction of semantic rules or constraints such as height balance, weight balance, or heuristic organization schemes. The Pascal data definition and search algorithm for all binary search trees are given below.

Tree definition

```
type tree = ↑node;
    node = record
      k : typekey;
      left, right : tree
      end;
```

Binary tree search

```
  procedure search(key : typekey; t : tree);
begin
if t=nil then {*** Not Found ***}
      notfound(key)
else if t↑.k = key then {*** Found ***}
      found(t↑)
else if t↑.k < key then search(key, t↑.right)
    else                search(key, t↑.left)
end;
```

The number of different binary trees with n nodes is

$$t_n = \frac{1}{n+1}\binom{2n}{n}$$

and the associated generating function is:

$$T(z) = \sum_{n\geq 0} t_n z^n = 1 + zT^2(z) = \frac{1-\sqrt{1-4z}}{2z}$$

The internal path length, I_n, of a tree with n nodes is defined as the sum of the depths of all its nodes. The external path length, E_n, of a tree is the sum of the depths of all its external nodes. For any binary tree

$$E_n = I_n + 2n\ .$$

We have

$$n(\log_2 n + 1 + \theta - 2^\theta) \leq E_n \leq \frac{n(n+1)}{2} - 1$$

where $\theta = \lceil \log_2 n \rceil - \log_2 n$ $(0 \leq \theta \leq 1)$. If Δ is the maximal path length difference in the tree (that is, the number of levels between the deepest and shallowest external node), then

$$E_n \leq n(\log_2 n + \Delta - \log_2 \Delta - \Psi(\Delta))$$

where

$$\Psi(\Delta) = \log_2 \mathrm{e} - \log_2 \log_2 \mathrm{e} - o(1) \geq 0.6622 \ .$$

This bound is tight to an $O(n)$ term for $\Delta \leq \sqrt{n}$.

Let a_k be the expected number of nodes at depth k and let b_k be the expected number of external nodes at depth k in a binary tree with n nodes. Then we have the associated generating functions

$$A(z) \ = \ \sum_k a_k z^k$$

$$B(z) \ = \ \sum_k b_k z^k \ = \ (2z-1)A(z) + 1$$

and

$$A(1) \ = \ B(1) - 1 \ = \ n$$

$$A'(1) \ = \ E[I_n]$$

$$B'(1) \ = \ E[E_n]$$

For a successful search we have

$$C_n \ = \ E[A_n] \ = \ \frac{E[I_n]}{n} + 1 \ = \ \frac{A'(1)}{n} + 1 \ = \ (1 + 1/n)C_n' - 1$$

$$\sigma^2(A_n) \ = \ 3C_n - 2 + \frac{A'(1)}{n} - C_n^2$$

$$1 \ \leq \ A_n \ \leq \ h(n)$$

and for an unsuccessful search

$$C_n' \ = \ E[A_n'] \ = \ \frac{E[E_n]}{n+1} \ = \ \frac{B'(1)}{n+1}$$

$$\sigma^2(C_n') \ = \ \frac{B'(1)}{n+1} + C_n'(1 - C_n')$$

$$1 \ \leq \ A_n' \ \leq \ h(n)$$

The ordered binary tree is a structure which allows us to perform many operations efficiently: inserting takes a time of $O(h(n))$; deleting a record also takes $O(h(n))$; finding the maximum or minimum key requires $O(h(n))$ comparisons; and retrieving all the elements in ascending or descending order can be done in a time of $O(n)$. With small changes, it permits the retrieval of the kth ordered record in the tree in $O(h(n))$.

General references:
[Hibbard, T.N., 62], [Batson, A., 65], [Bell, C., 65], [Lynch, W.C., 65], [Arora, S.R. *et al.*, 69], [Coffman, E.G. *et al.*, 70], [Stanfel, L., 70], [Nievergelt, J. *et al.*, 71], [Price, C.E., 71], [Knuth, D.E., 73], [Nievergelt, J. *et al.*, 73], [Robson, J.M., 73], [Aho, A.V. *et al.*, 74], [Nievergelt, J., 74], [Burkhard, W.A., 75], [Burge, W.H., 76], [Horowitz, E. *et al.*, 76], [Wilson, L.B., 76], [Wirth, N., 76], [Knott, G.D., 77], [Payne, H.J. *et al.*, 77], [Ruskey, F. *et al.*, 77], [Snyder, L., 77], [Soule, S., 77], [Driscoll, J.R. *et al.*, 78], [Gotlieb, C.C. *et al.*, 78], [Rotem, D. *et al.*, 78], [Flajolet, P. *et al.*, 79], [Flajolet, P. *et al.*, 79], [Kronsjo, L., 79], [Rosenberg, A.L., 79], [Strong, H.R. *et al.*, 79], [Yongjin, Z. *et al.*, 79], [Dasarathy, B. *et al.*, 80], [Flajolet, P. *et al.*, 80], [Gill, A., 80], [Kleitman, D.J. *et al.*, 80], [Lee, K.P., 80], [Proskurowski, A., 80], [Solomon, M. *et al.*, 80], [Standish, T.A., 80], [Stephenson, C.J., 80], [Fisher, M.T.R., 81], [Cesarini, F. *et al.*, 82], [Ottmann, T. *et al.*, 82], [Aho, A.V. *et al.*, 83], [Andersson, A. *et al.*, 83], [Kirschenhofer, P., 83], [Lescarne, P. *et al.*, 83], [Munro, J.I. *et al.*, 83], [Reingold, E.M. *et al.*, 83], [Sleator, D.D. *et al.*, 83], [van Leeuwen, J. *et al.*, 83], [Brown, G.G. *et al.*, 84], [Mehlhorn, K., 84], [Munro, J.I. *et al.*, 84], [Ottmann, T. *et al.*, 84], [Brinck, K., 85], [Ottmann, T. *et al.*, 85], [Pittel, B., 85], [Zerling, D., 85], [Brinck, K., 86], [Culberson, J.C., 86], [Gordon, D., 86], [Langenhop, C.E. *et al.*, 86], [Lee, C.C. *et al.*, 86], [Stout, Q.F. *et al.*, 86], [Wirth, N., 86], [Burgdorff, H.A. *et al.*, 87], [Levcopoulos, C. *et al.*, 88], [Sedgewick, R., 88], [Andersson, A., 89], [Aragon, C. *et al.*, 89], [Klein, R. *et al.*, 89], [Lentfert, P. *et al.*, 89], [Makinen, E., 89], [Manber, U., 89], [Slough, W. *et al.*, 89], [Andersson, A. *et al.*, 90], [Cormen, T.H. *et al.*, 90], [Francon, J. *et al.*, 90], [Fredman, M.L. *et al.*, 90], [Ottmann, T. *et al.*, 90], [Papadakis, T. *et al.*, 90], [Pugh, W., 90].

3.4.1.1 Randomly generated binary trees

These structures are also known as random search trees. Such trees are generated by taking elements in a random order and inserting them into an empty tree using the algorithm described below. Ordered binary search trees are normally considered to be created in this way. The efficiency measures for searching such trees are

$$B(z) = \prod_{i=1}^{n} \frac{i-1+2z}{i}$$

$$A(z) = \frac{B(z)-1}{2z-1}$$

$$C_n = 1 + n^{-1} \sum_{i=0}^{n-1} C_i'$$

$$E[A_n] = C_n = 2(1+1/n)H_n - 3 \approx 1.3863 \log_2 n - 1.8456$$

$$\sigma^2(A_n) = (2 + 10/n)H_n - 4(1 + 1/n)(H_n^2/n + H_n^{(2)}) + 4$$
$$\approx 1.3863 \log_2 n - 1.4253$$

$$E[A_n'] = C_n' = 2H_{n+1} - 2 \approx 1.3863 \log_2 n - 0.8456$$

$$\sigma^2(A_n') = 2H_{n+1} - 4H_{n+1}^{(2)} + 2 \approx 1.3863 \log_2 n - 3.4253$$

where $H_n = \sum_{i=1}^n 1/i$ is the nth harmonic number, and $H_n^{(2)} = \sum_{i=1}^n 1/i^2$ is the nth biharmonic number.

$$E[h(n)^k] = (4.31107...)^k \ln^k n + o(\ln^k n)$$
$$E[h(n)] = 4.31107... \ln n + O(\sqrt{\log n \log\log n})$$
$$\leq 4.31107... \ln n - 2.80654... \ln \ln n + O(1)$$

for any positive k, where the constant 4.31107... is a solution of the equation $c \ln(2e/c) = 1$.

Binary tree insertion

```
    procedure insert(key : typekey; var t : tree);
begin
if t = nil then
        t := NewNode(key, nil, nil)
else if t↑.k = key then
        Error {*** Key already in table ***}
else if t↑.k < key then insert(key, t↑.right)
    else                insert(key, t↑.left)
end;
```

At the cost of two extra pointers per element, randomly generated binary trees display excellent behaviour in searches. Unfortunately, the worst case can be generated when the elements are sorted before they are put into the tree. In particular, if any subset of the input records is sorted, it will cause the tree to degenerate badly. Compared to the random binary trees of the next section, however, ordered binary trees generated from random input are exceptionally well behaved.

Table 3.21 gives numerical values for several efficiency measures in trees of various sizes.

References:

[Knuth, D.E., 73], [Knuth, D.E., 74], [Palmer, E.M. *et al.*, 74], [Guibas, L.J., 75], [Wilson, L.B., 76], [Francon, J., 77], [Reingold, E.M. *et al.*, 77], [Meir, A. *et al.*, 78], [Robson, J.M., 79], [Brinck, K. *et al.*, 81], [Sprugnoli, R., 81],

Table 3.21: Exact complexity measures for binary search trees.

n	C_n	$\sigma^2(A_n)$	C'_n	$\sigma^2(A'_n)$	$E[h(n)]$
5	2.4800	1.1029	2.900	0.9344	3.8000
10	3.4437	2.1932	4.0398	1.8076	5.6411
50	6.1784	5.6159	7.0376	4.5356	10.8103
100	7.4785	7.2010	8.3946	5.8542	13.2858
500	10.6128	10.7667	11.5896	9.0179	19.3359
1000	11.9859	12.2391	12.9729	10.3972	22.0362
5000	15.1927	15.5608	16.1894	13.6105	28.4283
10000	16.5772	16.9667	17.5754	14.9961	31.2216

[Wright, W.E., 81], [Bagchi, A. *et al.*, 82], [Knott, G.D., 82], [Robson, J.M., 82], [Ziviani, N., 82], [Eppinger, J.L., 83], [Devroye, L., 84], [Mahmoud, H.M. *et al.*, 84], [Pittel, B., 84], [Flajolet, P. *et al.*, 85], [Devroye, L., 86], [Mahmoud, H.M., 86], [Cunto, W. *et al.*, 87], [Devroye, L., 87], [Devroye, L., 88].

3.4.1.2 Random binary trees

When discussing random binary trees, we consider the situation where all possible trees with the same number of nodes are equally likely to occur. In this case,

$$E[A_n] = \frac{4^n - \frac{3n+1}{n+1}\binom{2n}{n}}{\frac{n}{n+1}\binom{2n}{n}} = \sqrt{\pi n}\left(1 + \frac{9}{8n} + \frac{17}{128n^2} + O(n^{-3})\right) - 3 - \frac{1}{n}$$

$$E[A'_n] = \frac{4^n - \frac{n-1}{n+1}\binom{2n}{n}}{\frac{n}{n+1}\binom{2n}{n}} = \sqrt{\pi n}\left(1 + \frac{1}{8n} + \frac{1}{128n^2} + O(n^{-3})\right) - \frac{n-1}{n+1}$$

$$E[h(n)] = 2\sqrt{\pi n} + O(n^{1/4+\delta}) \qquad \text{(for any } \delta > 0\text{)}$$

$$\sigma^2(I_n) = \left(\frac{10}{3} - \pi\right)n^3 - \frac{n^2\sqrt{\pi n}}{2} + 9(1-\pi/4)n^2 - \frac{25n\sqrt{\pi n}}{16} + O(n)$$

If $t_{n,h}$ is the number of trees of height h and size n, then the associated generating function is

$$B_h(z) = \sum_{n=0}^{\infty} t_{n,h} z^n = z B_{h-1}^2(z) + 1$$

When all trees of height h are considered equally likely to occur, then

$$E[nodes] = (0.62896...)2^h - 1 + O(\delta^{-2^h}) \qquad (\delta > 1)$$

This situation is primarily a theoretical model. In practice, very few situations give rise to random trees.

References:
[Knuth, D.E., 73], [Kemp, R., 79], [Flajolet, P. *et al.*, 80], [Kemp, R., 80], [Flajolet, P. *et al.*, 82], [Flajolet, P. *et al.*, 84], [Kirschenhofer, P. *et al.*, 87], [Kemp, R., 89].

3.4.1.3 Height-balanced trees

These are also known as **AVL trees**. Height-balanced trees have the property that any two subtrees at a common node differ in height by 1 at the most. This balance property can be efficiently maintained by means of a counter in each node, indicating the difference in height between the right and left subtrees, $h(right) - h(left)$. The data structure used by an AVL tree is defined by **bt-(int, KEY)-LEAF.**

Because of the height balancing, the total height of a tree with n nodes is bounded by

$$\lceil \log_2 n + 1 \rceil \leq h(n) \leq 1.44042... \log_2 (n+2) - 0.32772...$$

There are AVL trees for which

$$C_n \geq 1.4404...(\log_2 n - \log_2 \log_2 n) + O(1)$$

and this is also an upper bound.

Let R_n indicate the average number of rotations per insertion required during the insertion of n keys into a random AVL tree. Then

$$0.3784... \leq R_n \leq 0.7261...$$

Let B_n be the average number of AVL nodes that are completely height balanced. Then

$$0.5637...n + o(n) \leq B_n \leq 0.7799...n + o(n)$$

Let $t_{n,h}$ be the number of height-balanced trees of height h and size n. The associated generating function is

$$T_h(z) = \sum_{n \geq 0} t_{n,h} z^n = z T_{h-1}(z)(2T_{h-2}(z) + T_{h-1}(z))$$

If we assume that all trees of height h are equally likely to occur, the average number of nodes in a balanced tree of height h is

$$E[nodes] = (0.70118...)2^h$$

Below we give the description of the AVL insertion algorithm. The insertion algorithm uses an additional balance counter in each node of the tree, *bal*.

The range of this balance field is $-2...2$. The procedures *rrot*() and *lrot*() which perform right and left rotations are common to several algorithms and are described in Section 3.4.1.8.

Height-balanced tree (AVL) insertion

```
function insert(key : typekey; var t : tree) : integer;
var  incr : integer;
begin
insert := 0;
if t = nil then begin
          t := NewNode(key, nil, nil);
          t↑.bal := 0;
          insert := 1
          end
else if t↑.k = key then
          Error {*** Key already in table ***}
else  with t↑ do begin
     if k < key then incr := insert(key, right)
          else incr := -insert(key, left);
     bal := bal + incr;
     if (incr <> 0) and (bal <> 0) then
          if bal < -1 then
               {*** left subtree too tall: right rotation needed ***}
               if left↑.bal < 0 then rrot(t)
                    else begin lrot(left); rrot(t) end
          else if bal > 1 then
               {*** right subtree too tall: left rotation needed ***}
               if right↑.bal > 0 then lrot(t)
                    else begin rrot(right); lrot(t) end
          else insert := 1;
     end
end;
```

AVL trees are of practical and theoretical importance as they allow searches, insertions and deletions in $O(\log n)$ time in the worst case.

The balance information in an AVL tree needs to represent three cases (five cases for some particular implementations). This requires two (or three) bits per node. It is not likely that this unit of storage is available, and a larger amount will be allocated for this purpose. Although a lot of emphasis has been placed on reducing the amount of extra storage used, the storage required by balance information is of little practical significance. If enough space is available it is best to store the height of the subtree, which contains more useful information and leads to simpler algorithms. Note that using six bits for height information we could store trees with up to 0.66×10^{13} nodes.

The constraint on the height balance can be strengthened to require that either both subtrees be of the same height or the right-side one be taller by one. These trees are called **one-sided height balanced** (OSHB), trees. In this case only one bit per node is required to store the balance information. Insertions in OSHBs become more complicated though; in particular, insertions in $O(\log n)$ time are extremely complicated.

Similarly, the constraint on the balance may be relaxed. One option is to allow the height of subtrees to differ at most by k. These trees are called **k-height balanced**, $HB[k]$, trees.

Table 3.22 shows some simulation results for AVL trees. C_n indicates the average number of comparisons required in a successful search, R_n is the average number of rotations (single or double) required by an insertion, and $E[h(n)]$ indicates the average height of a tree of size n.

Table 3.22: Exact and simulation results for AVL trees.

n	C_n	$E[h(n)]$	R_n
5	2.2	3.0	0.21333
10	2.907143	4	0.318095
50	4.930346±0.000033	6.94667±0.00017	0.42731±0.00005
100	5.888611±0.000042	7.998905±0.000043	0.44439±0.00005
500	8.192021±0.000087	10.92515±0.00073	0.46103±0.00006
1000	9.20056±0.00012	11.99842±0.00020	0.46329±0.00006
5000	11.55409±0.00028	14.9213±0.0026	0.46529±0.00007
10000	12.57009±0.00041	15.99885±0.00072	0.46552±0.00007
50000	14.92963±0.00094	18.9165±0.0096	0.46573±0.00007

The values for C'_n can be calculated from the above, for example, for all binary trees $C'_n = (C_n + 1)/(1 + 1/n)$.

From the above results we can see that the value for C_n is close to the value of $\log_2 n$; in particular, under the arbitrary assumption that

$$C_n = \alpha \log_2 n + \beta$$

for $n \geq 500$, then

$$\alpha = 1.01228 \pm 0.00006 \; ; \text{ and } \beta = -0.8850 \pm 0.0006 \; .$$

References:

[Adel'son-Vel'skii, G.M. *et al.*, 62], [Foster, C.C., 65], [Knott, G.D., 71], [Tan, K.C., 72], [Foster, C.C., 73], [Knuth, D.E., 73], [Aho, A.V. *et al.*, 74], [Hirschberg, D.S., 76], [Karlton, P.L. *et al.*, 76], [Luccio, F. *et al.*, 76], [Baer,

J.L. *et al.*, 77], [Reingold, E.M. *et al.*, 77], [Brown, M.R., 78], [Guibas, L.J. *et al.*, 78], [Kosaraju, S.R., 78], [Luccio, F. *et al.*, 78], [Luccio, F. *et al.*, 78], [Ottmann, T. *et al.*, 78], [Zweben, S.H. *et al.*, 78], [Brown, M.R., 79], [Ottmann, T. *et al.*, 79], [Ottmann, T. *et al.*, 79], [Pagli, L., 79], [Raiha, K.J. *et al.*, 79], [Luccio, F. *et al.*, 80], [Ottmann, T. *et al.*, 80], [Wright, W.E., 81], [Mehlhorn, K., 82], [Ziviani, N. *et al.*, 82], [Ziviani, N., 82], [Gonnet, G.H. *et al.*, 83], [Richards, R.C., 83], [Zaki, A.S., 83], [Tsakalidis, A.K., 85], [Chen, L., 86], [Li, L., 86], [Mehlhorn, K. *et al.*, 86], [Klein, R. *et al.*, 87], [Wood, D., 88], [Manber, U., 89], [Baeza-Yates, R.A. *et al.*, 90], [Klein, R. *et al.*, 90].

3.4.1.4 Weight-balanced trees

These are also known as **BB**(α) trees. Weight-balanced trees are binary search trees which obey a balance criterion on the subtrees of every node. Each node of the tree has a weight attached to it. A tree is said to be of **weighted balance** α or of **bounded balance** α, or in the set $BB[\alpha]$, for $0 \leq \alpha \leq 1/2$, if every node in the tree has balance, $\rho(t)$, between α and $1-\alpha$. The balance of a node is defined as

$$\rho(t) \;=\; \frac{\text{number of leaves in } t \uparrow .\mathit{left}}{\text{number of leaves in } t}$$

The empty binary tree is in $BB[\alpha]$ by convention.

The set $BB[\alpha]$ becomes more restricted as α goes from 0 to 1/2. $BB[0]$ is the class of all binary search trees, and $BB[1/2]$ is the class of completely balanced binary search trees of $n = 2^h - 1$ nodes. Interesting $BB[\alpha]$ trees are the ones with $2/11 \leq \alpha \leq 1-\sqrt{2}/2$. For these α, a balanced tree which is updated by an insertion or a deletion can be rebalanced with at most one rotation per level.

For any value of α,

$$\lceil \log_2 (n+1) \rceil \;\leq\; h(n) \;\leq\; -\frac{\log_2 n}{\log_2 (1-\alpha)} + O(1)$$

$$C_n \;\leq\; -\frac{\log_2 n}{\alpha \log_2 \alpha + (1-\alpha)\log_2 (1-\alpha)} - 2\,.$$

For any sequence of n updates (insertions and/or deletions), the worst-case average number of rotations is bounded by a constant which depends only on α:

$$R_n \;\leq\; c(\alpha)$$

For the class of trees $BB[1-\sqrt{2}/2]$

$$\begin{aligned}\lceil \log_2 (n+1) \rceil \;\leq\; h(n) \;&\leq\; 2\log_2 (n+1) \\ &\leq\; 2\log_2 (n+3) \;-\; 2.44549...\end{aligned}$$

$$C_n \leq 1.14622... \log_2 n + O(1)$$

Let R_n be the average number of rotations per insertion in a $BB[1-\sqrt{2}/2]$ tree after the random insertion of n keys into the empty tree. Let $f(\beta)$ be the fraction of internal nodes with weight balance factor exactly β or $1-\beta$ in a random $BB[1-\sqrt{2}/2]$ tree with n keys. Then

$$R_n \geq 0.40238...$$

$$0.54291... \leq f(1/2) \leq 0.72593...$$

$$0.17231... \leq f(1/3) \leq 0.34801...$$

$$0.05405... \leq f(2/5) \leq 0.22975...$$

Below we give a description of the insertion algorithm for weight-balanced trees with $\alpha = 1-\sqrt{2}/2 = 0.292893...$ The procedures *rrot*() and *lrot*() which perform right and left rotations, are common to several algorithms and are described in Section 3.4.1.8. The insertion algorithm uses a weight counter in each node of the tree, *weight*. For any node t, $t\uparrow.weight$ contains the number of external nodes in the subtree rooted at t. We use for convenience the function $wt(t)$ which returns 1 if the tree t is nil or $t\uparrow.weight$ otherwise.

Weight-balanced tree insertion

```
procedure insert(key : typekey; var t : tree);
begin
if t = nil then begin
        t := NewNode(key, nil, nil);
        t↑.weight := 2
        end
else if t↑.k = key then
        Error {*** Key already in table ***}
else with t↑ do begin
   if k < key then insert(key, right)
              else insert(key, left);
   weight := wt(left) + wt(right);
   checkrots(t)
   end
end;
```

Although the insertion algorithm is coded using real arithmetic, this is not really needed. For example, $\sqrt{2}/2$ can be approximated by its convergents 2/3, 5/7, 12/17, 29/41, 70/99, ... In case integer arithmetic must be used, the first test can be rewritten, for example, as

if 99*$wt(t\uparrow.left)$ > 70*$wt(t)$ **then** ...

Table 3.23 shows some simulation results on weight-balanced trees for $\alpha = 1 - \sqrt{2}/2$. C_n indicates the average number of comparisons required in a successful search, R_n is the average number of rotations (single or double) required by an insertion and $E[h(n)]$ indicates the average height of a tree of size n.

Table 3.23: Exact and simulation results for weight-balanced trees.

n	C_n	$E[h(n)]$	R_n
5	2.2	3	0.21333
10	2.9	4	0.3252381
50	4.944142±0.000046	7.02363±0.00027	0.40861±0.00006
100	5.908038±0.000067	8.20895±0.00063	0.42139±0.00007
500	8.23015±0.00017	11.2552±0.0018	0.43204±0.00008
1000	9.24698±0.00025	12.6081±0.0031	0.43343±0.00009
5000	11.62148±0.00061	15.6991±0.0076	0.43455±0.00010
10000	12.64656±0.00091	17.0366±0.0089	0.43470±0.00010
50000	15.0300±0.0022	20.110±0.022	0.43476±0.00011

From the above results we can see that the value for C_n is close to the value of $\log_2 n$; in particular, under the arbitrary assumption that

$$C_n = \alpha \log_2 n + \beta$$

for $n \geq 500$, then

$$\alpha = 1.02107 \pm 0.00013 \; ; \text{ and } \beta = -0.9256 \pm 0.0012 \; .$$

References:
[Knuth, D.E., 73], [Nievergelt, J. *et al.*, 73], [Baer, J.L. *et al.*, 77], [Reingold, E.M. *et al.*, 77], [Unterauer, K., 79], [Blum, N. *et al.*, 80], [Bagchi, A. *et al.*, 82].

3.4.1.5 Balancing by internal path reduction

These are also known as weight-balanced or **path-balanced** trees. These trees are similar to the trees described in the previous section, except that rotations are made only when they can reduce the total internal path of the tree. For this reason these trees are also known as **path trees**. In summary, a single left rotation is performed whenever

$$wt(t\uparrow.left) < wt(t\uparrow.right\uparrow.right)$$

and a double left rotation when

$$wt(t\uparrow.left) < wt(t\uparrow.right\uparrow.left)$$

and right rotations for the symmetric cases. For these balance conditions we have:

$$\lceil \log_2 (n+1) \rceil \leq h(n) \leq 1.44042... \log_2 n - 0.32772...$$

$$C_n \leq \frac{5 \log_3 2}{3} \log_2 n + O(1) = 1.05155... \log_2 n + O(1)$$

The amortized worst-case number of rotations per insertion is bounded by

$$R_n \leq 0.44042... \log_2 n + O(1)$$

The amortized worst-case number of rotations per deletion is bounded by

$$R_n \leq 0.42062... \log_2 n + O(1)$$

In the worst case, for a single insertion or deletion,

$$R_n = O(n)$$

Below we give a description of the insertion algorithm. The insertion code uses the procedure *checkrot* which checks the balancing, performs any necessary rotation and checks whether further rotations may be needed. The procedures *rrot*() and *lrot*(), which perform right and left rotations, are common to several algorithms and are described in Section 3.4.1.8. For any node t, $t \uparrow .weight$ contains the number of external nodes in the subtree rooted at t. We use for convenience the function $wt(t)$ which returns 1 if the tree t is nil or $t \uparrow .weight$ otherwise.

Internal path reduction trees: insertion

```
procedure checkrots(var t : tree);
{*** check need for rotations ***}
var wl, wll, wr, wrr : integer;
begin
if t <> nil then with t↑ do begin
    wl := wt(left);
    wr := wt(right);
    if wr > wl then begin
        {*** left rotation needed ***}
        wrr := wt(right↑.right);
        if (wrr > wl) and (2*wrr >= wr) then
            begin lrot(t); checkrots(left) end
```

```
            else if wr-wrr > wl then begin
               rrot(right);  lrot(t);
               Rots := Rots-1;
                checkrots(left);  checkrots(right)
                end
            end
        else if wl > wr then begin
            {*** right rotation needed ***}
            wll := wt(left↑.left);
            if (wll > wr) and (2*wll >= wl) then
                begin rrot(t); checkrots(right) end
            else if wl-wll > wr then begin
               lrot(left);  rrot(t);
               Rots := Rots-1;
                checkrots(left);  checkrots(right)
                end
            end
        end
end;

procedure insert(key : typekey; var t : tree);
begin
if t = nil then begin
        t := NewNode(key, nil, nil);
        t↑.weight := 2
        end
else if t↑.k = key then
        i:=i-1
else with t↑ do begin
    if k < key then insert(key, right)
              else insert(key, left);
    weight := wt(left) + wt(right);
    checkrots(t)
    end
end;
```

Although these trees are in the class $BB(1/3)$, there are some important restrictions on the rotations. This makes their performance superior to the $BB(1/3)$ trees.

A natural extension of this algorithm is to perform rotations only when the difference in weights is k or larger. This extension is called **k-balancing**. For these trees the main complexity measures remain of the same order, while the number of rotations is expected to be reduced by a factor of k.

$$h^k(n) \leq 1.44042...\log_2(n-k+2) + k - 1.32772...$$

$$C_n^k \le 1.05155... \log_2 n + O(1)$$

Table 3.24 shows simulation results for these trees. C_n indicates the average number of comparisons required in a successful search, R_n is the average number of rotations (single or double) required by an insertion and $E[h(n)]$ indicates the average height of a tree of size n.

Table 3.24: Exact and simulation results for path-trees.

n	C_n	$E[h(n)]$	R_n
5	2.2	3	0.213333
10	2.9	4	0.33
50	4.904496±0.000027	6.93788±0.00026	0.469722±0.000078
100	5.857259±0.000038	8.00408±0.00015	0.494494±0.000090
500	8.151860±0.000090	10.9169±0.0012	0.51836±0.00011
1000	9.15670±0.00013	12.0191±0.0010	0.52177±0.00012
5000	11.50285±0.00032	14.9529±0.0039	0.52476±0.00014
10000	12.51640±0.00048	16.0477±0.0052	0.52521±0.00014
50000	14.8702±0.0011	18.995±0.011	0.52564±0.00016

From the above results we can see that the value for C_n is close to the value of $\log_2 n$; in particular, under the arbitrary assumption that

$$C_n = \alpha \log_2 n + \beta$$

for $n \ge 500$, then

$$\alpha = 1.00892 \pm 0.00007 \; ; \text{ and } \beta = -0.8963 \pm 0.0007 \; .$$

References:
[Baer, J.L., 75], [Robson, J.M., 80], [Gonnet, G.H., 83], [Gerash, T.E., 88].

3.4.1.6 Heuristic organization schemes on binary trees

When the keys in a binary tree have different accessing probabilities, a randomly generated tree or a balanced tree may not be fully satisfactory. The following heuristic organization schemes offer ways to build better trees when the accessing probabilities are known.

For all these heuristics we will denote by p_i the accessing probability of the ith key. We will denote by q_i the probability of an unsuccessful access, searching for a key with value in between the ith and $i+1$st keys. In all cases, $\sum_i p_i + \sum_i q_i = 1$. The entropy, or uncertainty of the set of $p_i s$ (or $p_i s$ and $q_i s$), is

$$H(\vec{p}) = -\sum_i p_i \log_2 p_i$$

$$H(\vec{p}, \vec{q}) = -\sum_i p_i \log_2 p_i - \sum_i q_i \log_2 q_i$$

Heuristics for known probabilities

The first four algorithms allow a top-down construction, and share the common pseudo-code construction:

Top-down binary tree construction

```
BuildTree(SetOfKeys) : tree;
begin
    K := select(SetOfKeys);
    A1 := Keys in SetOfKeys < K;
    A2 := Keys in SetOfKeys > K;
    return(NewNode(K, BuildTree(A1), BuildTree(A2)))
end;
```

(1) **Insert in decreasing probability order** In this way, the keys most likely to be sought are closer to the root and have shorter search paths. This method requires either a reordering of the keys before they are put into the tree or the selection of the maximum probability at each step. For this analysis, we will assume that the keys are numbered in decreasing probability order, that is, $(p_1 \geq p_2 \geq ... \geq p_n)$. Then for a random tree

$$C_n = \sum_{i=1}^{n} p_i H_i - 1$$

where $H_i = \sum_{j=1}^{i} 1/j$ is the ith harmonic number.

(2) **Median split** In this scheme we choose the root so that the total accessing probabilities of both the left and right subtrees are as close as possible to 1/2. This is repeated recursively on both subtrees. This arrangement gives the information theoretic optimum. For this heuristic

$$C_n^{Opt} \leq C_n^{MS} \leq 2 + 1.44042...H(\vec{p}, \vec{q})$$

(3) It is possible to mix approaches (1) and (2). We allow a tolerance δ, and examine the elements for which the accessing probabilities of the left and right subtrees fall into the range $1/2 \pm \delta$. From these elements,

we choose the one with the highest accessing probability to be the root. This selection procedure is repeated recursively for the nodes in each subtree. Experimental results indicate that these trees are within 2% to 3% from optimal.

(4) Another way of combining approaches (1) and (2) produces trees which are also called **median split trees**. At every node we store two keys; the first one, the 'owner' of the node, is the one with higher probability in the subtree, and the second one is the median of all the values in the subtree. The searching algorithm is almost identical to the normal algorithm:

Median split trees: search

```
procedure search(key : typekey; t : tree);
begin
if t=nil then              {*** Not Found *** }
        notfound(key)
else if t↑.OwnerKey = key then {*** Found *** }
        found(t ↑ )
else if t↑.SplitKey < key then search(key, t↑.right)
        else        search(key, t↑.left)
end;
```

Using this approach we benefit from the advantages of both (1) and (2) above, at the cost of one extra key per node. The 'median split' may be interpreted as the statistical median (a key which splits the tree into two subtrees in such a way that both halves are the closest possible to equiprobable) or as the counting median (a key which splits the tree in equal size halves). Known algorithms to construct optimal median split trees are not very efficient (at least $O(n^4)$).

(5) **Greedy trees** This is a heuristic which constructs trees bottom-up. The construction resembles the Huffman encoding algorithm. At each step we select the three consecutive external/internal/external nodes which add to the lowest accessing probability. A node is constructed with the two external nodes as direct descendants and the triplet is replaced by a single external node with the sum of the accessing probabilities. Under this heuristic

$$C_n^{GT} \leq 2 + 1.81335...H(\vec{p}, \vec{q})$$

Self-organizing heuristics

When we do not know the accessing probabilities we may try heuristic organization schemes similar to the transpose and move-to-front techniques in sequential searching.

(6) **Exchange with parent** or **simple exchange** The transpose method can be adapted for trees by exchanging a node with its parent each time the node is accessed. This is achieved by performing a single rotation on the parent node (a left rotation if the searched node is at its right, a right rotation otherwise). This is not a good heuristic, however, as it tends to be very unstable in some cases. For example, if the probability of accessing any key is uniform, $p_i = 1/n$, then this exchange-with-parent technique produces a random binary tree and

$$C_n^{EP} \approx \sqrt{\pi n} - 3 + O(n^{-1/2})$$

(7) **Move-to-root** Corresponding to the move-to-front scheme in linear searching, we have the technique of moving an accessed element to the root. This is achieved, while maintaining the lexicographical order of the tree, by several single rotations on the ancestors of the accessed element. With this move-to-root approach we have

$$C_n^{MR} = 1 + 2 \sum_{1 \leq i < j \leq n} \frac{p_i p_j}{p_i + \ldots + p_j} \leq 2 \ln (2) H(\vec{p}) + 1$$

$$\sigma^2(A_n^{MR}) \leq 2 \ln n + O(1)$$

(8) **Dynamic trees** (or **D-trees**) Dynamic trees use a self-organizing technique based on estimating the accessing probabilities by keeping counters for the number of successful/unsuccessful searches at each internal/external node. The tree is balanced with respect to these counters, like the balance done for $BB[\alpha]$ trees (see Section 3.4.1.4). If f_i denotes the relative accessing frequency of node i, then the number of access needed to locate node i is $O(\log (1/f_i))$.

(9) **Splay trees** This scheme is similar to the move-to-root technique (7). Splay trees are reorganized whenever they are accessed or updated. The basic reorganizing operation (**splaying**) moves the accessed node towards the root by a sequence of rotations. Therefore, frequently accessed keys tend to be near the root. For the worst sequence of splayings, the number of operations is $O(\log n)$ per node in the tree, where n is the number of nodes.

Shape heuristics

(10) **Fringe reorganization** This type of heuristics guarantees that any subtree with size k or smaller is of minimal height (or, equivalently, of minimal internal path). The simplest heuristic is for $k = 3$ which reorganizes any subtree with three nodes which is not in perfect balance. Under random insertions, a tree constructed using $k = 3$ will have

$$C_n' = \frac{12}{7}H_{n+1} - \frac{75}{49} \approx 1.18825\ldots\log_2 n - 0.54109\ldots \text{ for } n \geq 6$$

$$\sigma^2(A_n') = \frac{300}{343}H_{n+1} - \frac{144}{49}H_{n+1}^{(2)} + \frac{5056}{2401} + \frac{2304}{343(n+1)n\cdots(n-5)}$$

for $n \geq 13$.

In general, if $k = 2t - 1$ ($t \geq 1$) then

$$C_n' = \frac{H_n}{H_{2t} - H_t} + O(1)$$

$$\sigma^2(A_n') = \frac{H_{2t}^{(2)} - H_t^{(2)}}{(H_{2t} - H_t)^3}H_n + O(1)$$

References:
[Gotlieb, C.C. *et al.*, 72], [Martin, W.A. *et al.*, 72], [Knuth, D.E., 73], [Fredman, M.L., 75], [Mehlhorn, K., 75], [Walker, W.A. *et al.*, 76], [Baer, J.L. *et al.*, 77], [Mehlhorn, K., 77], [Allen, B. *et al.*, 78], [Sheil, B.A., 78], [Horibe, Y. *et al.*, 79], [Mehlhorn, K., 79], [Comer, D., 80], [Eades, P. *et al.*, 81], [Korsh, J.F., 81], [Allen, B., 82], [Korsh, J.F., 82], [Poblete, P.V., 82], [Greene, D.H., 83], [Huang, S-H.S. *et al.*, 83], [Chang, H. *et al.*, 84], [Huang, S-H.S. *et al.*, 84], [Huang, S-H.S. *et al.*, 84], [Huang, S-H.S. *et al.*, 84], [Perl, Y., 84], [Bent, S.W. *et al.*, 85], [Hermosilla, L. *et al.*, 85], [Poblete, P.V. *et al.*, 85], [Sleator, D.D. *et al.*, 85], [Hester, J.H. *et al.*, 86], [Huang, S-H.S., 87], [Levcopoulos, C. *et al.*, 87], [Makinen, E., 87], [Hester, J.H. *et al.*, 88], [Moffat, A. *et al.*, 89], [Sherk, M., 89], [Cole, R., 90].

3.4.1.7 Optimal binary tree search

When we want to minimize the average case search and all the nodes in the tree are equally probable, or when we want to minimize the worst case, it is easy to see that the optimal tree is the one with minimum height. Equivalently, such an optimal tree has all its leaves at a maximum of two consecutive levels.

When the nodes in the tree have different accessing probabilities, and these probabilities are known, we can construct an optimal (minave) search tree. For these optimal trees,

$$H(\vec{p}) - \log_2(eH(\vec{p})) + 1 \le C_n^{Opt} \le H(\vec{p}) + 1$$

$$H(\vec{q}) \le C_n^{Opt} \le 2 + H(\vec{q})$$

if $p_i = 0$.

The following algorithm constructs an optimal tree given the probabilities of successful searches (p_i) and the probabilities of unsuccessful searches (q_i). This algorithm due to Knuth uses a dynamic programming approach, computing the cost and root of every tree composed of contiguous keys. To store this information, the algorithm uses two upper triangular matrices dimensioned $n \times n$. Both its storage and time requirements are $O(n^2)$.

Optimal binary tree construction (Knuth)

```
function OptTree(keys : ArrayKeys; p : ArrayCost; q : ArrayCost) : tree;

var   wk, wki, min : cost;
      i, ik, indxmin, j, k : integer;
      {*** r[i,j] indicates the root of the optimal tree formed
            with keys from i to j ***}
      r : array[0..n,0..n] of integer;
      {*** c[i,j] indicates the optimal cost of the tree with
            keys from i to j ***}
      c : array[0..n,0..n] of cost;

      function CreateTree(i, j : integer) : tree;
      {*** Create optimal tree from information in r[i,j] ***}
      var  t : tree;
      begin
      if i=j then CreateTree := nil
      else begin
            new(t);
            t↑.k := keys[r[i,j]];
            t↑.left := CreateTree(i, r[i,j]-1);
            t↑.right := CreateTree(r[i,j], j);
            CreateTree := t
            end
      end;

begin
{*** Initializations ***}
c[0,0] := q[0];
for i:=1 to n do begin
      c[i,i] := q[i];
```

```
        c[i-1,i] := 2*(q[i-1] + q[i]) + p[i];
        r[i-1,i] := i
        end;

{*** Main loop to compute r[i,j] ***}
wk := q[0];
for k:=2 to n do begin
        wk := wk + q[k-1] + p[k-1];
        wki := wk;
        for i:=0 to n-k do begin
                ik := i+k;
                wki := wki + q[ik] + p[ik];
                min := maxint;
                {*** Select root with lowest cost ***}
                for j:=r[i,ik-1] to r[i+1,ik] do
                        if c[i,j-1]+c[j,ik] < min then begin
                                min := c[i,j-1]+c[j,ik];
                                indxmin := j
                                end;
                c[i,ik] := min + wki;
                r[i,ik] := indxmin;
                wki := wki - q[i] - p[i+1];
                end
        end;

OptTree := CreateTree(0, n);
end;
```

If we are interested in the unsuccessful probabilities alone ($p_i = 0$), the **Hu–Tucker algorithm** algorithm will construct an optimal tree in $O(n \log n)$ time and $O(n)$ space.

References:
[Bruno, J. *et al.*, 71], [Hu, T.C. *et al.*, 71], [Knuth, D.E., 71], [Hu, T.C. *et al.*, 72], [Kennedy, S., 72], [Hu, T.C., 73], [Knuth, D.E., 73], [Garey, M.R., 74], [Hosken, W.H., 75], [Itai, A., 76], [Wessner, R.L., 76], [Choy, D.M. *et al.*, 77], [Garsia, A.M. *et al.*, 77], [Horibe, Y., 77], [Reingold, E.M. *et al.*, 77], [Choy, D.M. *et al.*, 78], [Bagchi, A. *et al.*, 79], [Horibe, Y., 79], [Hu, T.C. *et al.*, 79], [Wikstrom, A., 79], [Kleitman, D.J. *et al.*, 81], [Allen, B., 82], [Hu, T.C., 82], [Akdag, H., 83], [Shirg, M., 83], [Bender, E.A. *et al.*, 87], [Larmore, L.L., 87], [Levcopoulos, C. *et al.*, 87], [Baase, S., 88], [Brassard, G. *et al.*, 88], [Kingston, J.H., 88], [Sedgewick, R., 88], [Levcopoulos, C. *et al.*, 89].

3.4.1.8 Rotations in binary trees

Rotations in binary trees are operations that modify the structure (shape) of the tree without altering the lexicographical ordering in the tree. These transformations are very useful in keeping the tree structure balanced.

The simplest rotation, which is usually called **single rotation**, is illustrated by Figure 3.1.

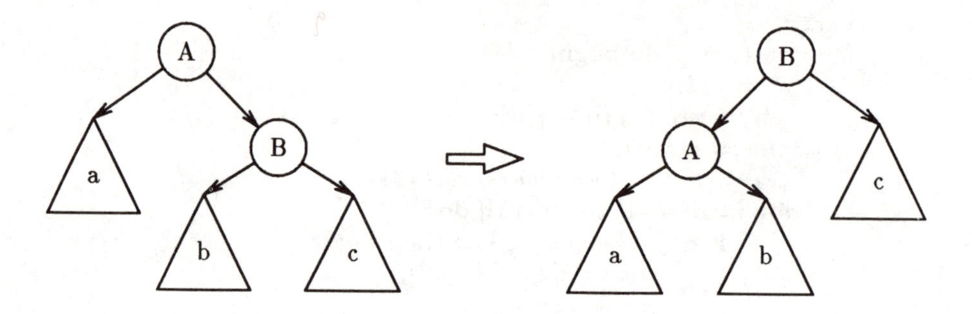

Figure 3.1: Single left rotation.

There are two possible such situations, the one shown in Figure 3.1 and its symmetric which are called *left* and *right* single rotations respectively. The procedures to perform these rotations are

Single left rotation

```
procedure lrot(var t : tree);
var  temp : tree;
begin
     temp := t;
     t := t↑.right;
     temp↑.right := t↑.left;
     t↑.left := temp;
end;
```

Single right rotation

```
procedure rrot(var t : tree);
```

```
var temp : tree;
begin
      temp := t;
      t := t↑.left;
      temp↑.left := t↑.right;
      t↑.right := temp;
end;
```

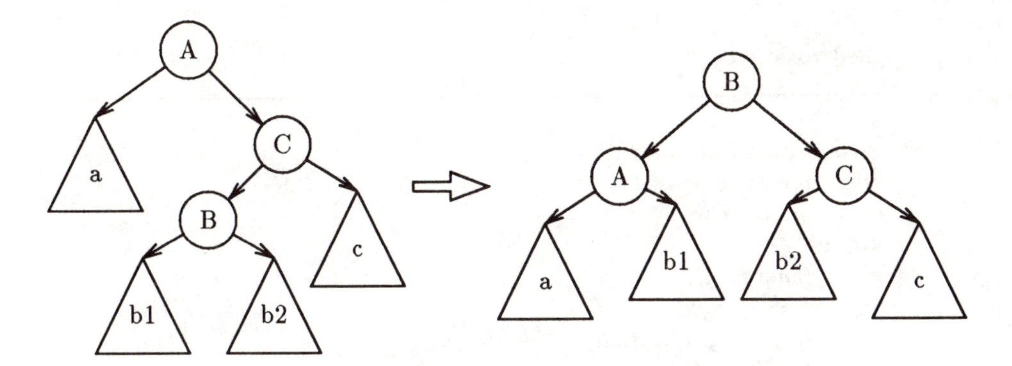

Figure 3.2: Double left rotation.

A double rotation is a more complicated transformation. Figure 3.2 illustrates a transformation called double *left* rotation. Its symmetric is called a double *right* rotation. Both rotations can be described in terms of two single rotations, for example a double left rotation at the node pointed by t is achieved by

Double left rotation

```
rrot(t↑.right);    lrot(t);
```

In many cases the nodes carry some information about the balance of their subtrees. For example, in AVL trees (see Section 3.4.1.3), each node contains the difference in height of its subtrees; in weight-balanced trees (see Section 3.4.1.4) each node contains the total number of nodes in its subtree. This information should be reconstructed by the single rotation, and consequently double rotations or more complicated rotations based on single rotations do not need to reconstruct any information.

Let *bal* contain the difference in height between the right subtree and the left subtree ($h(t \uparrow .right) - h(t \uparrow .left)$), as in AVL trees (see Section 3.4.1.3).

For example, after a single left rotation, the new balance of the nodes A and B (Figure 3.1) is given by:

$$NewBal(A) = OldBal(A) - 1 - \max(OldBal(B), 0)$$

$$NewBal(B) = \min(OldBal(A) - 2, OldBal(A) + OldBal(B) - 2, OldBal(B) - 1)$$

The complete code for a single left rotation becomes

Single left rotation

```
procedure lrot(var t : tree);
var  temp : tree;
     a : integer;
begin
     temp := t;
     t := t↑.right;
     temp↑.right := t↑.left;
     t↑.left := temp;
     {*** adjust balance ***}
     a := temp↑.bal;
     temp↑.bal := a - 1 - max(t↑.bal, 0);
     t↑.bal := min(a-2, a+t↑.bal-2, t↑.bal-1);
end;
```

References:
[Tarjan, R.E., 83], [Zerling, D., 85], [Sleator, D.D. *et al.*, 86], [Stout, Q.F. *et al.*, 86], [Wilber, R., 86], [Bent, S.W., 90], [Cormen, T.H. *et al.*, 90], [Ottmann, T. *et al.*, 90].

3.4.1.9 Deletions in binary trees

The operation of deleting a node in a binary tree is relatively simple if the node to be deleted has a null descendant. In this case the node is replaced by the other descendant. If both descendants are non-null the node has to be moved down the tree until it has a non-null descendant.

One way of moving the node to the fringe of the tree is to swap it with one of its lexicographically ordered neighbours. Experimental and theoretical evidence suggests that always choosing the successor (or the predecessor) may degenerate to a tree of $O(\sqrt{n})$ height after a big number of updates, for a random tree containing n keys (after the updates). On the other hand, using a random choice (or alternating) seems to maintain the height of the tree logarithmic. Another strategy, better suited for balanced trees, is to gradually move the node towards the fringe by the use of rotations.

The following procedure performs deletions on weight-balanced (see Section 3.4.1.4) or path-balanced trees (see Section 3.4.1.5).

Deletions on weight-balanced trees

```
procedure delete(key : typekey; var t : tree);

begin
if t = nil then Error {*** key not found ***}
else begin
    {*** search for key to be deleted ***}
    if      t↑.k < key then delete(key, t↑.right)
    else if t↑.k > key then delete(key, t↑.left)

    {*** key found, delete if a descendant is nil ***}
    else if t↑.left  = nil then t := t↑.right
    else if t↑.right = nil then t := t↑.left

    {*** no descendant is null, rotate on heavier side ***}
    else if wt(t↑.left) > wt(t↑.right) then
         begin rrot(t);  delete(key, t↑.right) end
    else begin lrot(t);  delete(key, t↑.left) end;

    {*** reconstruct weight information ***}
    if t <> nil then begin
        t↑.weight := wt(t↑.left) + wt(t↑.right);
        checkrots(t)
        end
    end
end;
```

For height balanced (AVL) trees (see Section 3.4.1.3) we simply replace the function $wt()$ by the height of the subtree.

References:
[Knuth, D.E., 73], [Knott, G.D., 75], [Knuth, D.E., 77], [Jonassen, A.T. *et al.*, 78], [Brinck, K., 86], [Baeza-Yates, R.A., 89], [Culberson, J.C. *et al.*, 89], [Cormen, T.H. *et al.*, 90], [Culberson, J.C. *et al.*, 90].

3.4.1.10 *m*-ary search trees

An m-ary search tree is a multiway tree where:

(1) every internal node has $m-1$ keys and m descendants;

(2) every external node has between 0 and $m-2$ keys.

The lexicographical order is given by the fact that, in each internal node, all the keys stored in the ith descendant are greater than the $i-1$th key and less than the ith key of the node. The relation between the internal path length, I_n, and the external path length, E_n, on a tree with n internal nodes, is

$$E_n = I_n + mn$$

The average internal path length of an m-ary search tree built from n random insertions is:

$$E[I_n] = \frac{(n+1)H_n}{H_m - 1} - \left(\frac{m}{m-1} + \frac{1}{H_m - 1} + O(1)\right) n + o(n)$$

with variance:

$$\sigma^2(I_n) = \frac{1}{(H_m - 1)^2}\left(\frac{(m+1)H_m^{(2)} - 2}{m-1} - \frac{\pi^2}{6}\right) n^2 \; + \; o(n^2)$$

For the expected height, we have the following limit (in probability)

$$\lim_{m\to\infty} \frac{h(n)}{\ln n} = \frac{1}{H_m - 1}$$

The average space utilization of an m-ary search tree is

$$\frac{n}{2(H_m - 1)} + o(n)$$

A surprising result is that the variance of the above complexity measure is linear in n for $3 \leq m \leq 26$, but superlinear for $m > 26$ (almost quadratic for large m).

There exist several variations that improve the storage utilization of these trees, making them suitable for use as external data structures.

References:
[Ruskey, F., 78], [Szwarcfiter, J.L. *et al.*, 78], [Pagli, L., 79], [Vaishnavi, V.K. *et al.*, 80], [Culik II, K. *et al.*, 81], [Arnow, D. *et al.*, 84], [Szwarcfiter, J.L., 84], [Mahmoud, H.M., 86], [Baeza-Yates, R.A., 87], [Huang, S-H.S., 87], [Cunto, W. *et al.*, 88], [Mahmoud, H.M. *et al.*, 89], [Sherk, M., 89].

3.4.2 B-trees

A B-tree is a balanced multiway tree with the following properties:

(1) Every node has at most $2m+1$ descendants.

(2) Every internal node except the root has at least $m+1$ descendants, the root either being a leaf or having at least two descendants.

(3) The leaves are null nodes which all appear at the same depth.

B-trees are usually named after their allowable branching factors, that is, $m+1$–$2m+1$. For example, 2–3 trees are B-trees with $m=1$; 6–11 trees are B-trees with $m=5$. B-trees are used mainly as a primary key access method for large databases which cannot be stored in internal memory. Recall the definition of multiway trees:

$$\mathbf{mt-N-D-LEAF}: \ [\mathbf{int}, \{\mathbf{D}\}_1^{\mathbf{N}}, \{\mathbf{mt-N-D-LEAF}\}_0^{\mathbf{N}}]; \mathbf{LEAF}.$$

Then the data structure for a general B-tree is **mt – 2m – D – nil**. For our C algorithms we will use the definition:

B-tree data structure

```
typedef struct btnode { /*** B-Tree Definition ***/
        int d;                /*** number of active entries ***/
        typekey k[2*M]; /*** Keys ***/
        struct btnode *p[2*M+1]; /*** Pointers to subtrees ***/
        } node, *btree;
```

Note that, in C, arrays always start with index 0, consequently the array containing the keys runs from 0 to $2M-1$. The lexicographical order is given by the fact that all the keys in the subtree pointed by $p[i]$ are greater than $k[i-1]$ and less than $k[i]$.

Let E_n and E_n' represent the number of nodes accessed in successful and unsuccessful searches respectively. Let $h(n)$ be the height of a B-tree with n keys. Then

$$1 \le E_n \le h(n)$$

$$E_n' = h(n)$$

$$E[E_n] = h(n) - \frac{1}{2m \ln 2} + O(m^{-2})$$

$$\lceil \log_{2m+1}(n+1) \rceil \le h \le 1 + \lfloor \log_{m+1}((n+1)/2) \rfloor$$

$$\sigma^2(E_n) \leq \frac{m+1}{m^2}$$

Let t_n be the number of different B-trees with n leaves. We have

$$B(z) = \sum_{n=0}^{\infty} t_n z^n = B(P(z)) + z$$

where

$$P(z) = \frac{z^{m+1}(z^{m+1}-1)}{z-1},$$

and

$$t_n = \frac{\phi_m^{-n}}{n} Q(\log n)(1 + O(n^{-1}))$$

where $0 < \phi_m < 1$ and ϕ_m is a root of $P(z) = z$ and $Q(x)$ is a periodic function in x with average value $\phi_m / \ln P'(\phi_m)$ and period $\ln P'(\phi_m)$. Table 3.25 shows some exact values.

Table 3.25: Parameters for counting different B-trees.

m	ϕ_m	$\ln P'(\phi_m)$	$\phi_m / \ln P'(\phi_m)$
1	0.61803...	0.86792...	0.71208...
2	0.68232...	1.01572...	0.67176...
5	0.77808...	1.21563...	0.64006...
10	0.84439...	1.34229...	0.62907...

$$\phi_m = 1 - \frac{w(m)}{m} + \frac{w(m)+2}{2w(m)+2}\left(\frac{w(m)}{m}\right)^2 + O((w(m)/m)^3)$$

where $w(m)e^{w(m)} = m$, and

$$t_n = O\left(\frac{e^{w(m)n/m}}{n}\right)$$

Let N_n be the expected number of nodes in a randomly generated B-tree with n keys. Then

$$\frac{2m+1}{4m(m+1)(H_{2m+2} - H_{m+1})} \leq \frac{N_n}{n} \leq \frac{1}{2m(H_{2m+2} - H_{m+1})}$$

$$\frac{N_n}{n} = \frac{1}{2m \ln 2} + O(m^{-2})$$

Let S_n be the average number of node-splits produced by an insertion into a randomly generated B-tree with n keys.

$$\frac{1}{2(m+1)(H_{2m+2} - H_{m+1})} \leq S_n$$

Below we present a description of the algorithm for searching B-trees. Note that in this case we can convert the 'tail recursion' into an iteration very easily.

B-tree search

```
search(key, t)
typekey key;
btree t;

{     int i;
while (t != NULL) {
        for (i=0; i<t->d && key>t->k[i]; i++);
        if (key == t->k[i])
                { found(t, i);     return; }
        t = t->p[i];
        }
notfound(key);
};
```

B-tree insertion

```
btree insert(key, t)
typekey key;
btree t;

{
typekey ins;
extern btree NewTree;
typekey InternalInsert();
        ins = InternalInsert(key, t);
        /*** check for growth at the root ***/
        if (ins != NoKey) return(NewNode(ins, t, NewTree));
        return(t);
};

   typekey InternalInsert(key, t)
   typekey key;
```

```
btree t;
{int i, j;
typekey ins;
btree tempr;
extern btree NewTree;
if (t == NULL) {  /*** the bottom of the tree has been reached:
                       indicate insertion to be done ***/
   NewTree = NULL;
   return(key);
   }
else {
   for (i=0; i<t->d && key>t->k[i]; i++);
   if (i<t->d && key == t->k[i])
      Error; /*** Key already in table ***/
   else {
    ins = InternalInsert(key, t->p[i]);
    if (ins != NoKey)
    /*** the key in "ins" has to be inserted in present node ***/
       if (t->d < 2*M) InsInNode(t, ins, NewTree);
       else    /*** present node has to be split ***/
         {/*** create new node ***/
         if (i<=M) {
            tempr = NewNode(t->k[2*M-1], NULL, t->p[2*M]);
            t->d--;
            InsInNode(t, ins, NewTree);
            }
         else tempr = NewNode(ins, NULL, NewTree);
         /*** move keys and pointers ***/
         for (j=M+2; j<=2*M; j++)
            InsInNode(tempr, t->k[j-1], t->p[j]);
         t->d = M;
         tempr->p[0] = t->p[M+1];
         NewTree = tempr;
         return(t->k[M]);
         }
    }
 return(NoKey);
 }
};
```

The above algorithm is structured as a main function *insert* and a subordinate function *InternalInsert*. The main function handles the growth at the root, while the internal one handles the recursive insertion in the tree.

The insertion function returns a pointer to the resulting tree. This pointer may point to a new node when the B-tree grows at the root.

The insertion algorithm uses the global variable *NewNode* to keep track of newly allocated nodes in the case of node splitting. The function *InsInNode* inserts a key and its associated pointer in lexicographical order in a given node. The function *CreateNode* allocates storage for a new node and inserts one key and its left and right descendant pointers. The value *NoKey* is an impossible value for a key and it is used to signal that there is no propagation of splittings during an insertion.

Although B-trees can be used for internal memory dictionaries, this structure is most suitable for external searching. For external dictionaries, each node can be made large enough to fit exactly into a physical record, thus yielding, in general, high branching factors. This produces trees with very small height.

B-trees are well suited to searches which look for a range of keys rather than a unique key. Furthermore, since the B-tree structure is kept balanced during insertions and deletions, there is no need for periodic reorganizations.

Several variations have been proposed for general B-trees with the intention of improving the utilization factor of the internal nodes. Note that a better storage utilization will result in a higher effective branching factor, shorter height and less complexity. The variations can be loosely grouped in three different classes.

Overflow techniques

There are several overflow techniques for B-trees. The most important are **B*-trees** and solutions based on multiple bucket sizes. Both cases are variations which try to prevent the splitting of nodes.

In B*-trees, when an overflow occurs during an insertion, instead of splitting the node we can:

(1) scan a right or left brother of the node to see if there is any room, and, if there is, we can transfer one key–pointer pair (the leftmost or rightmost respectively) to make room in the overflowed node;

(2) scan both left and right siblings of a node;

(3) scan all the descendants of the parent of the node.

If splitting is still necessary, the new nodes may take some keys from their siblings to achieve a more even distribution of keys in nodes. In the worst-case a 67% node storage utilization is achieved, with an average value of approximately 81%.

When we have multiple bucket sizes, instead of splitting the node, we expand it. This is called a partial expansion. When the bucket reaches the maximum size, we split it into two buckets of minimum size. The simplest case is having two bucket sizes of relative size ratio 2/3. This also gives a 67% worst-case storage utilization and around 80% average storage utilization (including external fragmentation owing to two bucket sizes). There are also adaptive overflow techniques that perform well for sorted or non-uniformly distributed inputs based on multiple bucket sizes.

Variable-length array implementations

These variations replace the arrays used to store keys and pointers at every node for some other structure which allows variable length, and may save space when the node is not full. For example, we could use a linked list where each node in the list contains a key and a pointer to the subtree at its left and the last pointer of the list points to the rightmost subtree. The sequence, in this case, is defined by:

$$\mathbf{s - D} : [\mathbf{KEY}, [\mathbf{D}], \mathbf{s - D}] ; [\mathbf{D}]$$

Each node in the B-tree contains one of these sequences. These sequences can be viewed as restricted binary trees, with two types of pointers: vertical pointers (those which point to nodes down the tree) and horizontal pointers (those pointing at the next link of the list). This type of tree is called **symmetric binary tree** (see Section 3.4.2.2).

When the keys are themselves of variable length, we can slightly relax the conditions on B-trees and require that each node be between 50% and 100% full, without any explicit reference to the actual number of keys stored.

Let m be the total number of characters that can be stored in a node, and let k be the maximum size of a key. Then we can guarantee that the number of characters per node will be between $\lfloor (m+1)/2 \rfloor - k$ and m.

Index B-trees, B+-trees or B*-trees

The idea behind these trees is to move all the data which is normally associated with a record to the leaves of the tree. The internal nodes contain only keys which are used to direct the searching; the complete records reside at the leaves. The keys in the internal nodes may not even belong to the file. Typically the leaves are pointers to external buckets of uniform size b. The data structure is now represented as:

$$\mathbf{mt - N - D - LEAF} \rightarrow \mathbf{mt - 2m - KEY} - [\mathbf{D}_1^b].$$

The above variations are somewhat orthogonal, in the sense that these can be applied simultaneously to achieve varying degrees of optimization. Note that the limits of the range for any gain in efficiency are from about 70% occupation (for randomly generated trees) to 100% occupation (optimal trees). The coding complexity of some of these implementations may not justify the gains.

Table 3.26 presents simulation results of 6–11 trees for several sizes, and Table 3.27 shows simulation results for various branching factors and a constant size. In both cases, E_n indicates the number of nodes accessed, $h(n)$ indicates the height of the tree, N_n is the average number of nodes in the tree, and S_n is the average number of splits that the $n+1$th insertion will require.

The simulation results indicate that the variance on the number of nodes accessed is very small. Induced by the formula for the upper bound on the variance, and with the arbitrary assumption that

Table 3.26: Simulation results for 6–11 trees.

n	$E[E_n]$	$E[h(n)]$	N_n/n	S_n
5	1	1	0.2	0
10	1	1	0.1	1
50	1.889599±0.000007	2±0.0000003	0.150401±0.000007	0.12718±0.00009
100	2.83386±0.00016	2.9623±0.0002	0.158109±0.000009	0.13922±0.00013
500	2.860087±0.000008	3±0.000003	0.145913±0.000008	0.13623±0.00012
1000	3.857201±0.000009	4±0.000007	0.146799±0.000009	0.13972±0.00013
5000	3.8792±0.0011	4.0243±0.0011	0.145827±0.000011	0.14724±0.00015
10000	4.854505±0.000011	5±0.000089	0.145995±0.000011	0.14704±0.00016
50000	5.85293±0.00079	5.9990±0.0008	0.146199±0.000012	0.14651±0.00016

Table 3.27: Simulation results for B-trees with 10000 keys.

type	$E[E_n]$	$E[h(n)]$	N_n/n	S_n
2–3	10.25436±0.00032	10.9993±0.0003	0.746064±0.000039	0.74588±0.00029
6–11	4.854505±0.000011	5.00000±0.00009	0.145995±0.000011	0.14704±0.00016
11–21	3.927589±0.000008	4.00000±0.00009	0.072811±0.000008	0.07636±0.00016
21–41	2.963877±0.000006	3.00000±0.00010	0.036423±0.000006	0.03806±0.00016
51–101	2.986036±0.000005	3.00000±0.00016	0.014264±0.000005	0.01278±0.00016

$$\sigma^2(E_n) = \alpha \frac{m+1}{m^2} + \beta$$

for $n = 10000$ we find that

$$\alpha = 0.6414 \pm 0.0005; \text{ and } \beta = 0.0053 \pm 0.0005 .$$

General references:
[Bayer, R., 71], [Bayer, R. *et al.*, 72], [Knuth, D.E., 73], [Wagner, R.E., 73], [Wong, C.K. *et al.*, 73], [Bayer, R., 74], [Bayer, R. *et al.*, 76], [Horowitz, E. *et al.*, 76], [Samadi, B., 76], [Shneiderman, B. *et al.*, 76], [Wirth, N., 76], [Bayer, R. *et al.*, 77], [Guibas, L.J. *et al.*, 77], [McCreight, E.M., 77], [Reingold, E.M. *et al.*, 77], [Gotlieb, C.C. *et al.*, 78], [Held, G. *et al.*, 78], [Maly, K., 78], [Snyder, L., 78], [Comer, D., 79], [Frederickson, G.N., 79], [Strong, H.R. *et al.*, 79], [Quitzow, K.H. *et al.*, 80], [Standish, T.A., 80], [Wright, W.E., 80], [Batory, D.S., 81], [Culik II, K. *et al.*, 81], [Gotlieb, L.R., 81], [Hansen, W.J., 81], [Huddleston, S. *et al.*, 81], [Ouksel, M. *et al.*, 81], [Robinson, J.T., 81],

[Rosenberg, A.L. *et al.*, 81], [Eisenbarth, B. *et al.*, 82], [Ottmann, T. *et al.*, 82], [Ziviani, N., 82], [Aho, A.V. *et al.*, 83], [Cesarini, F. *et al.*, 83], [Gupta, U.I. *et al.*, 83], [Kuspert, K., 83], [Tamminen, M., 83], [van Leeuwen, J. *et al.*, 83], [Arnow, D. *et al.*, 84], [Bell, D.A. *et al.*, 84], [Diehr, G. *et al.*, 84], [Leung, H.C., 84], [Mehlhorn, K., 84], [Bagchi, A. *et al.*, 85], [Huang, S-H.S., 85], [Langenhop, C.E. *et al.*, 85], [Wright, W.E., 85], [Gupta, G.K. *et al.*, 86], [Wirth, N., 86], [Driscoll, J.R. *et al.*, 87], [Litwin, W. *et al.*, 87], [Lomet, D.B., 87], [Aldous, D. *et al.*, 88], [Pramanik, S. *et al.*, 88], [Ramakrishna, M.V. *et al.*, 88], [Salzberg, B., 88], [Sedgewick, R., 88], [Veroy, B.S., 88], [Baeza-Yates, R.A. *et al.*, 89], [Baeza-Yates, R.A., 89], [Baeza-Yates, R.A., 89], [Burton, F.W. *et al.*, 89], [Johnson, T. *et al.*, 89], [Langenhop, C.E. *et al.*, 89], [Baeza-Yates, R.A. *et al.*, 90], [Baeza-Yates, R.A., 90], [Cormen, T.H. *et al.*, 90], [Huang, S-H.S. *et al.*, 90], [Odlyzko, A.M., to app.].

3.4.2.1 2–3 trees

2–3 trees are the special case of B-trees when $m = 1$. Each node has two or three descendants, and all the leaves are at the same depth.

$$\lceil \log_3 n + 1 \rceil \;\le\; h(n) \;\le\; \lfloor \log_2 n + 1 \rfloor$$

Let t_n be the number of different 2–3 trees with n leaves. Then

$$B(z) \;=\; \sum_{n=0}^{\infty} t_n z^n \;=\; B(z^2 + z^3) + z$$

$$t_n \;=\; \frac{\phi^n}{n} Q(\ln\, n)(1 + O(n^{-1}))$$

where $\phi = (1 + \sqrt{5})/2$ is the 'golden ratio', and $Q(x)$ is a periodic function with period $\ln(4 - \phi)$ and mean value $(\phi \ln(4 - \phi))^{-1}$.

Let N_n be the expected number of nodes in a 2–3 tree built by the insertion of a random permutation of n keys. Then

$$0.7377... + O(1/n) \;\le\; \frac{N_n}{n} \;\le\; 0.7543... + O(1/n)$$

Let S_n be the number of node-splits produced by an insertion into a random 2–3 tree with n keys, then

$$0.7212... + O(1/n) \;\le\; S_n \;\le\; 0.5585... + 0.03308... \log_2(n+1) + O(1/n)$$

If S_n converges when $n \to \infty$ then

$$S_\infty \;\le\; 0.7543...$$

If we assume all trees of height h are equally likely, then

$$N_n \;=\; (0.48061...)3^h$$

$$E[keys] = (0.72161...)3^h$$

The algorithm for searching and performing insertions in 2–3 trees is the same as the general algorithm for B-trees with $m = 1$.

As opposed to general B-trees, 2–3 trees are intended for use in main memory.

In Table 3.28, we give figures showing the performance of 2–3 trees constructed from random sets of keys.

Table 3.28: Exact and simulation results for 2–3 trees.

n	$E[E_n]$	$E[h(n)]$	N_n/n	S_n
5	1.68	2	0.72	0.40
10	2.528571	3	0.771429	0.522078
50	4.18710±0.00023	4.84606±0.00025	0.755878±0.000032	0.71874±0.00021
100	4.71396±0.00047	5.40699±0.00049	0.747097±0.000035	0.75062±0.00023
500	6.46226±0.00093	7.19371±0.00094	0.745831±0.000035	0.74726±0.00025
1000	7.27715±0.00042	8.01493±0.00042	0.745800±0.000035	0.74550±0.00025
5000	9.25824±0.00040	10.0023±0.0004	0.746027±0.000038	0.74591±0.00028
10000	10.25436±0.00032	10.9993±0.0003	0.746064±0.000039	0.74588±0.00029
50000	12.2518±0.0014	12.9977±0.0014	0.746090±0.000043	0.74610±0.00031

2–3 brother trees

2–3 brother trees are 2–3 trees with the additional constraint that a binary node has to have ternary brothers. With this restriction it is still possible, albeit complicated, to update a tree in $O(\log n)$ time. Let N_n^B be the number of nodes and $h^B(n)$ the height of a 2–3 brother tree with n keys. Then

$$\lceil \log_3 (n+1) \rceil \leq h^B(n) \leq \lfloor 0.78644... \log_2 n - 0.39321... \rfloor$$

$$\frac{1}{2} \leq \frac{N_n^B}{n} \leq \frac{1}{\sqrt{2}} = 0.70710...$$

$$1 \leq \frac{E[N_n^B]}{n} \leq 1.4142...$$

References:

[Aho, A.V. *et al.*, 74], [Brown, M.R. *et al.*, 78], [Brown, M.R., 78], [Kriegel, H.P. *et al.*, 78], [Rosenberg, A.L. *et al.*, 78], [Yao, A.C-C., 78], [Brown, M.R., 79], [Larson, J.A. *et al.*, 79], [Miller, R. *et al.*, 79], [Reingold, E.M., 79], [Vaishnavi, V.K. *et al.*, 79], [Bent, S.W. *et al.*, 80], [Brown, M.R. *et al.*, 80], [Olivie, H.J., 80], [Bitner, J.R. *et al.*, 81], [Kosaraju, S.R., 81], [Maier, D. *et al.*, 81], [Eisenbarth, B. *et al.*, 82], [Gupta, U.I. *et al.*, 82], [Huddleston, S. *et al.*, 82], [Mehlhorn, K., 82], [Ziviani, N., 82], [Kriegel, H.P. *et al.*, 83], [Murthy, Y.D. *et al.*, 83], [Zaki, A.S., 83], [Zaki, A.S., 84], [Baeza-Yates, R.A. *et al.*, 85], [Bagchi, A. *et al.*, 85], [Klein, R. *et al.*, 87], [Aldous, D. *et al.*, 88], [Wood, D., 88].

3.4.2.2 Symmetric binary B-trees

Symmetric binary B-trees (SBB trees) are implementations of 2–3 trees. A 2–3 tree node with a single key is mapped into a binary tree node directly; a 2–3 node with two keys is mapped into two nodes as indicated in Figure 3.3.

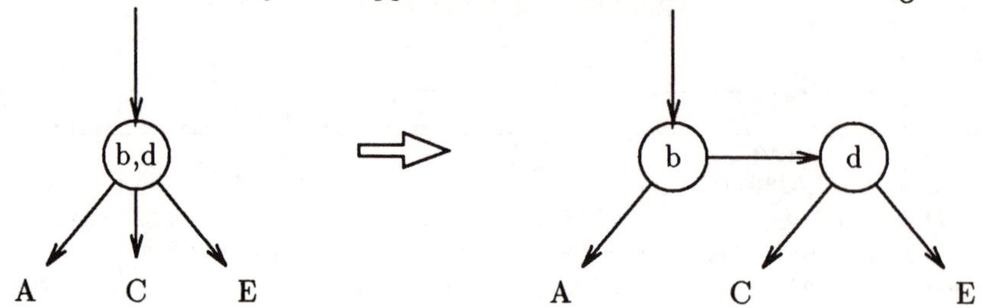

Figure 3.3: Transformation of 2–3 node into an SBB node.

SBB trees are binary search trees in which the right and left pointers may be either **vertical pointers** (normal pointers) or **horizontal pointers**. In an SBB tree all paths have the same number of vertical pointers (as in a true B-tree). All nodes except the leaves have two descendants and there are no two consecutive horizontal pointers in any path. In order to maintain the SBB tree property one bit per node is needed to indicate whether the incoming pointer is horizontal or vertical.

Random retrievals, insertions and deletions of keys in an SBB tree with n keys can be done in time of $O(\log n)$. If we let $k(n)$ be the maximum number of keys in any path and $h(n)$ be the height of the SBB tree (calculated by counting only vertical pointers plus one), we have

$$h(n) \leq k(n) \leq 2h(n)$$

$$\lceil \log_2 (n+1) \rceil \leq k(n) \leq 2\lfloor \log_2 (n+2) \rfloor - 2$$

Let S_n indicate the number of split transformations (a split transformation for SBB trees is similar to a rotation for AVL trees, see Section 3.4.1.3) required during the insertion of the nth key into a random tree. Let HI_n indicate the number of local height increase transformations required during the insertion of the nth key into the tree. Then

$$0.35921... + O(n^{-5}) \leq E[S_n] \leq 0.55672... + O(n^{-5})$$

$$E[HI_n] = \frac{23}{35} + O(n^{-5})$$

Let V_n be the number of nodes visited to process n random insertions/deletions into the empty tree. Then

$$E[V_n] \leq \frac{5}{2}n + O(n^{-5})$$

Table 3.29 shows some simulation results for SBB trees. C_n is the average number of nodes visited during a successful search and S_n, V_n and $h(n)$ have the meaning described earlier.

Table 3.29: Simulation results for SBB trees.

n	C_n	S_n	V_n/n	$E[h(n)]$
5	2.2000±0.0003	0.213±0.023	1.213±0.023	3.000±0.020
10	2.9057±0.0035	0.293±0.015	1.663±0.021	4.023±0.022
50	4.9720±0.0051	0.3594±0.0050	2.1692±0.0073	7.009±0.016
100	5.9307±0.0054	0.3733±0.0046	2.2757±0.0072	8.093±0.033
500	8.2419±0.0059	0.3868±0.0027	2.3801±0.0047	11.027±0.026
1000	9.2537±0.0062	0.3872±0.0023	2.3975±0.0042	12.140±0.068
5000	11.6081±0.0073	0.3876±0.0013	2.4088±0.0023	15.014±0.028
10000	12.6287±0.0083	0.3880±0.0011	2.4109±0.0019	16.180±0.108

From the simulation results we can see that the value for C_n is close to the value of $\log_2 n$; in particular, under the arbitrary assumption that

$$C_n = \alpha \log_2 n + \beta$$

then

$$\alpha = 1.0186 \pm 0.0010; \text{ and } \beta = -0.909 \pm 0.011 .$$

While every AVL tree (see Section 3.4.1.3) can be transformed into an SBB tree, the converse is not true. Thus the class of AVL trees is a proper subclass of the SBB trees. Experimental results show that, on the average, SBB trees perform approximately as well as AVL trees. Indeed SBB trees require less work than AVL trees to maintain balance, but this is at the expense of search time. (The search time is only slightly longer and the maintenance time is in some areas significantly less.) As a practical structure SBB trees should be considered as an option for representing dictionaries.

References:
[Bayer, R., 72], [Olivie, H.J., 80], [Ziviani, N. *et al.*, 82], [Ziviani, N., 82], [Tarjan, R.E., 83], [Ziviani, N. *et al.*, 85].

3.4.2.3 1–2 trees

1–2 trees are a special case of B-trees in which every node can have either one key or no keys. Consequently, every node has either two or one descendants. A node with no keys and one descendant is called a **unary node**. Since we allow nodes without keys, some additional restrictions are usually imposed so that a tree containing n keys is of bounded size (number of nodes).

1–2 brother trees

1–2 brother trees are 1–2 trees with the additional constraint that every unary node has a binary brother.

There is a close correspondence between 1–2 brother trees and AVL trees (see Section 3.4.1.3), as any 1–2 brother tree can be easily converted into an AVL tree and vice versa. This correspondence is a very natural one and consists in deleting the unary nodes (brother→avl) or inserting a unary node on the shorter subtree (avl→brother) of every node. Moreover, for some methods of insertion and deletion, any sequence of operations (insertions/deletions) on AVL trees and 1–2 brother trees will produce equivalent trees.

Let N_n^B be the number of nodes in a 1–2 brother tree with n keys, then for a tree constructed from a set of random keys:

$$n \leq N_n^B \leq 1.61803...n$$

$$\frac{40n-5}{35} \leq E[N_n^B] \leq \frac{156n-19}{105}$$

for $n \geq 6$.

1–2 son trees

1–2 son trees are 1–2 trees with the additional constraint that in no parent–descendant pair are both nodes unary. There is a close correspondence between 1–2 son trees and SBB trees (see Section 3.4.2.2) as any son tree can be converted to an SBB tree and vice versa. With this restriction, letting N_n^S denote the number of nodes used by a tree with n keys

$$n \leq N_n^S \leq 3n+1$$

$$\frac{48n+13}{35} \leq E[N_n^S] \leq \frac{72n-33}{35}$$

for $n \geq 6$.

1–2 neighbour trees

Neighbour trees of order k are 1–2 trees with the additional constraint that every unary node has at least one right neighbour and its first k right neighbours, if these exist, are binary. For these trees the height is bounded by

$$\lfloor \log_2(n+1) \rfloor \leq h(n) \leq \frac{\log_2 n}{\log_2(2-1/(k+1))} + 1$$

References:
[Maurer, H.A. *et al.*, 76], [Ottmann, T. *et al.*, 78], [Ottmann, T. *et al.*, 79], [Olivie, H.J., 80], [Ottmann, T. *et al.*, 80], [Ottmann, T. *et al.*, 80], [Olivie, H.J., 81], [Ottmann, T. *et al.*, 81], [Mehlhorn, K., 82], [Ottmann, T. *et al.*, 84], [Klein, R. *et al.*, 87], [Wood, D., 88].

3.4.2.4 2-3-4 trees

2-3-4 trees are similar to B-trees. We allow nodes having two, three, or four children. As for B-trees, all the leaves are at the same level, and this property is maintained through node splitting when we perform an insertion.

It is possible to represent 2-3-4 trees as binary trees. These are called **red-black trees**. A red-black tree is a binary search tree where every node has a colour, which can be either red or black. The correspondence with 2-3-4 trees is as follows:

(1) A black node with two red children is equivalent to a four children node;

(2) a black node with one red child (the other must be black) corresponds to a three children node;

(3) a black node with no red children is a two-child node (both children are black).

According to the above, the colouring of the nodes satisfies the following properties:

(1) Every leaf (external node) is black.

(2) A red node must have black children.

(3) Every path from a node to a leaf contains the same number of black nodes.

With these restrictions, we have

$$\lfloor \log_2(n+1) \rfloor \leq h(n) \leq 2\log_2(n+1)$$

Maintaining the colouring properties (that is, balancing the tree) of red-black trees, during an insertion or a deletion, is done through rotations (Section 3.4.1.8).

References:
[Guibas, L.J. *et al.*, 78], [Sedgewick, R., 88], [Cormen, T.H. *et al.*, 90].

3.4.2.5 B-tree variations

As the B-tree is one of the most popular external data structures, many variations have been devised. Of particular interest has been the combination of the fast access time of hashing, with the lexicographic order of B-trees. The most important variations are:

(1) *Prefix B-trees* this is a B-tree oriented to store strings (or variable length keys). Every internal node has a variable number of keys, with each key being the minimal length string that satisfies the lexicographical order condition of the B-tree.

(2) *Bounded disorder* this is an index B-tree where each bucket is organized as a multipage hashing table. Inside each page the keys are maintained in sorted order.

(3) *Digital B-trees* in this case, the access to the buckets of an index B-tree is done using the digital decomposition of the keys.

References:
[Lomet, D.B., 81], [Scheurmann, P. *et al.*, 82], [Lomet, D.B., 83], [Litwin, W. *et al.*, 86], [Hsiao, Y-S. *et al.*, 88], [Baeza-Yates, R.A., 89], [Christodoulakis, S. *et al.*, 89], [Lomet, D.B. *et al.*, 89], [Baeza-Yates, R.A., 90], [Lomet, D.B. *et al.*, 90].

3.4.3 Index and indexed sequential files

An **indexed file** is a superimposition of a dictionary structure called the **main file** upon another dictionary structure called the **index file**. The index file is constructed on a subset of the keys of the main file. Using our notation for data structures, a single level index is defined by:

$$\mathbf{main-file(KEY)} \;:\; \mathbf{SET(\,bucket(KEY)\,)}$$

$$\mathbf{index(KEY)} \;:\; \mathbf{DICT1(\,KEY, [bucket(KEY)]\,)}$$

$$\mathbf{bucket(KEY)} \;:\; \mathbf{DICT2(KEY)};$$

In the above definition, **DICT1** stands for the organization of the index file and **DICT2** for the organization of each individual bucket (both mapping to **DICT**), while the collection of all the **bucket(KEY)** forms the main file.

Indexed files can be organized in several levels. By adding an index of the index we increase the number of levels by one. This is formally described by mapping the **bucket(KEY)** to

$$\mathbf{bucket(KEY)} \;:\; \mathbf{index(KEY)}$$

instead. If the same **DICT** structures for each level of indexing are chosen, the file has **homogeneous indexing**. In practice, the number of levels is very small and homogeneous (typically one or two levels).

The typical choices for the **DICT** structure in the index file are arrays and trees. The typical choice for the **bucket** is a sequential array. An indexed file can, however, be implemented using any selection for the **DICT** structures in the index file and bucket and the **SET** representation for the main file. Normally the following constraints are imposed on the structure:

(1) each index entry contains as key the maximum key appearing in the pointed **bucket(KEY)**.

(2) the index file structure should perform range searches, or nearest-neighbour searches efficiently, the type of search of most interest being 'search the smallest key $\geq X$'.

(3) the **bucket(KEY)** should allow some type of dynamic growth (overflow records, chaining, and so on), which should not be of bounded size.

(4) it should be possible to scan all the components in a bucket sequentially and all the components of the set sequentially, or, in other words, it should be possible to scan all the main file sequentially.

(5) the index contains an artificial key (∞) which is larger than any other key in the file.

Searching an array index

```
function SearchIndex(key : typekey) : BucketAddress;
var  low, high, j : integer;

begin
low := 0;
high := n; {*** highest index entry ***}
while high-low > 1 do begin
      j := (high+low) div 2;
      if key <= index[j].k then high := j
            else        low := j
      end;
SearchIndex := index[high].BuckAddr
end;
```

Searching array buckets with overflow

```
procedure SearchBucket(key : typekey; p : BucketAddress);
label 999;
var   i : integer;

begin
while p <> nil do begin
      ReadBucket(p) into bucket;
      i := B;
      while (i>1) and (bucket.r[i].k>key) do i := i-1;
      if bucket.r[i].k = key then goto 999  {*** break ***}
      else if i=B then  p := bucket.next
                  else  p := nil
      end;
999:
if p <> nil then  found(bucket.r[i])
      else        notfound(key)
end;
```

The goal of indexed files is to have an index small enough to keep in main memory, and buckets small enough to read with a single access. In this ideal situation, only one external access per random request is needed.

B*-trees (see Section 3.4.2) are a generalization of a special implementation of index files.

Searching a single-level indexed file

```
SearchBucket(key, SearchIndex(key));
```

Typically the index part of the file is considered to be a fixed structure and no updates are performed on it. In case the file grows or shrinks or alters its distribution significantly, it is easier to reconstruct the index entirely.

3.4.3.1 Index sequential access method

A particular implementation of indexed files are the **index sequential access method** (ISAM) files. For these files the index file and set are both arrays. The buckets are composed of an array of records of fixed maximum size and an additional pointer to 'overflow' buckets. Since the index file and main file are both arrays, there is no need to keep pointers in the index. The array index in the index file corresponds to the array index (bucket index) on the main file.

$$\mathbf{index(KEY)} : \{\mathbf{KEY}\}_1^{\mathbf{N}}$$

$$\mathbf{main-file} : \{\mathbf{bucket(KEY)}\}_1^{\mathbf{N+W}};$$

$$\mathbf{bucket(KEY)} : (\{\mathbf{KEY, D}\}_1^{\mathbf{B}}, \mathbf{int});$$

In the above definition, **B** is the bucket size, **N** denotes the number of buckets in the main file, and **W** denotes the number of buckets reserved for overflow. The integer in the **bucket(KEY)** is the index of the corresponding overflow bucket.

The buckets are designed to match closely the physical characteristics of devices, for example, typically a bucket fully occupies a track in a disk. In some cases the index is organized as an indexed file itself, in which case the ISAM becomes a two-level index. For two-level indices the same array structures are used. The top level index is made to match a physical characteristic of the device, for example, a cylinder in a disk.

General references:
[Chapin, N., 69], [Chapin, N., 69], [Ghosh, S.P. *et al.*, 69], [Senko, M.E. *et al.*, 69], [Collmeyer, A.J. *et al.*, 70], [Lum, V.Y., 70], [Mullin, J.K., 71], [Nijssen, G.M., 71], [Mullin, J.K., 72], [Cardenas, A.F., 73], [Casey, R.G., 73], [Wagner, R.E., 73], [Behymer, J.A. *et al.*, 74], [Grimson, J.B. *et al.*, 74], [Keehn, D.G. *et al.*, 74], [Shneiderman, B., 74], [Schkolnick, M., 75], [Schkolnick, M., 75], [Whitt, J.D. *et al.*, 75], [Wong, K.F. *et al.*, 75], [Yue, P.C. *et al.*, 75], [Gairola, B.K. *et al.*, 76], [Shneiderman, B. *et al.*, 76], [Anderson, H.D. *et al.*, 77], [Cardenas, A.F. *et al.*, 77], [Maruyama, K. *et al.*, 77], [Schkolnick, M., 77], [Senko, M.E., 77], [Severance, D.G. *et al.*, 77], [Gotlieb, C.C. *et al.*, 78], [Kollias, J.G., 78], [Nakamura, T. *et al.*, 78], [Mizoguchi, T., 79], [Strong, H.R. *et al.*, 79], [Zvegintzov, N., 80], [Batory, D.S., 81], [Larson, P., 81], [Leipala, T., 81], [Leipala, T., 82], [Willard, D.E., 82], [Burkhard, W.A., 83], [Cooper, R.B. *et al.*, 84], [Manolopoulos, Y.P., 86], [Willard, D.E., 86], [Ramakrishna, M.V. *et al.*, 88], [Rao, V.N.S. *et al.*, 88].

3.4.4 Digital trees

Digital trees or **tries** are recursive tree structures which use the characters, or digital decomposition of the key, to direct the branching. The name trie comes from the word re*trie*val. A node in a trie is either an external node and contains one record, or it is an internal node and contains an array of pointers to nodes or null pointers. The selection of the subtries of a node (entries of the array) is done by the ordering of the ith character of each key, where i is the depth of the node. The root node uses the first character of the key, the direct descendants of the root use the second character, and so on. At any level where the remaining subtrie has only one record, the branching is suspended. A trie of order **M** is defined by

$$\mathbf{tr - M - D} \; : \; [\{\mathbf{tr - M - D}\}_1^{\mathbf{M}}] \; ; \; [\mathbf{D}] \; ; \; \mathbf{nil}$$

The basic trie tree, if the underlying alphabet is ordered, is a lexicographically ordered tree. The character set is usually the alphabet or the decimal digits or both. Typically the character set has to include a string-terminator character (blank). If a string terminator character is available, tries can store variable length keys. In particular, as we use the smallest prefix of the key which makes the key unique, digital trees are well suited for handling unbounded or semi-infinite keys.

Let C_n and C'_n denote the average number of internal nodes inspected during a successful search and an unsuccessful search respectively. Let N_n denote the number of internal nodes in a trie with n keys, and let $h(n)$ denote its height. The **digital cardinality** will be denoted by m; this is the size of the alphabet and coincides with the dimension of the internal-node arrays.

In all the following formulas, $P(x)$ denotes complicated periodic (or convergent to periodic) functions with average value 0 and very small absolute value. These functions should be ignored for any practical purposes. Although we use $P(x)$ for all such functions, these may be different.

For tries built from random keys, uniformly distributed in $U(0,1)$ (or keys composed of random-uniform digits) we have:

$$E[N_n] = 1 + m^{1-n} \sum_{i=0}^{n} \binom{n}{i} (m-1)^{n-i} E[N_i] \qquad (N_0 = N_1 = 0)$$

$$= \frac{n}{\ln m}(1 + P(\log_m n)) + O(1)$$

$$C_n = 1 + m^{1-n} \sum_{i=1}^{n} \binom{n-1}{i-1} (m-1)^{n-i} C_i \,, \qquad (C_0 = C_1 = 0)$$

$$= \frac{H_{n-1}}{\ln m} + \frac{1}{2} + P(\log_m n) + O(n^{-1})$$

$$C'_n = 1 + m^{-n} \sum_{i=2}^{n} \binom{n}{i} (m-1)^{n-i} C'_i \qquad (C'_0 = C'_1 = 0)$$

$$= \frac{H_n - 1}{\ln m} + \frac{1}{2} + P(\log_m n) + O(n^{-1})$$

$$E[h(n)] = 2 \log_m n + o(\log n)$$

where $H_n = \sum_{i=1}^{n} 1/i$ denote the harmonic numbers. Table 3.30 shows some exact values.

Digital tree (trie) search

```
search(key, t)
typekey key;
trie t;

{
int depth;
for(depth=1; t!=NULL && !IsData(t); depth++)
      t = t->p[charac(depth,key)];
if(t != NULL && key == t->k)
      found(t);
else  notfound(key);
}
```

Digital tree (trie) insertion

```
trie insert(key, t, depth)
typekey key;
trie t;
int depth;

{
int j;
trie t1;
if (t==NULL) return(NewDataNode(key));
if (IsData(t))
      if (t->k == key)
            Error /*** Key already in table ***/;
      else { t1 = NewIntNode();
            t1->p[charac(depth,t->k)] = t;
            t = insert(key, t1, depth);
            }
else  { j = charac(depth,key);
      t->p[j] = insert(key, t->p[j], depth+1);
      }
return(t);
}
```

The function $charac(i, key)$ returns the ith character of a key. It is expected that the result is an integer in the range 0 to $m-1$. The function *insert* uses the level indicator *depth* to facilitate the search. The user should call this function with depth 1; for example, *insert(key, trie*, 1). The function *IsData(t)* tests whether a pointer points to an internal node or to a data node. The functions *NewIntNode* and *NewDataNode* create new nodes of the corresponding types.

In cases where there is no value associated with the key, we can avoid the data records completely with a special terminator (such as nil*) which indicates that a string key terminates there. The key, if desired, can be reconstructed from the path in the tree.

There is a very close correspondence between a trie tree and top-down radix sort, as the trie structure reflects the execution pattern of the sort, each node corresponds to one call to the sorting routine.

Table 3.30: Exact results for general tries.

	$m = 2$			
n	$E[N_n]$	C_n	C'_n	$E[h(n)]$
10	13.42660	4.58131	3.28307	6.92605±0.00068
50	71.13458	6.96212	5.54827	11.6105±0.0017
100	143.26928	7.96937	6.54110	13.6108±0.0025
500	720.34810	10.29709	8.85727	18.2517±0.0060
1000	1441.69617	11.29781	9.85655	20.2566±0.0087
5000	7212.47792	13.62031	12.17792	24.877±0.020
10000	14425.95582	14.62039	13.17785	26.769±0.027
50000	72133.67421	16.94237	15.49970	30.246±0.031
	$m = 10$			
10	4.11539	1.70903	1.26821	2.42065±0.00022
50	20.92787	2.43643	2.05685	3.84110±0.00059
100	42.60540	2.73549	2.26860	4.43724±0.00082
500	210.60300	3.44059	3.05159	5.8418±0.0021
1000	427.45740	3.73802	3.26849	6.4373±0.0029
5000	2107.33593	4.44100	4.05106	7.8286±0.0071
10000	4275.97176	4.73827	4.26847	8.3965±0.0091
50000	21074.66351	5.44104	5.05100	9.494±0.020

When the cardinality of the alphabet is large and consequently internal nodes are of significant size compared to a record, the trie becomes inefficient in its use of storage. For example, if only two keys reach a given internal node, we have to include a complete internal node which will be mostly under-utilized. In some sense, tries are efficient close to the root where the branching is dense, but inefficient close to the leaves.

General references:
[de la Brandais, R., 59], [Fredkin, E., 60], [Sussenguth, E.H., 63], [Patt, Y.N., 69], [Knuth, D.E., 73], [Burkhard, W.A., 76], [Horowitz, E. *et al.*, 76], [Maly, K., 76], [Stanfel, L., 76], [Burkhard, W.A., 77], [Comer, D. *et al.*, 77], [Miyakawa, M. *et al.*, 77], [Nicklas, B.M. *et al.*, 77], [Reingold, E.M. *et al.*, 77], [Gotlieb, C.C. *et al.*, 78], [Comer, D., 79], [Mehlhorn, K., 79], [Tarjan, R.E. *et al.*, 79], [Comer, D., 81], [Litwin, W., 81], [Lomet, D.B., 81], [Regnier, M., 81], [Tamminen, M., 81], [Devroye, L., 82], [Flajolet, P. *et al.*, 82], [Knott, G.D., 82], [Orenstein, J.A., 82], [Comer, D., 83], [Flajolet, P. *et al.*, 83], [Flajolet, P., 83], [Devroye, L., 84], [Mehlhorn, K., 84], [Flajolet, P. *et al.*, 85], [Flajolet, P. *et al.*, 86], [Jacquet, P. *et al.*, 86], [Kirschenhofer, P. *et al.*, 86], [Litwin, W. *et al.*, 86], [Pittel, B., 86], [Szpankowski, W., 87], [de la Torre, P., 87], [Kirschenhofer, P. *et al.*, 88], [Lomet, D.B., 88], [Sedgewick, R., 88], [Szpankowski, W., 88], [Szpankowski, W., 88], [Luccio, F. *et al.*, 89], [Szpankowski, W., 89], [Murphy, O.J., 90].

3.4.4.1 Hybrid tries

It is for the above reason that tries are usually composed with some other structure to allow for their efficient behaviour at the root but to switch to some other data structure closer to the leaves. All these compositions have the common definition:

$$\mathbf{tr-M-D} \; : \; [\{\mathbf{tr-M-D}\}_1^{\mathbf{M}}] \, ; \, [\mathbf{D}] \, ; \, \mathbf{DICT(D)} \, ; \, \mathbf{nil}$$

Common compositions are with external buckets ($\mathbf{DICT(D)} \rightarrow \{\mathbf{D}\}_{\mathbf{n}}^{\mathbf{b}}$), called **bucket tries**, and with binary search trees ($\mathbf{DICT(D)} \rightarrow \mathbf{bt-D-nil}$, see Section 3.4.1).

For bucket tries, after the insertion of n random keys uniformly distributed in $U(0,1)$, we have

$$N_n = \frac{n}{b \ln m}(1 + P(\log_m n)) + O(1)$$

$$C_n = \frac{H_{n-1} - H_{b-1}}{\ln m} + \frac{1}{2} + P(\log_m n) + O(n^{-1})$$

$$C_n' = \frac{H_n - H_b}{\ln m} + \frac{1}{2} + P(\log_m n) + O(n^{-1})$$

The exact formulas for the above quantities are the same as the ones for general tries but with the extended initial condition: $N_0 = N_1 = ... = N_b = 0$. For bucket binary tries, that is, when $m = 2$ we have

$$E[h(n)] = (1 + 1/b)\log_2 n + \frac{\gamma}{\ln 2} - \log_2((b+1)!) + P((1 + 1/b)\log_2 n) + o(1)$$

Bucket binary tries are used as the collision resolution mechanism for dynamic hashing (see Section 3.3.14).

A different type of hybrid trie is obtained by implementing the array in the internal nodes with a structure which takes advantage of its possible sparsity: for example, a linked list consisting of links only for non-empty subtries. Almost any technique of those used for economizing storage in B-tree nodes can be applied to the internal nodes in the tries (see Section 3.4.2).

3.4.4.2 Tries for word-dictionaries

Digital trees seem very appropriate to implement language dictionaries. The most important reason, besides their efficiency, is that tries allow for efficient prefix searching. **Prefix search** is searching for any word which matches a given prefix, for example, searching for *comput** where the asterisk can be matched by any string (see Section 7.2.2).

There are some problems associated with this particular application though: long common prefixes tend to create unnecessary additional levels with very little (maybe unary) branching. For example, the words *computation, computational, computations* will force 11 levels of branching before these words can be separated. If prefix searching is not needed, this problem can be remedied by organizing the scan of the characters of the key in reverse order (as suffixes are shorter and less common than prefixes).

More generally and much better, if we are prepared to lose the lexicographical ordering of the keys, is to consider the function $charac(i, key)$ as a hashing function which operates on the key and returns an integer value with a rather uniform distribution. This option may be particularly appealing when the cardinality of the alphabet is large and the usage distribution is uneven (as would be the case for a full ASCII set under normal circumstances). In this latter case the hashing function can be applied to the characters individually.

3.4.4.3 Digital search trees

Digital search trees are a particular implementation of tries where a record is stored in each internal node. The hyperrule which defines these trees is

$$\mathbf{dst - M - D} \ : \ [\mathbf{D}, \{\mathbf{dst - M - D}\}_1^{\mathbf{M}}] \ ; \ \mathbf{nil}$$

The binary digital search trees use the same structure as the binary search trees (see Section 3.4.1); the only difference between them is that the selection of subtrees is not based on comparisons with the key in the node, but on bit inspections.

Let C_n and C_n' be the average number of nodes inspected during a successful and an unsuccessful search respectively. Then for digital search trees constructed from random uniform keys (or keys composed of random digits) we have:

$$N_n = n$$

$$C_n = 1 + \frac{(n-1)m^{1-n}}{n} \sum_{i=1}^{n} \binom{n-2}{i-1} (m-1)^{n-i-1} C_i \qquad (C_0 = 0)$$

$$C_n = \log_m n + \frac{\gamma - 1}{\ln m} + \frac{3}{2} - \alpha_m + P(\log_m n) + O\left(\frac{\ln n}{n}\right)$$

$$\begin{aligned} C'_n &= (n+1)C_{n+1} - nC_n - 1 \\ &= \log_m n + \frac{\gamma}{\ln m} + \frac{1}{2} - \alpha_m + P(\log_m n) + O(n^{-1}) \end{aligned}$$

$$\lim_{n \to \infty} E[h(n)] = \log_m n \qquad \text{(in probability)}$$

where

$$\alpha_m = \sum_{k \geq 1} \frac{1}{m^k - 1}$$

$$\alpha_2 = 1.60669...$$

Table 3.31 shows some exact values.

The selection of which key is placed in each node is arbitrary among all the keys of its subtree. As the selected key does not affect the branching (other than by not being in the subtree), any choice will give almost equivalent subtrees. This fact leaves room for optimizing the trees. The most common, and possibly the best, strategy is to choose the key with highest probability. This is equivalent to building the tree by inserting keys in descending probability order.

Table 3.31: Exact results for digital search trees.

	$m = 2$		$m = 10$	
n	C_n	C'_n	C_n	C'_n
10	3.04816	3.24647	2.19458	1.64068
50	5.06061	5.41239	2.90096	2.32270
100	6.00381	6.39134	3.19015	2.61841
500	8.26909	8.69616	3.89782	3.31913
1000	9.26011	9.69400	4.18865	3.61622
5000	11.57373	12.01420	4.89731	4.31876
10000	12.57250	13.01398	5.18840	4.61600

3.4.4.4 Compressed tries

A compressed trie is a static tree for which the array of pointers at each internal node is represented by one base address and a bit array. Each bit indicates whether the corresponding entry points to a non-null subtrie or not. All non-null subtries are stored consecutively starting at the base address.

The easiest way of guaranteeing contiguity is by storing the trie as an array of records. The base address is an integer used as an index into the array. The hyperrule which defines the compressed trie is:

$$\mathbf{tr-M-D} \ : \ \{\mathbf{int}, \{\mathbf{bool}\}_1^{\mathbf{M}}\}_1^{\mathbf{N}}$$

By convention the root of the trie is at location 1. Given an internal node, its ith descendant will be found by adding the base integer plus the count of '1' bits in the array at the left of location i.

Compressed tries have the same complexity measures as the basic tries. Compressed tries achieve a good efficiency in searching and a very compact representation at the cost of being static structures.

3.4.4.5 Patricia trees

A Patricia tree is a particular implementation of a binary trie. The Patricia tree uses an index at each node to indicate the bit used for that node's branching. By using this index, we can avoid empty subtrees and hence guarantee that every internal node will have non-null descendants, except for the totally empty tree. A Patricia tree is defined by

$$\mathbf{pat-D} \ : \ [\mathbf{int}, \mathbf{pat-D}, \mathbf{pat-D}] \ ; \ [\mathbf{D}] \ \equiv \ \mathbf{bt-int-[D]}$$

As a binary tree, the Patricia tree stores all its data at the external nodes and keeps one integer, the bit index, in each internal node.

Let C_n be the average number of internal node inspections during a successful search and C_n' for an unsuccessful search. Then for trees constructed from n randomly distributed keys in $U(0,1)$ we have:

$$N_n = n - 1$$

$$C_n = 1 + \frac{1}{2^{n-1}-1}\sum_{i=1}^{n-1}\binom{n-1}{i-1}C_i \qquad (C_0 = C_1 = 0)$$

$$= \log_2 n + \frac{\gamma}{\ln 2} - \frac{1}{2} + P(\log_2 n) + O(n^{-1})$$

$$C_n' = 1 + \frac{1}{2^n-2}\sum_{i=1}^{n-1}\binom{n}{i}C_i' \qquad (C_0' = C_1' = 0)$$

$$= \log_2 n + \frac{\gamma - \ln \pi}{\ln 2} + \frac{1}{2} + P(\log_2 n) + O(n^{-1})$$

$$\lim_{n\to\infty} E[h(n)] = \log_2 n \qquad \text{(in probability)}$$

Table 3.32 shows some exact values.

Patricia tree search

```
search(key, t)
typekey key;
patricia t;

{
if (t==NULL) notfound(key);
else { while (!IsData(t))
            t = bit(t->level,key) ? t->right : t->left;
       if (key == t->k) found(t);
            else        notfound(key);
     }
};
```

Patricia tree insertion

```
patricia insert(key, t)
typekey key;
patricia t;

{patricia p;
 patricia InsBetween();
int i;
if (t==NULL) return(NewDataNode(key));

for(p=t; !IsData(p);)
      p = bit(p->level, key) ? p->right : p->left ;

/* find first different bit */
for (i=1; i<=D && bit(i,key)==bit(i,p->k); i++);
if (i>D) { Error /* Key already in table */;
         return(t); }
else return(InsBetween(key, t, i));
}

patricia InsBetween(key, t, i)
```

```
typekey key;
patricia t;
int i;

{patricia p;
if (IsData(t) || i < t->level) {
      /* create a new internal node */
      p = NewDataNode(key);
      return(bit(i,key) ? NewIntNode(i,t,p) : NewIntNode(i,p,t));
      }
if (bit(t->level,key)==1)
      t->right = InsBetween(key, t->right, i);
else  t->left  = InsBetween(key, t->left, i);
return(t);
};
```

The function $bit(i, key)$ returns the ith bit of a key. The functions *IsData*, *NewIntNode* and *NewDataNode* have the same functionality as the ones for tries.

Some implementations keep the number of bits skipped between the bit inspected by a node and the bit inspected by its parent, instead of the bit index. This approach may save some space, but complicates the calling sequence and the algorithms.

Patricia trees are a practical and efficient solution for handling variable length or very long keys; they are particularly well suited for text searching. Note that the problem generated by very long common prefixes virtually disappears for Patricia trees.

The structure generated by building a Patricia tree over all the semi-infinite strings resulting from a base string (or base text) is called a PAT tree and has several important uses in text searching (see Section 7.2.2).

Given a set of keys, the shape of the tree is determined, so there cannot be any conformation or reorganization algorithm.

In summary, digital trees provide a convenient implementation for several database applications. The most important reasons are:

(1) short searching time (successful or unsuccessful);

(2) they allow searching on very long or unbounded keys very efficiently;

(3) flexibility, as they allow composition with many other structures;

(4) they allow search of interleaved keys and hence they are amenable to multidimensional search.

Table 3.32: Exact and simulation results for Patricia trees.

n	C_n	C_n'	$E[h(n)]$
10	3.58131	3.07425	4.63400±0.00023
50	5.96212	5.33950	7.88927±0.00060
100	6.96937	6.33232	9.21029±0.00080
500	9.29709	8.64847	12.1681±0.0018
1000	10.29781	9.64775	13.3669±0.0029
5000	12.62031	11.96910	16.2120±0.0059
10000	13.62039	12.96903	17.382±0.010
50000	15.94237	15.29091	20.147±0.018

References:
[Morrison, D.R., 68], [Knuth, D.E., 73], [Merrett, T.H. *et al.*, 85], [Szpankowski, W., 86], [Kirschenhofer, P. *et al.*, 88], [Sedgewick, R., 88], [Kirschenhofer, P. *et al.*, 89].

3.5 Multidimensional search

The algorithms which allow non-atomic search keys, or keys composed of several subkeys, are called **multidimensional search** algorithms.

Any searching algorithm could, in principle, deal with composite keys, just by considering the composed key as a single block. For this reason only those search algorithms which treat the subkeys individually are called multidimensional search algorithms. In particular, the most important property of multidimensional search is to allow searching when only some of the subkeys are specified. This problem is called **partial-match searching** or **partial-match retrieval**. Retrieval on ranges of subkeys also requires special multidimensional searching algorithms.

Partial-match queries may have multiple answers, that is, more than one record may match part of the key. We will define two types of searches: **positive search**, when we search an element which is in the tree and we stop as soon as the record is found (denoted by C_n); **negative search**, when we do not know how many matches there will be and we search all of them (the *rsearch* function searches for all possible matches), denoted by C_n'.

Partial-match queries can be treated as a special case of range queries; for a specified subkey, the range is defined by a single value (upper bound = lower bound), and for an unspecified key the range is infinite (or sufficiently large to include all keys).

Partial-match query using range searching

```
lowk[0] = uppk[0] = value;   /*** specified value ***/
lowk[1] = -infinity;
uppk[1] = infinity;          /*** unspecified value ***/
. . . .
rsearch(lowk, uppk, t);
```

General references:
[Lum, V.Y., 70], [Dobkin, D. *et al.*, 74], [Rothnie, J.B. *et al.*, 74], [Dobkin, D. *et al.*, 76], [Raghavan, V.V. *et al.*, 77], [Bentley, J.L. *et al.*, 79], [Kosaraju, S.R., 79], [Ladi, E. *et al.*, 79], [Lipski, Jr., W. *et al.*, 79], [Bentley, J.L., 80], [Guting, R.H. *et al.*, 80], [Hirschberg, D.S., 80], [Lee, D.T. *et al.*, 80], [Guting, R.H. *et al.*, 81], [Ouksel, M. *et al.*, 81], [Eastman, C.M. *et al.*, 82], [Orenstein, J.A., 82], [Scheurmann, P. *et al.*, 82], [Willard, D.E., 82], [Guttman, A., 84], [Madhavan, C.E.V., 84], [Mehlhorn, K., 84], [Kent, P., 85], [Cole, R., 86], [Faloutsos, C. *et al.*, 87], [Karlsson, R.G. *et al.*, 87], [Munro, J.I., 87], [Sacks-Davis, R. *et al.*, 87], [Sellis, T. *et al.*, 87], [Willard, D.E., 87], [Fiat, A. *et al.*, 88], [Seeger, B. *et al.*, 88], [Henrich, A. *et al.*, 89], [Lomet, D.B. *et al.*, 89].

3.5.1 Quad trees

A quad search tree is an extension of the concept of binary tree search in which every node in the tree has 2^k descendants. While searching for a k-dimensional key, the corresponding descendant is selected based on the result of k comparisons. Each internal node of the quad tree contains one k-dimensional key and associated data. The hyperrule defining the quad trees is:

$$\mathbf{qt-N-D} \; : \; \mathbf{nil} \, ; \, [\mathbf{D}, \{\mathbf{qt-N-D}\}_0^{2^\mathbf{N}-1}]$$

The descendants of a quad tree node are numbered from 0 to $2^k - 1$. Let $b_0 b_1 ... b_{k-1}$ be the binary representation of a descendant number. If b_i is 1 then the ith subkeys in the descendant subtree are all larger than the ith key at the node; if $b_i = 0$ the subkeys are less or equal. For example, in two dimensions, say x and y, descendant $2 = 10_2$ contains the south-east sector of the plane.

$$E_n^k = (2^k - 1)I_n^k + 2^k n$$

$$C_n = \left(1 + \frac{1}{3n}\right) H_n - \frac{n+1}{6n} \qquad \text{(for } k = 2\text{)}$$

$$C_n' = H_n - \frac{n-1}{6(n+1)} \qquad \text{(for } k = 2\text{)}$$

$$Var[C_n'] = H_n^{(2)} + \frac{H_n}{2} + \frac{5n}{9} - \frac{4}{9n^2} - \frac{13}{6} \qquad \text{(for } k = 2\text{)}$$

$$C_n = 1 + \frac{4}{n^2}\sum_{i=1}^{n-1} i\,C_i\left((H_n - H_i)^2 + H_n^{(2)} - H_i^{(2)}\right) \qquad \text{(for } k = 3\text{)}$$

$$= \left(\frac{2}{3} + \frac{2}{21n}\right) H_n + 0.588226... + O(1/n)$$

$$C_n = \frac{2}{k}\ln n + \gamma_k + O\left(\frac{\log n}{n} + \log n\, n^{-2+2\cos\frac{2\pi}{k}}\right) \qquad \text{(for any } k\text{)}$$

where γ_k is independent of n.

For partial matches, for $k = 2$ when only one key is specified,

$$C_n' = \frac{\Gamma(2\alpha)}{2\Gamma(\alpha)^3} n^{\alpha-1} - 1 + o(1)$$

$$= 1.595099...n^{0.561552...} - 1 + o(1)$$

where $\alpha = \frac{\sqrt{17}-1}{2}$.

Quad tree search

```
search(key, t)
typekey key[K];
tree t;
{
int i, indx, noteq;
while(t != NULL) {
    indx = noteq = 0;
    for (i=0; i<K; i++) {
        indx = indx << 1;
        if (key[i] > t->k[i]) indx++;
        if (key[i] != t->k[i]) noteq++;
        }
    if (noteq)  t = t->p[indx];
        else { found(t); return; }
    }
notfound(key);
};
```

Quad tree insertion

```
tree insert(key, t)
typekey key[ ];
tree t;
{
int i, indx, noteq;
if (t==NULL) t = NewNode(key);

else {   indx = noteq = 0;
     for (i=0; i<K; i++) {
          indx = indx << 1;
          if (key[i] > t->k[i]) indx++;
          if (key[i] != t->k[i]) noteq++;
          }
     if (noteq)   t->p[indx] = insert(key, t->p[indx]);
     else Error; /*** Key already in table ***/
     }
return(t);
};
```

There are no efficient or simple methods for performing 'rotations' in quad trees. Consequently it is difficult to maintain a quad tree balanced.

There are no simple methods for performing deletions either. The best method for deletions is to mark the nodes as deleted, and reconstruct the tree whenever too many nodes have been deleted.

Quad trees with dimension three or higher become excessively expensive in terms of storage used by pointers. A quad tree has $(2^k - 1)n + 1$ null pointers.

Table 3.33 displays simulation results on randomly generated quad trees. C_n denotes the average successful search and $E[h(n)]$ the average height of a quad tree with n nodes.

3.5.1.1 Quad tries

Quad tries are similar to quad trees, but instead of using comparisons to select the descendant, they use the bits of the keys, as in a digital trie or a Patricia tree. Quad tries are usually called quad trees. The quad trie has no data in the internal nodes, these are used just for branching, the record information is stored in the external nodes.

Quad tries are generated by the hyperrule:

$$\mathbf{qt-N-D} \; : \; \mathbf{nil} \, ; \, [\mathbf{D}] \, ; \, [\{\mathbf{qt-N-D}\}_0^{2^\mathbf{N}-1}]$$

In all the following formulas, $P(x)$ denotes complicated periodic (or con-

Table 3.33: Exact and simulation results for quad trees of two and three dimensions.

	$k = 2$		$k = 3$	
n	C_n	$E[h(n)]$	C_n	$E[h(n)]$
5	2.23556	3.28455±0.00014	2.09307	2.97251±0.00013
10	2.84327	4.41439±0.00025	2.53845	3.78007±0.00022
50	4.35920	7.30033±0.00075	3.59019	5.81713±0.00058
100	5.03634	8.6134±0.0011	4.04838	6.72123±0.00086
500	6.63035	11.7547±0.0029	5.11746	8.8586±0.0021
1000	7.32113	13.1337±0.0043	5.57895	9.7953±0.0031
5000	8.92842	16.382±0.011	6.65135	11.9847±0.0076
10000	9.62125	17.784±0.015	7.11336	12.942±0.011
50000	11.2304	21.106±0.038	8.18624	15.140±0.027

vergent to periodic) functions with average value 0 and very small absolute value. These functions should be ignored for any practical purposes. Although we use $P(x)$ for all such functions, these may be different. The behaviour of quad tries is identical to those of digital tries of order 2^k:

$$E[N_n] = 1 + 2^{k(1-n)} \sum_{i=0}^{n} \binom{n}{i} (2^k - 1)^{n-i} E[N_i] \quad (N_0 = N_1 = 0)$$
$$= \frac{n}{k \ln 2}(1 + P((\log_2 n)/k)) + O(1)$$

$$C_n = 1 + 2^{k(1-n)} \sum_{i=1}^{n} \binom{n-i}{i-1} (2^k - 1)^{n-i} C_i, \quad (C_0 = C_1 = 0)$$
$$= \frac{H_{n-1}}{k \ln 2} + \frac{1}{2} + P((\log_2 n)/k) + O(n^{-1})$$

$$C'_n = 1 + 2^{-kn} \sum_{i=2}^{n} \binom{n}{i} (2^k - 1)^{n-i} C'_i \quad (C'_0 = C'_1 = 0)$$
$$= \frac{H_n - 1}{k \ln 2} + \frac{1}{2} + P((\log_2 n)/k) + O(n^{-1})$$

Quad trie search

```
search(key, t)
typekey key[K];
tree t;

{int bn, i, indx;
for (bn=1; t != NULL && !IsData(t); bn++) {
    indx = 0;
    for (i=0; i<K; i++) indx = 2*indx + bit(bn,key[i]);
    t = t ->p[indx];
    }
if (t != NULL) for (i=0; i<K && key[i]==t ->k[i]; i++);
if (t==NULL || i < K) notfound(key);
    else found(t);
};
```

Quad trie insertion

```
tree insert(key, t)
typekey key[K];
tree t;
{ tree InsertIndx();
return(InsertIndx(key,t,1));
}

tree InsertIndx(key, t, lev)
typekey key[K];
tree t;
int lev;

{ int i, indx;
  tree t1;
if (t == NULL) return(NewDataNode(key));
if (IsData(t)) {
    for(i=0; i<K && key[i] == t ->k[i]; i++);
    if (i >= K) {
        Error /*** Key already in table ***/;
        return(t);
        }
    else {   t1 = NewIntNode();
```

```
        indx = 0;
        for (i=0; i<K; i++) indx = 2*indx + bit(lev,t->k[i]);
        t1->p[indx] = t;
        t = t1;
        }
    }
indx = 0;
for (i=0; i<K; i++) indx = 2*indx + bit(lev,key[i]);
t->p[indx] = InsertIndx(key, t->p[indx], lev+1);
return(t);
};
```

Quad tries have been successfully used to represent data associated with planar coordinates such as maps, graphics, and bit-map displays. For example, in describing a planar surface, if all the surface is homogeneous, then it can be described by an external node, if not, the surface is divided into four equal-size quadrants and the description process continues recursively.

References:
[Finkel, R.A. *et al.*, 74], [Bentley, J.L. *et al.*, 75], [Lee, D.T. *et al.*, 77], [Overmars, M.H. *et al.*, 82], [Flajolet, P. *et al.*, 83], [Beckley, D.A. *et al.*, 85], [Flajolet, P. *et al.*, 85], [Fabbrini, F. *et al.*, 86], [Nelson, R.C. *et al.*, 87], [Cunto, W. *et al.*, 89], [Flajolet, P. *et al.*, 91].

3.5.2 K-dimensional trees

A **k-d tree** is a binary tree which stores k-dimensional keys in its nodes. The subkeys are used to direct the searching in the same way they are used in a binary search tree. The only difference is that the subkeys are used cyclically, one subkey per level. In our algorithm we use the first subkey at the root, the second subkey for the direct descendants of the root, and so on.

For k-d trees built from random insertions, the complexity measures are the same as for binary search trees (see Section 3.4.1):

$$E[A_n] = C_n = 2(1+1/n)H_n - 3 \approx 1.3863 \log_2 n - 1.8456$$

$$\sigma^2(A_n) = (2+10/n)H_n - 4(1+1/n)(H_n^2/n + H_n^{(2)}) + 4 \approx 1.3863 \log_2 n - 1.4253$$

$$E[A_n'] = C_n' = 2H_{n+1} - 2 \approx 1.3863 \log_2 n - 0.8456$$

$$\sigma^2(A_n') = 2H_{n+1} - 4H_{n+1}^{(2)} + 2 \approx 1.3863 \log_2 n - 3.4253$$

K-d tree search

```
search(key, t)
typekey key[K];
tree t;
{
int lev, i;
for (lev=0; t != NULL; lev=(lev+1)%K) {
    for (i=0; i<K && key[i]==t->k[i]; i++);
    if (i==K)    { found(t); return; }
    if (key[lev] > t->k[lev]) t = t->right;
        else      t = t->left;
    }
notfound(key);
};
```

K-d tree insertion

```
tree insert(key, t, lev)
typekey key[ ];
tree t;
int lev;
{
int i;
if (t==NULL) t = NewNode(key);

else {    for (i=0; i<K && key[i]==t->k[i]; i++);
    if (i==K)    Error /*** Key already in table ***/;
    else if (key[lev] > t->k[lev])
        t->right = insert(key, t->right, (lev+1)%K);
    else t->left  = insert(key, t->left, (lev+1)%K);
    }
return(t);
};
```

For a k-d tree grown from random keys, a partial-match query which involves p of the k subkeys will require

$$E[C_n] = O(n^{\lambda})$$

where λ is the only positive root of

$$(2+\lambda)^p(1+\lambda)^{k-p} = 2^k$$

We have

$$\lambda = 1 - \frac{p}{k} + \theta$$

with $0 < \theta < 0.07$. Table 3.34 shows some values for λ.

The constant which multiplies the n^λ term depends on which subkeys are used in the partial-match query. This constant is lowest when the subkeys used for the search are the first subkeys of the key.

Table 3.34: Order of magnitude of partial-match queries in k-d trees.

	λ		
k	$p=1$	$p=2$	$p=3$
2	0.56155		
3	0.71618	0.39485	
4	0.78995	0.56155	0.30555

K-d trees allow range searches; the following algorithm searches a k-d tree for values contained between *lowk* and *uppk*. The function *found*() is called for each value in the tree within the range.

Range search in k-d trees

```
rsearch(lowk, uppk, t, lev)
typekey lowk[ ], uppk[ ];
tree t;
int lev;

{int j;
if (t==NULL) return;
if (lowk[lev] <= t->k[lev])
    rsearch(lowk, uppk, t->left, (lev+1)%K);

for (j=0; j<K && lowk[j]<=t->k[j] && uppk[j]>=t->k[j]; j++);
if (j==K) found(t);

if (uppk[lev] > t->k[lev])
    rsearch(lowk, uppk, t->right, (lev+1)%K);
};
```

There are no efficient or simple methods for performing 'rotations' in k-d trees. Consequently it is difficult to maintain a k-d tree balanced.

There are no efficient or simple methods for performing deletions either. The best method for deletions is to mark the nodes as deleted, and reconstruct the tree whenever too many nodes have been deleted.

It is possible to construct a perfectly balanced k-d tree in $O(n \log n)$ time. This is done by a divide-and-conquer approach:

Construction of perfectly balanced k-d tree

```
function MakeBalTree(S : SetOfKeys; lev : integer) : tree;
var med : typekey;
    median : KDKey;
    A : SetOfKeys;
    i, n : integer;
    SubKey : array [1..Max] of typekey;
begin
if S=[ ] then MakeBalTree := nil
else begin
    n := SizeOf(S);
    {*** Select subkeys to find median ***}
    for i:=1 to n do SubKey[i] := element(i,S)[lev];
    {*** find median of subkeys ***}
    med := select(n div 2 + 1, SubKey, 1, n);
    A := [ ];
    for i:=1 to n do
        if element(i,S)[lev] > med then
            A := A + element(i,S)
        else if element(i,S)[lev] = med then
            median := element(i,S);
    MakeBalTree := NewNode(median,
            MakeBalTree(S-A-[median], (lev+1) mod K),
            MakeBalTree(A, (lev+1) mod K))
    end
end;
```

References:

[Bentley, J.L., 75], [Friedman, J.H. *et al.*, 77], [Lee, D.T. *et al.*, 77], [Bentley, J.L., 79], [Silva-Filho, Y.V., 79], [Eastman, C.M., 81], [Robinson, J.T., 81], [Silva-Filho, Y.V., 81], [Eastman, C.M. *et al.*, 82], [Hoshi, M. *et al.*, 82], [Overmars, M.H. *et al.*, 82], [Flajolet, P. *et al.*, 83], [Beckley, D.A. *et al.*, 85], [Flajolet, P. *et al.*, 86], [Murphy, O.J. *et al.*, 86], [Lea, D., 88].

Sorting Algorithms

4.1 Techniques for sorting arrays

The typical definition for procedures to sort arrays in place is, in Pascal:

Procedure definition for sorting arrays

```
procedure sort(var r : ArrayToSort; lo, up : integer);
```

and in C:

Procedure definition for sorting arrays

```
sort(r, lo, up)
ArrayToSort r;
int lo, up;
```

where r is the array to be sorted between $r[lo]$ and $r[up]$. The sorting is done 'in place', in other words, the array is modified by permuting its components into ascending order.

4.1.1 Bubble sort

The bubble sort algorithm sorts an array by interchanging adjacent records that are in the wrong order. The algorithm makes repeated passes through the array probing all adjacent pairs until the file is completely in order. Every complete pass sets at least one element into its final location (in an upward pass the maximum is settled, in a downward the minimum). In this way, every pass is at least one element shorter than the previous pass.

Let C_n be the number of comparisons needed to sort a file of size n using the bubble sort, and let I_n be the number of interchanges performed in the process. Then

$$n - 1 \le C_n \le \frac{n(n-1)}{2}$$

$$E[C_n] = \frac{n^2 - n \ln n - (\gamma + \ln 2 - 1)n}{2} + O(n^{1/2})$$

$$0 \le I_n \le \frac{n(n-1)}{2}$$

$$E[I_n] = \frac{n(n-1)}{4}$$

$$E[passes] = n - \sqrt{\pi n/2} + 5/3 + O\left(\frac{1}{\sqrt{n}}\right)$$

The simplest form of the bubble sort always makes its passes from the top of the array to the bottom.

Bubble sort

```
procedure sort(var r : ArrayToSort; lo, up : integer);

var i, j : integer;
    tempr : ArrayEntry;
begin
while up>lo do begin
    j := lo;
    for i:=lo to up-1 do
        if r[i].k > r[i+1].k then begin
            tempr := r[i];
            r[i] := r[i+1];
            r[i+1] := tempr;
            j := i
            end;
```

```
        up := j
        end
    end;
```

A slightly more complicated algorithm passes from the bottom to the top, then makes a return pass from top to bottom.

Bubble sort (double direction)

```
sort(r, lo, up)
ArrayToSort r;
int lo, up;

{int i, j;
while (up>lo) {
    j = lo;
    for (i=lo; i<up; i++)
        if (r[i].k > r[i+1].k) {
            exchange(r, i, i+1);
            j = i;}
    up = j;
    for (i=up; i>lo; i--)
        if (r[i].k < r[i-1].k) {
            exchange(r, i, i-1);
            j = i;}
    lo = j;
    }
}
```

The bubble sort is a simple sorting algorithm, but it is inefficient. Its running time is $O(n^2)$, unacceptable even for medium-sized files. Perhaps for very small files its simplicity may justify its use, but the linear insertion sort (see Section 4.1.2) is just as simple to code and more efficient to run.

For files with very few elements out of place, the double-direction bubble sort (or cocktail shaker sort) can be very efficient. If only k of the n elements are out of order, the running time of the double-direction sort is $O(kn)$. One advantage of the bubble sort is that it is stable: records with equal keys remain in the same relative order after the sort as before.

References:
[Knuth, D.E., 73], [Reingold, E.M. *et al.*, 77], [Dobosiewicz, W., 80], [Meijer, H. *et al.*, 80], [Sedgewick, R., 88], [Weiss, M.A. *et al.*, 88].

4.1.2 Linear insertion sort

The linear insertion sort is one of the simplest sorting algorithms. With a portion of the array already sorted, the remaining records are moved into their proper places one by one. This algorithm uses sequential search to find the final location of each element. Linear insertion sort can be viewed as the result of the iterative application of inserting an element in an ordered array. Let C_n be the number of comparisons needed to sort an array of size n using linear insertion sort. Then sorting a randomly ordered file requires

$$E[C_n] = \frac{n(n+3)}{4} - H_n$$

$$\sigma^2(C_n) = \frac{(2n-11)n(n+7)}{72} + 2H_n - H_n^{(2)}$$

Linear insertion sort

```
sort(r, lo, up)
ArrayToSort r;
int lo, up;

{int i, j;
ArrayEntry tempr;
for (i=up-1; i>=lo; i--) {
    tempr = r[i];
    for (j=i+1; j<=up && (tempr.k>r[j].k); j++)
        r[j-1] = r[j];
    r[j-1] = tempr;
    }
};
```

If the table can be extended to add one **sentinel** record at its end (a record with the largest possible key), linear insertion sort will improve its efficiency by having a simpler inner loop.

Linear insertion sort with sentinel

```
sort(r, lo, up)
ArrayToSort r;
int lo, up;
```

```
{int i, j;
ArrayEntry tempr;

r[up+1].k = MaximumKey;
for (i=up-1; i>=lo; i--) {
    tempr = r[i];
    for (j=i+1; tempr.k>r[j].k; j++)
        r[j-1] = r[j];
    r[j-1] = tempr;
    }
}
```

The running time for sorting a file of size n with the linear insertion sort is $O(n^2)$. For this reason, the use of the algorithm is justifiable only for sorting very small files. For files of this size (say $n < 10$), however, the linear insertion sort may be more efficient than algorithms which perform better asymptotically. The main advantage of the algorithm is the simplicity of its code.

Like the bubble sort (see Section 4.1.1), the linear insertion sort is stable: records with equal keys remain in the same relative order after the sort as before.

A common variation of linear insertion sort is to do the searching of the final position of each key with binary search. This variation, called **binary insertion sort**, uses an almost optimal number of comparisons but does not reduce the number of interchanges needed to make space for the inserted key. The total running time still remains $O(n^2)$.

Binary insertion sort

```
/* Binary insertion sort */
sort(r, lo, up)
ArrayToSort r;
int lo, up;

{int i, j, h, l;
ArrayEntry tempr;
for (i=lo+1; i<=up; i++) {
    tempr = r[i];
    for (l=lo-1, h=i; h-l > 1;) {
        j = (h+l)/2;
        if (tempr.k < r[j].k) h = j; else l = j;
        }
    for (j=i; j>h; j--) r[j] = r[j-1];
```

```
        r[h] = tempr;
        }
    }
```

References:
[Knuth, D.E., 73], [Horowitz, E. *et al.*, 76], [Janko, W., 76], [Reingold, E.M. *et al.*, 77], [Gotlieb, C.C. *et al.*, 78], [Melville, R. *et al.*, 80], [Dijkstra, E.W. *et al.*, 82], [Doberkat, E.E., 82], [Panny, W., 86], [Baase, S., 88], [Sedgewick, R., 88].

4.1.3 Quicksort

Quicksort is a sorting algorithm which uses the divide-and-conquer technique. To begin each iteration an element is selected from the file. The file is then split into two subfiles, those elements with keys smaller than the selected one and those elements whose keys are larger. In this way, the selected element is placed in its proper final location between the two resulting subfiles. This procedure is repeated recursively on the two subfiles and so on.

Let C_n be the number of comparisons needed to sort a random array of size n, let I_n be the number of interchanges performed in the process (for the present algorithm I_n will be taken as the number of record assignments), and let $k = \lfloor \log_2 n \rfloor$. Then

$$(n+1)k - 2^{k+1} + 2 \;\le\; C_n \;\le\; \frac{n(n-1)}{2}$$

$$E[C_n] \;=\; n - 1 + \frac{2}{n}\sum_{k=1}^{n-1} E[C_k] \;=\; 2(n+1)H_n - 4n$$

$$\sigma^2(C_n) \;=\; n\sqrt{7 - 2\pi^2/3} \;+\; o(n)$$

$$E[I_n] \;=\; \frac{n+3}{2} + \frac{2}{n}\sum_{k=1}^{n-1} E[I_k] \;=\; (n+1)(H_n - 2/3)$$

Table 4.1 shows some exact results.

We now present the Pascal code for Quicksort. Note that one of the two recursions of the divide-and-conquer method has been transformed into a **while** loop, like the transformations for tail recursions.

Table 4.1: Exact average results for Quicksort.

n	$E[C_n]$	$E[I_n]$	n	$E[C_n]$	$E[I_n]$
10	24.437	24.885	50	258.92	195.46
100	647.85	456.59	500	4806.41	3069.20
1000	10985.9	6825.6	5000	70963.3	42147.6
10000	155771.7	91218.5	50000	939723.2	536527.6

Quicksort algorithm

```
procedure sort(var r : ArrayToSort; lo, up : integer);

var i, j : integer;
    tempr : ArrayEntry;
begin
while up>lo do begin
    i := lo;
    j := up;
    tempr := r[lo];
    {*** Split file in two ***}
    while i<j do begin
        while r[j].k > tempr.k do
            j := j-1;
        r[i] := r[j];
        while (i<j) and (r[i].k<=tempr.k) do
            i := i+1;
        r[j] := r[i]
        end;
    r[i] := tempr;
    {*** Sort recursively ***}
    sort(r,lo,i-1);
    lo := i+1
    end
end;
```

The above algorithm uses the same technique even for very small files. As it turns out, very small subfiles can be sorted more efficiently with other techniques, such as, linear insertion sort or binary insertion sort (see Section 4.1.2). It is relatively simple to build a hybrid algorithm which uses Quicksort for large files and switches to a simpler, more efficient, algorithm for small files.

Composition of Quicksort

```
....
begin
while up-lo > M do begin
    .... body of quicksort; ....
    end;
if up > lo then begin
    .... simpler-sort ....
    end
end;
```

Quicksort is a very popular sorting algorithm; although its worst case is $O(n^2)$, its average performance is excellent.

Unfortunately, this worst case occurs when the given file is in order already, a situation which may well happen in practice. Any portion of the file that is nearly in order will significantly deteriorate Quicksort's efficiency. To compensate for this, small tricks in the code of the algorithm can be used to ensure that these worst cases occur only with exponentially small probability.

It should be noted that for the worst case, Quicksort may also use $O(n)$ levels of recursion. This is undesirable, as it really implies $O(n)$ additional storage. Moreover, most systems will have a limited stack capacity. The above algorithm can be protected to force it to use a $O(\log n)$ stack (see Appendix IV). In its present form, it will not use $O(n)$ levels of recursion for a file in increasing order.

Quicksort allows several variations, improvements, or mechanisms to protect from its worst case. Most of these variations rely on different methods for selecting the 'splitting' element.

(1) The **standard (Quicksort, Quickersort)** algorithms select the splitting element from a fixed location (as in the algorithm above: the first element of the array). Selecting the element in the middle of the array does not deteriorate the random case and improves significantly for partially ordered files.

(2) The variation called **Samplesort** selects a small sample (for example, size 3) and determines the median of this sample. The median of the sample is used as a splitting element.

(3) The selection of the splitting element can be replaced by a pair of values which determine the range of the median. As the array is scanned, every time an element falls in between the pair, one of the values is updated to maintain the range as close to the median as possible. At the end of the splitting phase we have two elements in their final locations, dividing the interval.

(4) Arithmetic averages, or any other method which selects a value that is not part of the array, produce algorithms that may loop on equal keys. Arithmetic operations on keys significantly restrict the applicability of sorting algorithms.

References:
[Hoare, C.A.R., 61], [Hoare, C.A.R., 62], [Scowen, R.S., 65], [Singleton, R.C., 69], [Frazer, W.D. *et al.*, 70], [van Emden, M.H., 70], [van Emden, M.H., 70], [Knuth, D.E., 73], [Aho, A.V. *et al.*, 74], [Knuth, D.E., 74], [Loeser, R., 74], [Peters, J.G. *et al.*, 75], [Sedgewick, R., 75], [Horowitz, E. *et al.*, 76], [Reingold, E.M. *et al.*, 77], [Sedgewick, R., 77], [Sedgewick, R., 77], [Apers, P.M., 78], [Gotlieb, C.C. *et al.*, 78], [Sedgewick, R., 78], [Standish, T.A., 80], [Rohrich, J., 82], [Motzkin, D., 83], [Erkio, H., 84], [Wainwright, R.L., 85], [Bing-Chao, H. *et al.*, 86], [Wilf, H., 86], [Verkamo, A.I., 87], [Wegner, L.M., 87], [Baase, S., 88], [Brassard, G. *et al.*, 88], [Sedgewick, R., 88], [Manber, U., 89], [Cormen, T.H. *et al.*, 90].

4.1.4 Shellsort

Shellsort (or diminishing increment sort) sorts a file by repetitive application of linear insertion sort (see Section 4.1.2). For these iterations the file is seen as a collection of d files interlaced, that is, the first file is the one in locations $1, d+1, 2d+1, \ldots$, the second in locations $2, d+2, 2d+2, \ldots$, and so on. Linear insertion sort is applied to each of these files for several values of d. For example d may take the values in the sequence $\{n/3, n/9, \ldots, 1\}$. It is crucial that the sequence of increment values ends with 1 (simple linear insertion) to guarantee that the file is sorted.

Different sequences of increments give different performances for the algorithm.

Let C_n be the number of comparisons and I_n the number of interchanges used by Shellsort to sort n numbers.

For $d = \{h, k, 1\}$

$$E[I_n] = \frac{n^2}{4h} + \frac{\sqrt{\pi}}{8}\left(\frac{h^{1/2}}{k} - \frac{ch^{-1/2}}{k} + \frac{c-1}{\sqrt{c}}\right) n^{3/2} + O(n)$$

where $c = \gcd(h, k)$.

For $d = \{2^k - 1, 2^{k-1} - 1, \ldots 7, 3, 1\}$

$$E[I_n] = O(n^{3/2})$$

For $d = \{2^k, 2^{k-1}, \ldots, 8, 4, 2, 1\}$ and $n = 2^k$,

$$\begin{aligned} E[I_n] &= \frac{n}{16} \sum_{i=1}^{\log_2 n} \frac{\Gamma(2^{i-1})}{2^i \Gamma(2^i)} \sum_{r=1}^{2^{i-1}} r(r+3)2^r \frac{\Gamma(2^i - r + 1)}{\Gamma(2^{i-1} - r + 1)} \\ &= 0.534885\ldots n\sqrt{n} - 0.4387\ldots n - 0.097\ldots\sqrt{n} + O(1) \end{aligned}$$

$$E[C_n] = E[I_n] + n \log_2 n - \frac{3(n-1)}{2}$$

For $d = \{4^{k+1} + 3 \cdot 2^j + 1, ..., 77, 23, 8, 1\}$

$$E[I_n] = O(n^{4/3})$$

For $d = \{2^p 3^q, ..., 9, 8, 6, 4, 3, 2, 1\}$

$$E[I_n] = O(n(\log n)^2)$$

For $d = \{\alpha^p(\alpha - 1)^q, ..., \alpha, \alpha - 1\}$ when $\alpha = 2^{\sqrt{\log_2 n}}$

$$E[I_n] = O(n^{1+(2+\epsilon)/\sqrt{\log_2 n}})$$

for any $\epsilon > 0$. There exist sequences of increments that achieve

$$E[I_n] = O(n^{1+1/(c+1)}) \quad \text{and} \quad O(n^{1+\epsilon/\sqrt{\log n}})$$

for any $c > 0$ and $\epsilon > 0$.

The version we present here is closer to the original algorithm suggested by Shell; the increments are $\lfloor n\alpha \rfloor$, $\lfloor \lfloor n\alpha \rfloor \alpha \rfloor$,... . Extensive experiments indicate that the sequence defined by $\alpha = 0.45454 < 5/11$ performs significantly better than other sequences. The easiest way to compute $\lfloor 0.45454n \rfloor$ is by $(5 * n - 1)/11$ using integer arithmetic. Note that if $\alpha < 1/2$, some sequences will not terminate in 1, but in 0; this has to be corrected as a special case.

Shellsort

```
sort(r, lo, up)
ArrayToSort r;
int lo, up;

{int d, i, j;
ArrayEntry tempr;
for (d=up-lo+1; d>1;) {
     if (d<5)  d = 1;
         else  d = (5*d-1)/11;
     /*** Do linear insertion sort in steps size d ***/
     for (i=up-d; i>=lo; i--) {
          tempr = r[i];
          for (j=i+d; j<=up && (tempr.k>r[j].k); j+=d)
               r[j-d] = r[j];
          r[j-d] = tempr;
          }
     }
}
```

Table 4.2 presents simulation results of the sequence of increments $d = \{\frac{3^k-1}{2}..., 40, 13, 4, 1\}$ $(d_{k+1} = 3d_k + 1)$ and for the sequence $d = \{\lfloor n\alpha \rfloor, ..., 1\}$ $(\alpha = 0.45454)$.

Table 4.2: Exact and simulation results for Shellsort.

	$d_{k+1} = 3d_k + 1$		$\alpha = 0.45454$	
n	$E[C_n]$	$E[I_n]$	$E[C_n]$	$E[I_n]$
5	7.71667	4.0	8.86667	3.6
10	25.5133	14.1333	25.5133	14.1333
50	287.489±0.006	164.495±0.007	292.768±0.006	151.492±0.006
100	731.950±0.017	432.625±0.018	738.589±0.013	365.939±0.013
500	5862.64±0.24	3609.33±0.25	5674.38±0.11	2832.92±0.12
1000	13916.92±0.88	8897.19±0.88	13231.61±0.30	6556.54±0.31
5000	101080±16	68159±16	89350.7±3.4	46014.1±3.4
10000	235619±56	164720±56	194063.8±6.7	97404.5±6.7
50000	1671130±1163	1238247±1163	1203224±58	619996±58
100000	3892524±4336	2966745±4336	2579761±113	1313319±113

The simulation results indicate that the performance of both algorithms is rather discontinuous in the size of the file. Consequently, any approximation formula is applicable only in the computed range and will not reflect any discontinuities. For the above simulations, selecting the results with $n \geq 500$ we find the empirical formulas:

$$E[I_n] \approx 0.41\, n \ln(n) (\ln(\ln n) + 1/6) \qquad (\text{for } \alpha = 0.45454)$$

$$E[I_n] \approx 1.54 n^{1.257} - 190 \qquad (\text{for } d_{k+1} = 3d_k + 1)$$

Shellsort is not a stable sorting algorithm since equal keys may not preserve their relative ordering.

Shellsort seems a very attractive algorithm for internal sorting. Its coding is straightforward and usually results in a very short program. It does not have a bad worst case and, furthermore, it does less work when the file is partially ordered. These arguments make it a good choice for a library sorting routine.

References:
[Shell, D.L., 59], [Boothroyd, J., 63], [Espelid, T.O., 73], [Knuth, D.E., 73], [Ghoshdastidar, D. *et al.*, 75], [Erkio, H., 80], [Yao, A.C-C., 80], [Incerpi, J. *et al.*, 85], [Sedgewick, R., 86], [Incerpi, J. *et al.*, 87], [Baase, S., 88], [Sedgewick, R., 88], [Weiss, M.A. *et al.*, 88], [Selmer, E.S., 89], [Weiss, M.A. *et al.*, 90].

4.1.5 Heapsort

Heapsort (or **Treesort** III) is a sorting algorithm that sorts by building a priority queue and then repeatedly extracting the maximum of the queue until it is empty. The priority queue used is a **heap** (see Section 5.1.3) that shares the space in the array to be sorted. The heap is constructed using all the elements in the array and is located in the lower part of the array. The sorted array is constructed from top to bottom using the locations vacated by the heap as it shrinks. Consequently we organize the priority queue to extract the maximum element.

$$C_n \leq 2n\lfloor \log_2 n \rfloor + 3n$$

$$I_n \leq n\lfloor \log_2 n \rfloor + 2.5n$$

The complexity results for the heap-creation phase can be found in Section 5.1.3.

Heapsort

```
procedure sort(var r : ArrayToSort; lo, up : integer);

var i : integer;
    tempr : ArrayEntry;
begin
{*** construct heap ***}
for i := (up div 2) downto 2 do   siftup(r,i,up);
{*** repeatedly extract maximum ***}
for i := up downto 2 do begin
    siftup(r,1,i);
    tempr := r[1];
    r[1] := r[i];
    r[i] := tempr
    end
end;
```

The above algorithm uses the function *siftup* (defined in Section 5.1.3). A call to $siftup(r, i, n)$ constructs a subheap in the array r at location i not beyond location n assuming that there are subheaps rooted at $2i$ and $2i+1$. Although the above procedure accepts the parameter lo for conformity with other sorting routines, Heapsort assumes that $lo = 1$.

Heapsort is not a stable sorting algorithm since equal keys may be transposed.

Heapsort is guaranteed to execute in $O(n \log n)$ time even in the worst case. Heapsort does not benefit from a sorted array, nor is its efficiency significantly affected by any initial ordering. As indicated by simulation, its running time has a very small variance.

This algorithm does not use any extra storage or require languages supporting recursion. Although its average performance is not as good as some other sorting algorithms, the advantages noted indicate that Heapsort is an excellent choice for an internal sorting algorithm.

Heapsort can be modified to take advantage of a partially ordered table. This variation is called **Smoothsort**, and has an $O(n)$ performance for an ordered table and an $O(n \log n)$ performance for the worst case.

Table 4.3 shows simulation results on the total number of comparisons used by Heapsort (C_n) and the total number of interchanges (I_n).

Table 4.3: Exact and simulation results for Heapsort.

n	$E[C_n]$	$\sigma^2(C_n)$	$E[I_n]$
5	10.95	1.1475	8.86667
10	38.6310	3.84698	26.6893
50	414.7498±0.0027	36.664±0.023	241.9939±0.0022
100	1027.6566±0.0060	81.281±0.077	581.5611±0.0049
500	7426.236±0.034	431.7±1.0	4042.502±0.028
1000	16852.652±0.070	876.3±3.0	9081.915±0.058
5000	107686.13±0.38	4320±36	57105.41±0.31
10000	235372.42±0.81	8624±106	124205.77±0.66
50000	1409803.8±4.5	45628±1363	737476.2±3.7
100000	3019621.8±9.5	94640±4175	1574953.6±7.6

The following are approximate formulas computed from the simulation results.

$$E[C_n] \approx 2n \log_2 n - 3.0233n$$

$$E[I_n] \approx n \log_2 n - 0.8602n$$

References:

[Floyd, R.W., 64], [Williams, J.W.J., 64], [Knuth, D.E., 73], [Aho, A.V. *et al.*, 74], [Horowitz, E. *et al.*, 76], [Reingold, E.M. *et al.*, 77], [Doberkat, E.E., 80], [Standish, T.A., 80], [Dijkstra, E.W. *et al.*, 82], [Dijkstra, E.W., 82], [Hertel, S., 83], [Doberkat, E.E., 84], [Carlsson, S., 87], [Baase, S., 88], [Sedgewick, R., 88], [Manber, U., 89], [Cormen, T.H. *et al.*, 90], [Xunuang, G. *et al.*, 90].

4.1.6 Interpolation sort

This sorting algorithm is similar in concept to the bucket sort (see Section 4.2.3). An interpolation function is used to estimate where records should appear in the file. Records with the same interpolation address are grouped together in contiguous locations in the array and later linear insertion sorted (see Section 4.1.2). The main difference between this algorithm and the bucket sort is that the interpolation sort is implemented in an array, using only one auxiliary index array and with no pointers.

Let C_n be the number of comparisons needed to sort an array of size n using the interpolation sort, and let F_n be the total number of interpolation function evaluations made in the process. Then

$$F_n = 2n$$

$$n-1 \leq C_n \leq \frac{n(n-1)}{2}$$

$$E[C_n] = \frac{5(n-1)}{4}$$

$$\sigma^2(C_n) = \frac{(20n-13)(n-1)}{72n}$$

The algorithm below uses the interpolation function *phi*(*key*, *lo*, *up*) to sort records of the array. A good interpolation formula for uniformly distributed keys is

General interpolation formula

```
phi(key, lo, up)
typekey key;
int lo, up;

{int i;
i = (key-MinKey) * (up-lo+1.0) / (MaxKey-MinKey) + lo;
return(i>up ? up : i<lo ? lo : i);
};
```

Note that if the above multiplication is done with integers, this operation is likely to cause overflow.

The array *iwk* is an auxiliary array with the same dimensions as the array to be sorted and is used to store the indices to the working array.

The array *iwk* does not need to be as big as the array to be sorted. If we make it smaller, the total number of comparisons during the final linear insertion phase will increase. In particular, if *iwk* has m entries and $m \leq n$ then

$$E[C_n] = 2n - m - 1 + \frac{n(n-1)}{4m}$$

Interpolation sort

```
sort(r, lo, up)
ArrayToSort r;
int lo, up;

{ArrayIndices iwk;
ArrayToSort out;
ArrayEntry tempr;
int i, j;

for (i=lo+1; i<=up; i++) iwk[i] = 0;
iwk[lo] = lo-1;
for (i=lo; i<=up; i++)  iwk[phi(r[i].k,lo,up)]++;
for (i=lo; i<up;  i++)  iwk[i+1] += iwk[i];
for (i=up; i>=lo; i--)  out[iwk[phi(r[i].k,lo,up)]--] = r[i];
for (i=lo; i<=up; i++)  r[i] = out[i];
for (i=up-1; i>=lo; i--) {
    tempr = r[i];
    for (j=i+1; j<=up && (tempr.k>r[j].k); j++)
        r[j-1] = r[j];
    r[j-1] = tempr;
    }
};
```

The above implementation uses the array *out* to copy the sorted elements. This array can be avoided completely if we can add a flag to each location indicating whether the record has been moved or not.

Because the standard deviation of C_n is $\approx 0.53n^{1/2}$, the total number of comparisons used by the interpolation sort is very stable around its average.

One of the restrictions of the interpolation sort is that it can only be used when records have numerical keys which can be handled by the interpolation function. Even in this case, if the distribution of the record key values departs significantly from the uniform distribution, it may mean a dramatic difference in running time. If, however, the key distribution is suitable and we can afford the extra storage required, the interpolation sort is remarkably fast, with a running time of $O(n)$.

The above implementation of interpolation sort is stable since equal keys are not transposed.

References:
[Isaac, E.J. *et al.*, 56], [Flores, I., 60], [Kronmal, R.A. *et al.*, 65], [Tarter, M.E. *et al.*, 66], [Gamzon, E. *et al.*, 69], [Jones, B., 70], [Ducoin, F., 79], [Ehrlich, G., 81], [Gonnet, G.H., 84], [Lang, S.D., 90].

4.1.7 Linear probing sort

This is an interpolation sort (see Section 4.1.6) based on a collision resolution technique similar to that of linear probing hashing. Each key is interpolated into one of the first m positions in an array. (Note that m will be taken to be greater than n unlike most other interpolation sort methods.) If a collision arises, then the smaller element takes the location in question and the larger element moves forward to the next location, and the process repeats until we find an empty location. (This may, ultimately, cause elements to overflow beyond position m.) After insertion of all elements, a single pass through the array compresses the file to the first n locations. The sorting process can be described as creating a table with linear probing hashing, using an interpolation function as a hashing function and using the technique of ordered hashing.

Let the size of our table be $m + w$; we will use the first m locations to interpolate the keys and the last w locations as an overflow area. We will let n denote the total number of keys to be sorted and $\alpha = n/m$ be the load factor. Let C_n be the number of comparisons needed to sort the n keys using the linear probing sort, and let F_n be the total number of interpolation function evaluations performed in the process. Then

$$F_n = n$$

$$E[C_n] = \frac{n}{2}\left(\frac{2-\alpha}{1-\alpha}\right) - \frac{\alpha(\alpha^4 - 4\alpha^3 + 6\alpha^2 + 6)}{12(1-\alpha)^3} + O(m^{-1})$$

Let W_n be the number of keys in the overflow section beyond the location m in the table. We have

$$E[W_n] = \frac{1}{2}\sum_{i=2}^{n} \frac{n^{\underline{i}}}{m^i} = \frac{\alpha^2}{2(1-\alpha)} - \frac{\alpha}{2(1-\alpha)^3 m} + O(m^{-2})$$

where $n^{\underline{i}} = n(n-1)\cdots(n-i+1)$ denotes the descending factorial

$$E[W_m] = \sqrt{m\pi/8} - 2/3 + O(m^{-1/2})$$

$$\sigma^2(W_n) = \frac{6\alpha^2 - 2\alpha^3 - \alpha^4}{12(1-\alpha)^2} + O(m^{-1})$$

$$\sigma^2(W_m) = \frac{(4-\pi)m}{8} + \frac{1}{9} - \frac{\pi}{48} + O(m^{-1/2})$$

$$Pr\{W_n > k\} \leq (1-\alpha)\sum_{j\geq 0}\frac{e^{-j\alpha}(j\alpha)^{j+k+1}}{(j+k+1)!}$$

$$\ln(\Pr\{W_n > k\}) \approx -2k(1-\alpha)$$

The expected value of the total number of table probes to sort n elements using linear probing sort is minimized when $n/m = 2-\sqrt{2} = 0.5857....$ At this point the expected number of probes is

$$C_n + m + W_n = (2+\sqrt{2})n + O(1)$$

Below we describe the linear probing sort using the interpolation function *phi(key, lo, up)*. This sorting function depends on two additional global parameters: m, which is the size of the interpolation area, and *UppBoundr*, which is the upper bound of the input array ($UppBoundr \geq m+w$). Selecting $m \approx \sqrt{n \times UppBoundr}$ minimizes the probability of failure due to exceeding the overflow area.

Linear probing sort

```
procedure sort(var r : ArrayToSort; lo, up : integer);

var  i, j : integer;
     r1 : ArrayToSort;
begin
r1 := r;
for j:=lo to UppBoundr do r[j].k := NoKey;
for j:=lo to up do begin
     i := phi(r1[j].k,lo,m);
     while r[i].k <> NoKey do begin
          if r1[j].k < r[i].k then begin
               r1[j-1] := r[i];
               r[i] := r1[j];
               r1[j] := r1[j-1]
               end;
          i := i+1;
          if i > UppBoundr then Error
          end;
     r[i] := r1[j]
     end;
i := lo-1;
for j:=lo to UppBoundr do
     if r[j].k <> NoKey then begin
          i := i+1;
```

```
            r[i] := r[j]
            end;
    for j:=i+1 to UppBoundr do    r[j].k := NoKey;
    end;
```

With a good interpolation formula, this algorithm can rank among the most efficient interpolation sort (see Section 4.1.6) algorithms.

The application of this algorithm to external storage appears to be promising; its performance, however, cannot be improved by using larger buckets. Letting E_n be the number of external accesses required to sort n records, we have

$$E[E_n] \;=\; n\left(1+\frac{\alpha}{2b(1-\alpha)}\right)+\frac{m}{b}$$

Table 4.4 gives the efficiency measures for two table sizes with various load factors. I_n denotes the number of interchanges performed owing to collisions while building the table.

Table 4.4: Exact and simulation results for linear probing sort.

	$m = 100$			$m = 5000$		
α	$E[C_n]$	$E[W_n]$	$E[I_n]$	$E[C_n]$	$E[W_n]$	$E[I_n]$
50%	72.908	.23173	13.785±0.003	3747.65	.24960	765.29±0.18
80%	200.696	1.27870	84.795±0.019	11932.90	1.59014	5917.0±2.4
90%	310.184	2.47237	164.387±0.039	24149.77	3.96477	16168±12
95%	399.882	3.62330	234.827±0.056	45518.47	8.39737	35731±43
99%	499.135	5.10998	315.823±0.074	118134.0	26.46562	105444±236
100%	528.706	5.60498	340.260±0.080	169087.0	43.64542	154945±385

References:
[Melville, R. *et al.*, 80], [Gonnet, G.H. *et al.*, 81], [Gonnet, G.H. *et al.*, 84], [Poblete, P.V., 87].

4.1.8 Summary

Table 4.5 shows an example of real relative total times for sorting an array with 49998 random elements.

There are algorithms specially adapted to **partially sorted** inputs. That is, they run faster if the input is in order or almost in order. Several measures of presortedness have been defined, as well as optimal algorithms for each measure.

Table 4.5: Relative total times for sorting algorithms.

Algorithm	*C*	*Pascal*
Bubble sort		1254
Shaker sort	2370	
Linear insertion sort	544	541
Linear insertion sort with sentinel	450	366
Binary insertion sort	443	
Quicksort	1.0	1.0
Quicksort with bounded stack usage		1.0
Shellsort	1.9	2.0
Shellsort for fixed increments	1.9	
Heapsort	2.4	2.4
Interpolation sort	2.5	2.1
Interpolation sort (in-place, positive numbers)	2.6	
Linear probing sort	1.4	1.2

References:
[Warren, H.S., 73], [Meijer, H. *et al.*, 80], [Gonzalez, T.F. *et al.*, 82], [Mannila, H., 84], [Skiena, S.S., 88], [Estivill-Castro, V. *et al.*, 89], [Levcopoulos, C. *et al.*, 89], [Levcopoulos, C. *et al.*, 90].

General references:
[Friend, E.H., 56], [Flores, I., 61], [Boothroyd, J., 63], [Hibbard, T.N., 63], [Flores, I., 69], [Martin, W.A., 71], [Nozaki, A., 73], [Knuth, D.E., 74], [Lorin, H., 75], [Pohl, I., 75], [Preparata, F.P., 75], [Fredman, M.L., 76], [Wirth, N., 76], [Trabb Pardo, L., 77], [Horvath, E.C., 78], [Borodin, A. *et al.*, 79], [Kronsjo, L., 79], [Manacher, G.K., 79], [Mehlhorn, K., 79], [Cook, C.R. *et al.*, 80], [Erkio, H., 81], [Borodin, A. *et al.*, 82], [Aho, A.V. *et al.*, 83], [Reingold, E.M. *et al.*, 83], [Mehlhorn, K., 84], [Bui, T.D. *et al.*, 85], [Merritt, S.M., 85], [Wirth, N., 86], [Beck, I. *et al.*, 88], [Richards, D. *et al.*, 88], [Richards, D., 88], [Huang, B. *et al.*, 89], [Munro, J.I. *et al.*, 89], [Douglas, C.C. *et al.*, 90], [Fredman, M.L. *et al.*, 90], [Munro, J.I. *et al.*, 90].

4.2 Sorting other data structures

The second most popular data structure used to store sorted data is the linked list, or linear list. The corresponding data structure is described by the production:

$$\mathbf{s - KEY} \ : \ [\mathbf{KEY}, \mathbf{s - KEY}] \ ; \ \mathbf{nil}$$

A typical Pascal definition of a linked list, containing a key field *k*, is:

Linked list definition

```
type
    list = ↑ rec;
    rec  = record
           k : typekey;
           next : list
           end;
```

Linked lists can be implemented in arrays; in this case a pointer to a record is an integer indexing into the array. The only non-trivial operation when implementing lists in arrays is to reorder the array according to the order given by the list. This is particularly useful for the case of sorting. The following algorithm reorders the array *r* based on the list rooted at *root*.

Reordering of arrays

```
i := 1;
while root <> 0 do begin
    tempr := r[root];
    r[root] := r[i];
    r[i] := tempr;
    r[i].next := root;
    root := tempr.next;
    i := i+1;
    while (root<i) and (root>0) do root := r[root].next;
    end;
end;
```

General references:
[Friend, E.H., 56], [Flores, I., 69], [Tarjan, R.E., 72], [Harper, L.H. *et al.*, 75], [Munro, J.I. *et al.*, 76], [Wirth, N., 76], [Gotlieb, C.C. *et al.*, 78], [Sedgewick, R., 78], [Tanner, R.M., 78], [Borodin, A. *et al.*, 79], [Nozaki, A., 79], [Bentley, J.L. *et al.*, 80], [Chin, F.Y. *et al.*, 80], [Colin, A.J.T. *et al.*, 80], [Power, L.R., 80], [Borodin, A. *et al.*, 82], [Aho, A.V. *et al.*, 83], [Goodman, J.E. *et al.*, 83], [Reingold, E.M. *et al.*, 83], [Mehlhorn, K., 84], [Wirth, N., 86].

4.2.1 Merge sort

Merge sort is a natural way of sorting lists by repeated merging of sublists. By counting the total number of records in the list, each merging step can be as balanced as possible. At the deepest level of the recursion, single element lists are merged together to form two element lists and so on.

Let C_n be the total number of comparisons used by merge sort, then

$$\sum_{i=1}^{n-1} \nu(i) \leq C_n \leq kn - 2^k + 1$$

where $k = \lceil \log_2 n \rceil$ and $\nu(i)$ is the number of 1s in the binary representation of i.

$$k2^{k-1} \leq C_{2^k}$$

$$E[C_{2^k}] = (k-\alpha)2^k + 2 - \frac{4 \cdot 2^{-k}}{3} + \frac{8 \cdot 4^{-k}}{7} - \frac{16 \cdot 8^{-k}}{15} + O(16^{-k})$$

$$(\log_2 n - \alpha)n + 2 + O(n^{-1}) \leq E[C_n] \leq (\log_2 n - \beta)n + 2 + O(n^{-1})$$

where $\alpha = 1.26449... = 2 - \sum_{i \geq 0} \frac{1}{2^i(2^i+1)}$ and $\beta = 1.24075...$.

Merge sort

```
function sort(var r : list; n : integer) : list;
var  temp : list;

begin
if r = nil then sort := nil
else if n>1 then
        sort := merge(sort(r, n div 2),
                      sort(r, (n+1) div 2))
else begin
        temp := r;
        r := r↑.next;
        temp↑.next := nil;
        sort := temp
        end
end;
```

It is assumed that we know the number of elements in the list, which is given as the second parameter. If this number is not known, it can be overestimated without deteriorating significantly the performance of the algorithm.

The function *merge* merges two ordered lists into a single list and is described in Section 4.3.1.

If the merging routine is stable, that is, in the output of $merge(a, b)$ equal keys are not transposed and those from the list a precede those from the list b, merge sort will be a stable sorting algorithm and equal keys will not be transposed.

Merge sort uses extra storage: the pointers that are associated with the list.

Merge sort can take advantage of partially ordered lists (**Natural merge**) as described in Appendix IV. For this variation, the algorithm will do a single pass on totally ordered (or reversely ordered) files and will have a smooth transition between $O(n)$ and $O(n \log n)$ complexity for partially ordered files. Merge sort is guaranteed to execute in $O(n \log n)$ even in the worst case.

In view of the above, merge sort is one of the best alternatives to sorting lists.

Table 4.6 illustrates some exact counts of the number of comparisons for merge sort. The average values are computed for random permutations of the input file.

Table 4.6: Number of comparisons used by merge sort.

n	min C_n	$E[C_n]$	max C_n
5	5	7.1667	8
10	15	22.667	25
50	133	221.901	237
100	316	541.841	573
500	2216	3854.58	3989
1000	4932	8707.17	8977
5000	29804	55226.3	56809
10000	64608	120450.7	123617
50000	382512	718184.3	734465

References:
[Jones, B., 70], [Bron, C., 72], [Knuth, D.E., 73], [Aho, A.V. *et al.*, 74], [Dewar, R.B.K., 74], [Horowitz, E. *et al.*, 76], [Peltola, E. *et al.*, 78], [Todd, S., 78], [Erkio, H., 80], [Baase, S., 88], [Brassard, G. *et al.*, 88], [Manber, U., 89].

4.2.2 Quicksort for lists

A natural way of sorting a list is by the use of the divide-and-conquer technique. This will produce an algorithm similar to Quicksort (see Section 4.1.3); that is, pick an element of the list (the head of the list), split the remaining list according to elements being smaller or larger than the selected one, sort the two resulting lists recursively, and finally concatenate the lists.

The execution pattern (sizes of subfiles, and so on) of this algorithm is the same as for Quicksort for arrays. Let I_n be the number of times the inner loop is executed to sort a file with n elements. The inner loop involves one or two comparisons and a fixed number of pointer manipulations. Let C_n be the number of comparisons and $k = \lfloor \log_2 n \rfloor$, then

$$(n+1)k - 2^{k+1} + 2 \;\leq\; I_n \;\leq\; \frac{n(n-1)}{2}$$

$$E[I_n] \;=\; n - 1 + \frac{2}{n}\sum_{i=1}^{n-1} E[I_i] \;=\; 2(n+1)H_n \;-\; 4n$$

$$E[C_n] \;=\; \frac{3I_n}{2}$$

$$\sigma^2(C_n) \;=\; n\sqrt{7 - 2\pi^2/3} \;+\; o(n)$$

Quicksort for lists

```
function sort (r : list) : list;
var  lowf,lowl, midf,midl, highf,highl : list;

begin
if r = nil then begin Last := nil; sort := r end
else begin
     lowf := nil; midf := nil; highf := nil;
     {*** First key becomes splitter ***}
     tailins(r, midf, midl);
     r := r↑.next;
     while r<>nil do begin
          if r↑.k<midf↑.k then  tailins(r,lowf,lowl)
          else if r↑.k=midf↑.k then  tailins(r,midf,midl)
               else  tailins(r,highf,highl);
          r := r↑.next
          end;
     {*** Assemble resulting list ***}
     if lowf <> nil then begin
          lowl↑.next := nil;
          sort := sort(lowf);
          Last↑.next := midf
          end
     else sort := midf;
     if highf <> nil then highl↑.next := nil;
```

```
            midl↑.next := sort(highf);
            if Last = nil then Last := midl
            end
      end;
```

This algorithm keeps track of lists by keeping a pair of pointers to each list: one to the head and one to the tail. This is particularly useful for concatenating lists together. The global variable *Last* is used to return a pointer to the last element of a sorted list. The procedure *tailins* inserts a record at the end of a list given by a pair of pointers.

Insert a record at the end of a list

```
procedure tailins (rec : list; var first,last : list);
begin
if first = nil then first := rec
             else last↑.next := rec;
last := rec
end;
```

The worst case, $O(n^2)$ comparisons, happens, among other cases, when we sort a totally ordered or reverse-ordered list.

The above implementation of Quicksort keeps a list of all the elements that are equal to the splitting record. By doing this, and by growing the lists at the tail, Quicksort for lists becomes a stable sorting algorithm, that is, equal keys are not transposed.

When sorting lists with Quicksort we cannot easily prevent the worst case. Consequently, portions of the list that are already in order will deteriorate the algorithm's performance significantly.

References:
[Motzkin, D., 81], [Wegner, L.M., 82].

4.2.3 Bucket sort

Bucket sort (or **address-calculation sort**) uses an interpolation formula on the keys to split the records between m buckets. The buckets are sets of records, which we implement using lists. After the splitting pass, the records in the first bucket will have smaller keys than the records in the second bucket and so on. The buckets are then sorted recursively and finally all the buckets are concatenated together.

Let I_n denote the number of times that a key is placed into a bucket.

This measure counts the number of times the innermost loop is executed. I_n satisfies the recurrence equation:

$$I_n = n + \sum_{i=0}^{n-2} \binom{n-2}{2} \frac{(m-1)^{n-2-i}}{m^{n-2}} ((m-2)I_i + 2I_{i+1})$$

for fixed m, and $m < n$ the solution of the above is

$$I_n = n \log_m n + n\, Q(\log_m n, m) - \frac{1}{2 \ln m} + O(n^{-1})$$

where $Q(x, m)$ is a periodic function in x. For m proportional to n, $n = \alpha m$, (m remains fixed for the recursive calls) then

$$I_n = (2 - e^{-\alpha})n + \alpha^2 - 2 + e^{-\alpha}(\alpha^2/2 + \alpha + 2) + O(n^{-1})$$

For $m = n$, (and m is set equal to n for each recursive call)

$$I_n = n + \sum_{i=0}^{n-2} \binom{n-2}{2} \frac{(1-1/n)^{n-2}}{(n-1)^i} ((n-2)I_i + 2I_{i+1})$$

$$= 1.76108...n - 0.39125... + O(n^{-1})$$

Bucket sort

```
list sort(s, min, max)
list s;
typekey min, max;

{
int i;
typekey div, maxb[M], minb[M];
list head[M], t;
struct rec aux;
extern list Last;
if (s==NULL) return(s);
if (max==min) {
      for (Last=s; Last->next!=NULL; Last = Last->next);
      return(s);
      }
div = (max-min) / M; /* Find dividing factor */
if (div==0) div = 1;
for (i=0; i<M; i++) head[i] = NULL;
/* Place records in buckets */
while (s != NULL) {
```

```
        i = (s->k-min) / div;
        if (i<0) i = 0; else if (i>=M) i = M-1;
        t = s;
        s = s->next;
        t->next = head[i];
        if (head[i]==NULL) minb[i] = maxb[i] = t->k;
        head[i] = t;
        if (t->k > maxb[i]) maxb[i] = t->k;
        if (t->k < minb[i]) minb[i] = t->k;
        }
    /* sort recursively */
    t = &aux;
    for (i=0; i<M; i++) if (head[i]!=NULL) {
        t->next = sort(head[i], minb[i], maxb[i]);
        t = Last;
        }
    return(aux.next);
    }
```

The above algorithm computes the maximum and minimum key for each bucket. This is necessary and convenient as it allows correct sorting of files containing repeated keys and reduces the execution time. Bucket sort requires two additional parameters, the maximum and minimum key. Since these are recomputed for each pass, any estimates are acceptable; in the worst case, it will force bucket sort into one additional pass.

The above function sets the global variable *Last* to point to the last record of a sorted list. This allows easy concatenation of the resulting lists.

Bucket sort can be combined with other sorting techniques. If the number of buckets is significant compared to the number of records, most of the sorting work is done during the first pass. Consequently we can use a simpler (but quicker for small files) algorithm to sort the buckets.

Although the worst case for bucket sort is $O(n^2)$, this can only happen for particular sets of keys and only if the spread in values is $n!$. This is very unlikely. If we can perform arithmetic operations on keys, bucket sort is probably the most efficient alternative to sorting lists.

References:
[Isaac, E.J. *et al.*, 56], [Flores, I., 60], [Tarter, M.E. *et al.*, 66], [Knuth, D.E., 73], [Cooper, D. *et al.*, 80], [Devroye, L. *et al.*, 81], [Akl, S.G. *et al.*, 82], [Kirkpatrick, D.G. *et al.*, 84], [Suraweera, F. *et al.*, 88], [Manber, U., 89], [Cormen, T.H. *et al.*, 90].

4.2.4 Radix sort

Radix sort (or **distributions sort**), sorts records by examining the digital decomposition of the key. This algorithm does one pass of the file for each digit in the key. In each pass, the file is split according to the values of the corresponding digit. The sorting can be done top-down or bottom-up depending on the relative order of the splitting and recursive sorting of the subfiles.

If we split, sort recursively and concatenate, the resulting algorithm, which we will call **top-down radix sort**, resembles bucket sort (see Section 4.2.3), where instead of computing a 'bucket address' the bucket is selected based on a digit of the key.

The **bottom-up radix sort**, where we sort recursively, split and concatenate, is the most common version of radix sort. This method was at one time very popular in the data-processing field as it is the best method for sorting punched cards.

There is a close correspondence between the top-down radix sort and digital trees or tries (see Section 3.4.4). The number of times a given record is analyzed corresponds to the depth of the record in an equivalent trie tree. The total complexity, that is, total number of records passed, coincides with the internal path in an equivalent trie. These results can be found in Section 3.4.4.

For the bottom-up algorithm, let m be the base of the numbering system, let D be the number of digits in the key and let I_n be the number of times the innermost loop is repeated (number of records passed through). Then

$$I_n = nD$$

It is possible to group several digits together, in which case D and m could vary as long as

$$m^D = K_1$$

(where K_1 is a constant for a given file). Given this constraint, the tradeoffs are simple: the time complexity is linear in D and the additional storage is linear in m.

Bottom-up radix sort

```
function sort(r : list) : list;
var   head, tail : array[1..M] of list;
      i, j, h : integer;

begin
for i:=D downto 1 do begin
```

```
        for j:=1 to M do head[j] := nil;
        while r <> nil do begin
            h := charac(i, r↑.k);
            if head[h]=nil then head[h] := r
            else tail[h]↑.next := r;
            tail[h] := r;
            r := r↑.next;
            end;
        {*** Concatenate lists ***}
        r := nil;
        for j:=M downto 1 do
            if head[j] <> nil then begin
                tail[j]↑.next := r;
                r := head[j]
                end
        end;
    sort := r
    end;
```

The above sorting algorithm uses the function *charac(i, key)* which returns the ith digit from the key *key*. The top-down radix sorting function is described in Appendix IV.

If $D \log m$ is larger than $\log n$ then bottom-up radix sort is not very efficient. On the other hand, if $D \log m < \log n$ (some keys must be duplicated), radix sort is an excellent choice.

References:

[Hibbard, T.N., 63], [MacLaren, M.D., 66], [Knuth, D.E., 73], [Aho, A.V. *et al.*, 74], [Reingold, E.M. *et al.*, 77], [McCulloch, C.M., 82], [van der Nat, M., 83], [Devroye, L., 84], [Baase, S., 88], [Sedgewick, R., 88], [Manber, U., 89], [Cormen, T.H. *et al.*, 90].

4.2.5 Hybrid methods of sorting

Most of the sorting algorithms described so far are basic in the sense that their building blocks are more primitive operations rather than other sorting algorithms. In this section we describe algorithms which combine two or more sorting algorithms. The basic sortings usually have different properties and advantages and are combined in a way to exploit their most advantageous properties.

4.2.5.1 Recursion termination

This is a general technique which has been described for Quicksort (see Section 4.1.3) in particular. Many recursive sorting algorithms have good general performance, except that they may do an inordinate amount of work for a file with very few elements (such as Quicksort or bucket sort for two elements).

On the other hand, being efficient for the tail of the recursion is very important for the total complexity of the algorithm.

The general scheme for hybrid recursive sorts is then

Hybrid termination

```
function sort(keys);
begin
if size(keys) > M then
        < ...main sorting algorithm... >
else  simplersort(keys);
end;
```

The *simplersort*() part may be just an analysis of one, two, and three elements by brute force or another sorting algorithm which does well for small files. In the latter case, linear insertion sort (see Section 4.1.2) is a favourite candidate.

4.2.5.2 Distributive partitioning

Distributive partitioning sort is a composition of a balanced Quicksort with bucket sort (see Sections 4.1.3 and 4.2.3). The file is split by the median element (or an element chosen to be very close to the median, for example, median of medians) and then the lower and upper elements, separately, are bucket sorted. The procedure may be applied recursively, or we may use still another composition for sorting the individual buckets.

The motive for this composition is to profit from the good average performance of bucket sort, while guaranteeing an $O(n \log n)$ time by splitting the file by the median at each step.

$$C_n = O(n \log n)$$

If the median is too costly to compute we could split the file into two equal-size parts and apply bucket sort twice. We then sort the buckets recursively and finally merge the two halves. This has the same effect as computing the median for the worst case, but it is much more efficient.

4.2.5.3 Non-recursive bucket sort

When the number of buckets is relatively large, bucket sort achieves an excellent average performance ($O(n)$). Not only is the time linear, but the constant in the linear term is reasonably small; the first pass does most of the sorting.

However, the $O(n^2)$ worst case is clearly unwanted. A family of hybrid algorithms can be derived from compositions of a single pass of bucket sorting and a second algorithm. This second algorithm should: (a) sort small files efficiently, as this is what it will do most; (b) have an $O(n \log n)$ worst-case performance, in case bucket sort hits an 'unlucky' distribution.

Again, we could have a double composition, one algorithm good for case (a) and one good for case (b). For example we could use linear insertion sort for less than 10 elements and Heapsort (see Section 4.1.5) otherwise.

Another alternative is to use natural merge sort (see Section 4.2.1). The worst case for bucket sort (batches of equal keys) is almost the best case for natural merge sort.

References:
[Dobosiewicz, W., 78], [Peltola, E. *et al.*, 78], [Dobosiewicz, W., 79], [Huits, M. *et al.*, 79], [Jackowski, B.L. *et al.*, 79], [Meijer, H. *et al.*, 80], [van der Nat, M., 80], [Akl, S.G. *et al.*, 82], [Allison, D.C.S. *et al.*, 82], [Noga, M.T. *et al.*, 85], [Tamminen, M., 85], [Handley, C., 86].

4.2.6 Treesort

A Treesort sorting algorithm sorts by constructing a lexicographical search tree with all the keys. Traversing the tree in an infix order, all the nodes can be output in the desired order. Treesort algorithms are a composition of search tree insertion with infix tree traversal.

The number of comparisons required to sort n records is related to the specific type of search tree. Let C_n be the average number of comparisons in a successful search, then

$$C_n^{sort} = nC_n$$

Almost any of the tree structures described in Section 3.4 can be used for this purpose. The following algorithm is based on binary trees.

Binary treesort

```
tree := nil;
for i:=1 to n do insert(tree, <ith-key>);
output_infix(tree);
```

where the function *output_infix* is

Scan binary tree in infix order

```
procedure output_infix(t : tree);
begin
if t <> nil then begin
      output_infix(t↑.left);
      output(t↑.key);
      output_infix(t↑.right);
      end
end;
```

These algorithms require two pointers per record and consequently are significantly more expensive than other methods in terms of additional storage. There is one circumstance when this structure is desirable, and that is when the set of records may grow or shrink, and we want to be able to maintain it in order at low cost.

To guarantee an $O(n \log n)$ performance it is best to select some form of balanced tree (such as AVL, weight-balanced or B-trees).

References:
[Frazer, W.D. *et al.*, 70], [Woodall, A.D., 71], [Aho, A.V. *et al.*, 74], [Szwarcfiter, J.L. *et al.*, 78].

4.3 Merging

A special case of sorting is to build a single sorted file from several sorted files. This process is called **merging** of files and it is treated separately, as it normally requires simpler algorithms.

Merging a small number of files together is easily achieved by repeated use of a function which merges two files at a time. In most cases, an optimal strategy is to merge the two smallest files repeatedly until there is only one file left. For this reason, the merging of two ordered files is the main function which we will analyze in this section. Algorithms for merging large numbers of files are studied in conjunction with external sorting. In particular, the second phases of the merge sort algorithms are good merging strategies for many files.

A **stable merging** algorithm is one which preserves the relative orderings of equal elements from each of the sequences. The concept of stability can be extended to enforce that equal elements between sequences will maintain a consistent ordering; this is called **full stability**.

General references:
[Floyd, R.W. *et al.*, 73], [Schlumberger, M. *et al.*, 73], [Hyafil, L. *et al.*, 74], [Harper, L.H. *et al.*, 75], [Yao, A.C-C. *et al.*, 76], [Fabri, J., 77], [Reingold, E.M. *et al.*, 77], [Sedgewick, R., 78], [Tanner, R.M., 78], [Brown, M.R. *et al.*, 79], [van der Nat, M., 79], [Mehlhorn, K., 84], [Munro, J.I. *et al.*, 87], [Salowe, J.S. *et al.*, 87], [Huang, B. *et al.*, 88], [Sedgewick, R., 88], [Huang, B. *et al.*, 89].

4.3.1 List merging

Merging two ordered lists is achieved by repeatedly comparing the top elements and moving the smallest key one to the output list.

Assuming that all the possible orderings of the lists are equally likely, then:

$$E[C_{n_a,n_b}] = n_a + n_b - \frac{n_a}{n_b+1} - \frac{n_b}{n_a+1}$$

$$\sigma^2(C_{n_a,n_b}) = \frac{n_b(2n_b+n_a)}{(n_a+1)(n_a+2)} + \frac{n_a(2n_a+n_b)}{(n_b+1)(n_b+2)} - \left(\frac{n_a}{n_b+1} + \frac{n_b}{n_a+1}\right)^2$$

$$C^{\max}_{n_a,n_b} = n_a + n_b - 1$$

List merging

```
function merge (a, b : list) : list;

var first, last, temp : list;
begin
first := nil;
while b <> nil do
     if a = nil then begin a := b;  b := nil end
     else begin
          if b↑.k > a↑.k then
                    begin temp := a; a := a↑.next end
          else begin temp := b; b := b↑.next end;
          temp↑.next := nil;
          tailins(temp, first, last)
          end;
```

```
    tailins(a, first, last);
    merge := first
    end;
```

The above function uses the procedure *tailins* which inserts a node into a list defined by its first and last pointers. Such a procedure is useful in general for working with lists and is described in Section 4.2.

The above algorithm is stable but not fully stable.

References:
[Knuth, D.E., 73], [Horowitz, E. *et al.*, 76], [Huang, B. *et al.*, 88], [Huang, B. *et al.*, 89].

4.3.2 Array merging

Array merging is a simple operation if enough additional space is available. For example, merging two arrays with additional space amounting to the smallest of the arrays can be accomplished in $n_a + n_b - 1$ comparisons. The next algorithm merges arrays a and b of size n_a and n_b respectively into the array a.

Merging of arrays

```
merge(a, b, na, nb)
RecordArray a, b;
int na, nb;

{ /*** Merge array b (0...nb-1) into array a (0...na-1) ***/
while (nb > 0)
        if (na<=0 || a[na-1].k < b[nb-1].k)
                { nb--; a[na+nb] = b[nb]; }
        else { na--; a[na+nb] = a[na]; }
};
```

There are several algorithms to merge arrays with little or no additional storage. However, these are quite complex. The problem can be slightly rephrased, and in that case is usually referred to as **in-place merging**: given an array a which contains two sequences of ordered elements, one in locations 1 to n_a and the other in locations $n_a + 1$ to $n_a + n_b$, merge them into one sequence using only m units of additional storage.

Most of these algorithms, although asymptotically better, will not compete with an in-place sorting method for practical purposes. In particular, Shellsort (see Section 4.1.4) will do less work for the merging of two sequences than for sorting a random array, and is thus recommended.

Table 4.7: Characteristics of in-place merging algorithms.

Comparisons	*Extra space*	*Stable*	*Reference*
$O(n)$	$O(1)$	No	[Kronrod, 69]
$O(n)$	$O(\log n)$	Yes	[Horvarth, 74]
$O(n)$	$O(1)$	Yes	[Trabb Pardo, 77]
$O(kn)$	$O(n^{1/k})$	Yes	[Wong, 81]
$O(n_b \log(n_a/n_b + 1))$	$O(\log n_b)$	Yes	[Dudzinski & Dydek, 81]
$O(n)$	$O(1)$	No	[Huang & Langston, 88]
$O(n)$	$O(1)$	Yes	[Huang & Langston, 89]

Table 4.7 lists the properties and references for some in-place merging algorithms, where n_a and n_b denote the sizes of the two arrays to be merged, $n_a + n_b = n$, and without loss of generality we assume $n_a \geq n_b$.

References:
[Kronrod, M.A., 69], [Knuth, D.E., 73], [Horvath, E.C., 74], [Trabb Pardo, L., 77], [Horvath, E.C., 78], [Murphy, P.E. *et al.*, 79], [Dudzinski, K. *et al.*, 81], [Wong, J.K., 81], [Alagar, V.S. *et al.*, 83], [Mannila, H. *et al.*, 84], [Carlsson, S., 86], [Thanh, M. *et al.*, 86], [Dvorak, S. *et al.*, 87], [Salowe, J.S. *et al.*, 87], [Dvorak, S. *et al.*, 88], [Dvorak, S. *et al.*, 88], [Huang, B. *et al.*, 88], [Huang, B. *et al.*, 89], [Sprugnoli, R., 89].

4.3.3 Minimal-comparison merging

Let $C^{MM}_{n_a,n_b}$ denote the minimum–maximum, or the minimum worst-case number of comparisons required to merge two files of sizes n_a and n_b. It is known that

$$C^{MM}_{n,1} = \lceil \log_2(n+1) \rceil$$

$$C^{MM}_{n,2} = \left\lceil \log_2 \frac{7(n+1)}{6} \right\rceil + \left\lceil \log_2 \frac{7(n+1)}{17} \right\rceil$$

$$C^{MM}_{n,3} = \left\lceil \log_2 \frac{7n+13}{17} \right\rceil + \left\lceil \log_2 \frac{7(n+2)}{107} \right\rceil + \left\lceil \log_2 \frac{7(n+2)}{43} \right\rceil + 5 \quad (n \geq 9)$$

$$C^{MM}_{n_a,n_b} \geq \left\lceil \log_2 \binom{n_a+n_b}{n_a} \right\rceil$$

$$C^{MM}_{n_a,n_b} = n_a + n_b - 1 \qquad (\text{if } \min(n_a, n_b) \geq 2 \mid n_a - n_b \mid -2)$$

The Hwang and Lin merging algorithm, sometimes called **binary merging**, merges two files with an almost optimal number of comparisons. This algorithm is optimal for merging a single element into a sequence, two equal sequences and other cases. Compared to the standard algorithm, it reduces the number of comparisons significantly for files with very different sizes, however the number of movements will not be reduced, and hence this algorithm is mostly of theoretical interest.

The basic idea of binary merging is to compare the first element of the shorter file with the 1st or 2nd or 4th or 8th... element of the longer file depending on the ratio of the file sizes. If $n_a \geq n_b$ then we compare the first element of file b with the 2^t element of a, where $t = \lfloor \log_2 n_a/n_b \rfloor$. If the key from file b comes first, then a binary search between $2^t - 1$ elements is required; otherwise 2^t elements of file a are moved ahead. The procedure is repeated until one of the files is exhausted.

In its worst case, Hwang and Lin's algorithm requires

$$C^{HL}_{n_a,n_b} = (t+1)n_b + \lfloor n_b/2^t \rfloor - 1$$

where $t = \lfloor \log_2 n_a/n_b \rfloor$.

Manacher introduced an improvement to the Hwang and Lin algorithm when $n_a/n_b \geq 8$, which reduces the number of comparisons by $n_b/12 + O(1)$.

References:
[Hwang, F.K. *et al.*, 71], [Hwang, F.K. *et al.*, 72], [Knuth, D.E., 73], [Christen, C., 78], [Manacher, G.K., 79], [Hwang, F.K., 80], [Stockmeyer, P.K. *et al.*, 80], [Schulte Monting, J., 81], [Thanh, M. *et al.*, 82], [Manacher, G.K. *et al.*, 89].

4.4 External sorting

Sorting files that do not fit in internal memory, and are therefore stored in external memory, requires algorithms which are significantly different from those used for sorting internal files. The main differences are:

(1) the most expensive operation is accessing (or storing) a record;

(2) the intermediate files may not support direct (or random) access of elements, and even if they do support direct accesses, sequential accesses are more efficient.

Our main measure of complexity is the number of times that the file has been copied, or read and written. A complete copy of the file is called a **pass**.

The algorithms we will describe use the following interface with the file system:

Interface with file system

```
function ReadFile(i : integer) : record;
procedure WriteFile(i : integer; r : record);
procedure OpenWrite(i : integer);
function OpenRead(i : integer);
function Eof(i : integer) : boolean;
ReadDirect(i : integer) : record;
WriteDirect(i : integer; r : record);
```

In all cases the argument i refers to a unit number, an integer in the range $1...maxfiles$. The function $Eof(i)$ returns the value '*true*' when the last $ReadFile$ issued failed. The functions $OpenWrite$ and $OpenRead$ set the corresponding indicator to the letters 'o' (output unit) and 'i' (input unit) respectively in the global array $FilStat$. The direct access operations use an integer to select the record to be read/written. These operations use the input file only. Without loss of generality we will assume that the input file is in unit 1, which can be used later for the sorting process. Furthermore, the output file will be placed in any file whose index is returned by the sorting procedure. In the worst case, if this is not desired and cannot be predicted, a single copy is sufficient.

The **external merge sorting** algorithms are the most common algorithms and use two phases: distribution phase and merging phase. During the **distribution phase** or **dispersion phase** the input file is read and sorted into sequences, each sequence as long as possible. These sequences, sometimes called **strings** or **runs**, are distributed among the output files. The **merging phase** merges the ordered sequences together until the entire file is a single sequence; at this point the sorting is completed.

The options available for creating the initial sequences (runs), for distributing them and organizing the merging phase (which files to merge with which, and so on) give rise to many variations of external merge sorting.

The distribution phase's objective is to create as few sequences as possible, and at the same time distribute these sequences in a convenient way to start the merging phase. There are three main methods for constructing the ordered sequences: **replacement selection**, **natural selection** and **alternating selection**.

General references:
[Friend, E.H., 56], [Gotlieb, C.C., 63], [Flores, I., 69], [Martin, W.A., 71], [Frazer, W.D. *et al.*, 72], [Barnett, J.K.R., 73], [Schlumberger, M. *et al.*, 73], [Hyafil, L. *et al.*, 74], [Lorin, H., 75], [Kronsjo, L., 79], [Munro, J.I. *et al.*, 80], [McCulloch, C.M., 82], [Tan, K.C. *et al.*, 82], [Reingold, E.M. *et al.*, 83], [Mehlhorn, K., 84], [Six, H. *et al.*, 84], [Aggarwal, A. *et al.*, 88], [Baase, S., 88], [Sedgewick, R., 88], [Salzberg, B., 89].

4.4.1 Selection phase techniques

4.4.1.1 Replacement selection

The replacement selection algorithm keeps records in an internal buffer of size M. When the buffer is full, the smallest key record is output and a new record is read. Subsequently, the smallest key record in the buffer, whose key is larger than or equal to the last written, is output, and a new record is read. When this is no longer possible, that is, when all keys in the buffer are smaller than the last output, a new sequence is initiated.

The expected length of the ith run, denoted by n_i, is

$$E[n_i] = L_i M = 2M + O(8^{-i}M)$$

$$E[n_1] = (e-1)M$$

where the values L_i are given by the generating function

$$L(z) = \sum_{i \geq 0} L_i z^i = \frac{z(1-z)}{e^{z-1} - z} - z$$

The simplest way to manage the buffers is to keep a priority queue with the elements larger than the last output key, and a pool with the others. The following code describes the function *distribute* which uses a heap as a priority queue.

Replacement selection distribution

```
distribute()
{int i, hbot, s;
typekey lastout;

for (i=0; i<M; i++) {
      Buff[i] = ReadFile(1);
      if (Eof(1)) break;
      }
```

```
    i--;

    while (i>=0) {
        for (hbot=0; hbot<i;) insert(++hbot, Buff);
        /*** Start a new sequence ***/
        s = nextfile();
        while (hbot >= 0) {
            lastout = Buff[0].k;
            WriteFile(s, Buff[0]);
            Buff[0] = Buff[hbot];
            siftup(Buff, 0, hbot-1);
            if (!Eof(1)) Buff[hbot] = ReadFile(1);
            if (Eof(1))  Buff[hbot--] = Buff[i--];
            else if (Buff[hbot].k < lastout)  hbot--;
                else insert(hbot, Buff);
            }
        }
    };
```

The function *nextfile* returns the file number on which the next sequence or run should be placed. The functions *insert* and *siftup* are described in the priority queue Section 5.1.3.

4.4.1.2 Natural selection

Natural selection is a mechanism for producing runs, similar to replacement selection, which uses a **reservoir** of records to increase the efficiency of the internal buffer. Until the reservoir is full, new records with keys smaller than the last output record are written into the reservoir. Once the reservoir is full, the current sequence is completed as with replacement selection. When a new sequence is initiated, the records from the reservoir are read first. Table 4.8 shows the average run length on function of the reservoir size.

It is assumed that the reservoir is on secondary storage, as, if main memory is available, pure replacement selection with a bigger buffer is always better. If the reservoir is in secondary storage, there is a cost associated with its usage, and there is an interesting trade off: for a larger reservoir, more records will be passed through it, but longer sequences will result and fewer merging passes will be required.

If the number of passes in the sorting algorithm is

$$E[P_n] \;=\; \log_b n + O(1)$$

then the optimal reservoir size is the value r which minimizes

$$-\log_b L(r) + \frac{r}{L(r)}$$

Table 4.8: Average run lengths for natural selection.

Reservoir size	*Average run length*
$M/2$	$2.15553...M$
M	$2.71828...M$
$3M/2$	$3.16268...M$
$2M$	$3.53487...M$
$5M/2$	$3.86367...M$
$3M$	$4.16220...M$

where $L(r)$ is the average run length with reservoir size r. Table 4.9 shows some values for the optimal reservoir size. The above function is very 'flat' around its minimum, so large variations in the reservoir size do not depart significantly from the optimum.

Table 4.9: Optimum reservoir sizes for various sorting orders.

b	*Reservoir*	*Average run length*	*Passes saved*
2	$6.55M$	$5.81M$	1.409
3	$2.22M$	$3.68M$	0.584
4	$1.29M$	$2.99M$	0.358

4.4.1.3 Alternating selection

Some algorithms require that sequences be stored in ascending and descending order alternatively. The replacement selection algorithm can be used for this purpose with a single change: the last **if** statement should be

if (*Buff*[*hbot*].*k* < *lastout* ↑ *direction* == 'a')

where *direction* is a global variable which contains the letter a or the letter d. The priority queue functions should also use this global indicator.

The alternation between ascending and descending sequences should be commanded by the function *nextfile*. As a general rule, longer sequences are obtained when the direction is not changed, so the function *nextfile* should be designed to minimize the changes in direction. If the direction is changed for every run, the average length of run is

$$E[n_i] = \frac{3M}{2} + o(1)$$

4.4.1.4 Merging phase

During the merging phase, the only difference between the algorithms is the selection of the input and output units. The function *merge* merges one run from all the input units (files with the letter **i** in their corresponding *FilStat*[ ] entry) into the file given as parameter. This function will be used by all the external merge sorts.

Merge one ordered sequence

```
merge(out)
int out;

{
int i, isml;
typekey lastout;
extern struct rec LastRec[ ];
extern char FilStat[ ];

lastout = MinimumKey;
LastRec[0].k = MaximumKey;
while (TRUE) {
	isml = 0;
	for (i=1; i<=maxfiles; i++)
		if (FilStat[i]=='i' && !Eof(i) &&
			LastRec[i].k >= lastout &&
			LastRec[i].k < LastRec[isml].k)
			isml = i;
	if (isml==0) {
		for (i=1; i<=maxfiles; i++)
			if (FilStat[i]=='i' && !Eof(i)) return(0);
		return('done');
		}
	WriteFile(out, LastRec[isml]);
	lastout = LastRec[isml].k;
	LastRec[isml] = ReadFile(isml);
	}
};
```

Merge uses the global record array *LastRec*. This array contains the last record read from every input file. When all the input files are exhausted simultaneously, this function returns the word **done**.

References:
[Goetz, M.A., 63], [Knuth, D.E., 63], [Dinsmore, R.J., 65], [Gassner, B.J., 67], [Frazer, W.D. *et al.*, 72], [McKellar, A.C. *et al.*, 72], [Knuth, D.E., 73], [Espelid, T.O., 76], [Ting, T.C. *et al.*, 77], [Dobosiewicz, W., 85].

4.4.2 Balanced merge sort

Balanced merge sorting is perhaps the simplest scheme for sorting external files. The files are divided into two groups, and every pass merges the runs from one of the groups while distributing the output into the other group of files.

Let T be the number of sequential files available and let P_n^T denote the number of passes necessary to sort n runs or strings. Then we have:

$$P_n^T = \lceil 2 \log_{\lfloor T/2 \rfloor \lceil T/2 \rceil} n \rceil$$

Balanced merge sort

```
sort()
{
int i, runs;
extern int maxfiles, unit;
extern char FilStat[ ];
extern struct rec LastRec[ ];

/*** Initialize input/output files ***/
OpenRead(1);
for (i=2; i<=maxfiles; i++)
        if (i <= maxfiles/2) FilStat[i] = '-';
                else OpenWrite(i);
distribute();
do { /*** re-assign files ***/
        for (i=1; i<=maxfiles; i++)
                if (FilStat[i] == 'o') {
                        OpenRead(i);
                        LastRec[i] = ReadFile(i);
                        }
                else OpenWrite(i);
        for (runs=1; merge(nextfile()) != 'done'; runs++);
```

```
        } while (runs>1);
    return(unit);
    };
```

The function that performs the selection of the files to output the runs is very simple and just alternates between all possible output files.

Selection of next file for balanced merge sort

```
nextfile()
{extern int maxfiles, unit;
extern char FilStat[ ];

do unit = unit%maxfiles + 1;
        while (FilStat[unit] != 'o');
return(unit);
};
```

For simplicity, the current output unit number is kept in the global variable *unit*.

For some particular values of n and T, the balanced merge may not be optimal, for example $P_9^5 = 5$, but an unbalanced merge can do it in four passes. Also it is easy to see that $P_n^T = 2$ for $n \leq T-1$. The difference between the optimal and normal balanced merge is not significant.

Table 4.10 shows the maximum number of runs that can be sorted in a given number of passes for the optimal arrangement of balanced merge sort.

Table 4.10: Maximum number of runs sorted by balanced merge sort.

	Number of passes				
Files	3	4	5	6	7
3	2	4	4	8	8
4	4	9	16	32	64
5	6	18	36	108	216
6	9	32	81	256	729
7	12	50	144	576	1728
8	16	75	256	1125	4096
10	25	147	625	3456	15625

References:
[Knuth, D.E., 73], [Horowitz, E. *et al.*, 76].

4.4.3 Cascade merge sort

Cascade merge sort distributes the initial sequences or runs among the output files in such a way that during the merging phase the following pattern can be maintained: each **merging pass** consists of the merging of $T-1$ files into one, until one of the input files is exhausted, then the merging of the $T-2$ remaining files into the emptied one, and so on. The final pass finds one sequence in every file and merges them into a single file.

A **perfect distribution** is a set of numbers of runs which allow this process to be carried to the end without ever having two files exhausted at the same time, except when the process is completed. Perfect distributions depend on the number of files, T, and the number of merging steps, k. For example $\{0,3,5,6\}$ is a perfect distribution for $T=4$ and $k=3$.

Let $\{0, s_k^1, s_k^2, \ldots, s_k^{T-1}\}$ be a perfect distribution for k merging steps and T files, then

$$s^i(z) = \sum_k s_k^i z^k = \frac{4}{2T-1} \sum_{-T/2<k<\lfloor T/2 \rfloor} \frac{\cos\alpha_k \cos((T-i)\alpha_k)}{1 - z/(2\sin\alpha_k)}$$

where $\alpha_k = \frac{(4k+1)\pi}{4T-2}$

$$s_k^i = \sum_{j=T-i}^{T-1} s_{k-1}^j s_k^0 = 0$$

$$s_k^i \approx \frac{4}{2T-1}\cos\frac{\pi}{4T-2} \times \cos\frac{(T-i)\pi}{4T-2} \times \left(2\sin\frac{\pi}{4T-2}\right)^{-k}$$

$$\frac{1}{2\sin\pi/(4T-2)} = \frac{2T-1}{\pi} + \frac{\pi}{24(2T-1)} + O(T^{-3})$$

Let t_k be the total number of runs sorted by a T-file cascade merge sort in k merging passes or the size of the kth perfect distribution. Then

$$t(z) = \sum_k t_k z^k$$

$$= \frac{4}{2T-1} \sum_{-T/2<k<\lfloor T/2 \rfloor} \sin 2T\alpha_k \cos\alpha_k (2\sin\alpha_k)^{-(k+1)}$$

$$t_k \approx \frac{4}{2T-1}\sin\frac{2T\pi}{4T-2}\cos\frac{\pi}{4T-2}\left(2\sin\frac{\pi}{4T-2}\right)^{-(k+1)}$$

$$k = \log_{2T/\pi} n\pi/4 \left(1 + \frac{1}{2T\ln 2T/\pi}\right) + O(T^{-2}\log n)$$

Table 4.11 shows the maximum number of runs sorted by cascade merge sort for various values of T and k.

Table 4.11: Maximum number of runs sorted by cascade merge sort.

	Number of passes				
Files	3	4	5	6	7
3	3	7	13	23	54
4	6	14	32	97	261
5	10	30	85	257	802
6	15	55	190	677	2447
7	21	91	371	1547	6495
8	28	140	658	3164	15150
10	45	285	1695	10137	62349

References:
[Knuth, D.E., 73], [Kritzinger, P.S. *et al.*, 74].

4.4.4 Polyphase merge sort

Polyphase merge sort distributes the initial sequences or runs among the output files in such a way that during the merging phase all merges are done from $T-1$ files into 1. Once the proper distribution has been obtained, the merge proceeds from $T-1$ to 1 until one of the input files is exhausted. At this point the output file is rewound and the empty file is opened for output. The merging continues until the whole file is merged into a single sequence.

A perfect distribution is a set of numbers of runs which allow this process to be carried to the end without ever having two files exhausted at the same time, except when the process is completed. Perfect distributions depend on the number of files, T, and the number of merging steps, k. Perfect numbers are a generalization of Fibonacci numbers. For example $\{0, 2, 3, 4\}$ is a perfect distribution for $T = 4$ and $k = 3$.

Let $\{0, s_k^1, s_k^2, \ldots, s_k^{T-1}\}$ be a perfect distribution for k merging steps and T files, then

$$s^i(z) = \sum_k s_k^i z^k = \frac{(z^i - 1)z}{2z - 1 - z^T}$$

$$s_k^i = s_{k-1}^{i-1} + s_{k-1}^{T-1} s_k^0 = 0$$

$$s_k^i \approx \frac{1 - \alpha_T^{-i}}{2 - 2T + T\alpha_T} (\alpha_T)^k$$

where $1/\alpha_T$ is the smallest positive root of $2z - 1 - z^T = 0$ and

$$\alpha_T = 2 - \frac{2}{2^T - T + 1} + O(T^2 8^{-T})$$

Let t_k be the total number of runs sorted by a T-file polyphase merge in k merging steps, or the size of the kth perfect distribution, then

$$t(z) = \sum_k t_k z^k = \frac{(z^T - Tz + T - 1)z}{(2z - 1 - z^T)(z - 1)}$$

$$t_k \approx \frac{T-2}{2 - 2T + T\alpha_T}(\alpha_T)^k$$

The number of merging steps, M_n, for a perfect distribution with n sequences is then

$$\begin{aligned} M_n &= \log_{\alpha_T}\left(\frac{n(2 - 2T + T\alpha_T)}{T-2}\right) + o(1) \\ &\approx \left(1 + \frac{1}{2^T \ln 2}\right)\log_2 n + 1 - \log_2(T-2) + O(T2^{-T} + n^{-\epsilon}) \end{aligned}$$

for some positive ϵ.

Let r_k be the total number of runs passed (read and written) in a k-step merge with a perfect distribution. Then

$$r(z) = \sum_k r_k z^k = \frac{(z^T - Tz + T - 1)z}{(2z - 1 - z^T)^2}$$

$$r_k \approx \left((T-2)k + \frac{(\alpha_T - 2)T(T^2 - 2T + 2) + 2T}{2 - 2T + T\alpha_T}\right)\frac{(\alpha_T - 1)(\alpha_T)^k}{(2 - 2T + T\alpha_T)^2}$$

Let P_n be the total number of passes of the entire file required to sort n initial runs in k merging steps. Then

$$\begin{aligned} P_n &= \frac{\alpha_T - 1}{2 - 2T + T\alpha_T}k + O(1) \\ &\approx \frac{1}{2}\left(1 + \frac{T - 2 + 1/\ln 2}{2^T}\right)\log_2 n + 2 - \frac{\log_2(T-2)}{2} \\ &\quad + \frac{1}{T-2} + O(T^2 2^{-T}) \end{aligned}$$

When the actual number of sequences or runs is not a perfect distribution, the sequences can be increased by **dummy sequences** (empty sequences) arbitrarily inserted in the files. Since it is possible to predict how many times each sequence will be processed, we can insert the dummy sequences in those positions which are processed the largest number of times. Of course, the sequences are not literally 'inserted' in the files, since the files are assumed to be in sequential devices and no merging is done with these. The selection of the best placement of dummy sequences together with the selection of the best possible order (any number of merges larger or equal to the minimum required) gives rise to the **optimal polyphase merge**.

Polyphase merge sort

```
sort()
{
int i, j, some;
extern int maxfiles, maxruns[ ], actruns[ ];
extern struct rec LastRec[ ];

/*** Initialize input/output files ***/
OpenRead(1);
for (i=2; i<=maxfiles; i++) OpenWrite(i);

/*** Initialize maximum and actual count of runs ***/
for (i=0; i<=maxfiles; i++) maxruns[i] = actruns[i] = 0;
maxruns[0] = maxruns[maxfiles] = 1;
distribute();

/*** Initialize merging phase ***/
for (i=2; i<=maxfiles; i++)
      { OpenRead(i); LastRec[i] = ReadFile(i); }
for (i=1; maxruns[0]>1; i = (i%maxfiles)+1) {
      OpenWrite(i);
      while (maxruns[(i%maxfiles)+1] > 0) {
            for (j=1; j<=maxfiles; j++)
                  if (j!=i) {
                        if (maxruns[j]>actruns[j])
                              FilStat[j] = '-';
                        else { FilStat[j] = 'i';
                              actruns[j]--;
                              some = TRUE;
                              }
                        maxruns[j]--; maxruns[0]--;
                        }
            maxruns[i]++; maxruns[0]++;
            if (some) { merge(i); actruns[i]++; }
            }
      OpenRead(i); LastRec[i] = ReadFile(i);
      };
return(i==1 ? maxfiles : i-1);
};
```

Selection of next file for polyphase merge sort

```
nextfile()
{extern int maxfiles, maxruns[ ], actruns[ ];
int i, j, inc;

actruns[0]++;
if (actruns[0]>maxruns[0]) {
      /*** Find next perfect distribution ***/
      inc = maxruns[maxfiles];
      maxruns[0] += (maxfiles-2) * inc;
      for (i=maxfiles; i>1; i--)
            maxruns[i] = maxruns[i-1] + inc;
      }
j = 2;
/*** select file farthest from perfect ***/
for (i=3; i<=maxfiles; i++)
      if (maxruns[i]-actruns[i] > maxruns[j]-actruns[j]) j = i;
actruns[j]++;
return(j);
};
```

Table 4.12 shows the maximum number of runs sorted by polyphase merge sort for various numbers of files and passes.

Table 4.12: Maximum number of runs sorted by polyphase merge sort.

	Number of passes				
Files	3	4	5	6	7
3	3	7	13	26	54
4	7	17	55	149	355
5	11	40	118	378	1233
6	15	57	209	737	2510
7	19	74	291	1066	4109
8	23	90	355	1400	5446
10	31	122	487	1942	7737

References:
[Gilstad, R.L., 60], [Gilstad, R.L., 63], [Malcolm, W.D., 63], [Manker, H.H., 63], [McAllester, R.L., 64], [Shell, D.L., 71], [Knuth, D.E., 73], [MacCallum, I.R., 73], [Kritzinger, P.S. *et al.*, 74], [Horowitz, E. *et al.*, 76], [Zave, D.A.,

77], [Colin, A.J.T. *et al.*, 80], [Er, M.C. *et al.*, 82].

4.4.5 Oscillating merge sort

Oscillating sort interleaves the distribution or dispersion phase together with the merging phase. To do this it is required that the input/output devices be able to:

(1) read backwards;

(2) switch from writing to reading backwards;

(3) switch from reading backwards to writing, without rewinding and without destroying what is at the beginning of the file.

Oscillating sort will always do the mergings reading backwards from $T-2$ units into one. Furthermore, the merging steps are done with balanced files, in the sense that their expected number of records in each is the same. A sequence, ascending (or descending), with $(T-2)n$ initial runs is constructed by a $T-2$-way merge from $T-2$ sequences (each containing n runs) in descending (or ascending) order.

A perfect distribution for oscillating sort can be produced when $n = (T-2)^k$. The number of passes required to sort n initial runs is:

$$P_n = \lceil \log_{T-1} n \rceil + 1$$

Oscillating sort

```
procedure sort(n, unit, direction : integer);
var   i, r : integer;

begin
if n=0 then {*** Mark as dummy entry ***}
      FilStat[unit] := '-'
else if n=1 then
      ReadOneRun(unit, direction)

else for i:=1 to T-2 do begin
      r := n div (T-i-1);
      n := n-r;
      sort(r, (unit+i-2) mod T + 2, -direction);
      MergeOneRunInto(unit, -direction)
      end
end;
```

Table 4.13 shows the maximum number of runs sorted by oscillating sort or any of its modified versions, for various numbers of files and passes. Note that since the input unit remains open during most of the sorting process, it is not possible to sort with less than four units.

Table 4.13: Maximum number of runs sorted by oscillating merge sort.

	Number of passes				
Files	3	4	5	6	7
3	-	-	-	-	-
4	4	8	16	32	64
5	9	27	81	243	729
6	16	64	256	1024	4096
7	25	125	625	3125	15625
8	36	216	1296	7776	46656
10	64	512	4096	32768	262144

References:
[Sobel, S., 62], [Goetz, M.A. *et al.*, 63], [Knuth, D.E., 73], [Lowden, B.G.T., 77].

4.4.6 External Quicksort

External Quicksort is a completely different type of procedure for external sorting. The basic algorithm is the same as the internal Quicksort: the file is split into two parts, the lower and the upper, and the procedure is applied to these recursively. Instead of keeping one single record to do the splitting, this procedure keeps an array of size M of splitting elements. This array of records is maintained dynamically as the splitting phase progresses. Its goal is to produce an even split and to place as many records as possible in their final position (all records in the buffer will be placed in their final location).

For a file consisting of random records, assuming that each pass leaves the records in random order,

$$E[P_n] = \frac{\ln n}{H_{2M} - H_M} + O(1)$$

External Quicksort

```
sort(a, b)
int a, b;

{int i, j, rlow, rupp, wlow, wupp, InBuff;
typekey MaxLower, MinUpper;
struct rec LastRead;
extern struct rec Buff[ ];

while (b>a) {
      rupp = wupp = b;
      rlow = wlow = a;
      InBuff = 0;
      MaxLower = MinimumKey;
      MinUpper = MaximumKey;
      i = a-1;
      j = b+1;
      /*** Partition the file ***/
      while (rupp >= rlow) {
            if (rlow-wlow < wupp-rupp)
                  LastRead = ReadDirect(rlow++);
            else  LastRead = ReadDirect(rupp--);
            if (InBuff < M) {
                  Buff[InBuff++] = LastRead;
                  intsort(Buff, 0, InBuff-1);
                  }
            else {
                  if (LastRead.k > Buff[M-1].k) {
                        if (LastRead.k > MinUpper) j = wupp;
                                    else  MinUpper = LastRead.k;
                        WriteDirect(wupp--, LastRead);
                        }
                  else if (LastRead.k < Buff[0].k) {
                        if (LastRead.k < MaxLower) i = wlow;
                                    else  MaxLower = LastRead.k;
                        WriteDirect(wlow++, LastRead);
                        }
                  else if (wlow-a < b-wupp) {
                        WriteDirect(wlow++, Buff[0]);
                        MaxLower = Buff[0].k;
                        Buff[0] = LastRead;
                        intsort(Buff, 0, M-1);
                        }
```

```
                else { WriteDirect(wupp--, Buff[M-1]) ;
                       MinUpper = Buff[M-1].k;
                       Buff[M-1] = LastRead;
                       intsort(Buff, 0, M-1);
                       }
                }
        }
    while (InBuff>0) WriteDirect(wupp--, Buff[--InBuff]);

    /*** sort the shortest subfile first ***/
    if (i-a < b-j) { sort(a,i); a = j; }
        else { sort(j,b); b = i; }
    }
return(1);
};
```

The most noticeable differences between internal and external quicksort are:

(1) the records kept in the buffer are maintained as close to the centre as possible, that is, deletions are done on the left or on the right depending on how many records were already passed to the left or right.

(2) the reading of records is also done as balanced as possible with respect to the writing positions. This is done to improve the performance when the file is not random, but slightly out of order.

(3) two key values are carried during the splitting phase: *MaxLower* and *MinUpper*. These are used to determine the largest interval which can be guaranteed to be in order. By this mechanism it is possible to sort a totally ordered or reversely ordered file in a single pass.

The function *intsort* is any internal sorting function. Its complexity is not crucial as this function is called about $M \ln n$ times per pass of size n. An internal sorting function which does little work when the file is almost totally sorted is preferred (for example, the linear insertion sort of Section 4.1.2).

Table 4.14 shows simulation results on external Quicksort. From these results we find that the empirical formula

$$E[P_n] = \log_2(n/M) - 0.924$$

gives an excellent approximation for files with 1000 or more elements.

For very large internal buffers, a double-ended priority queue should be used, instead of the function *intsort*.

External Quicksort requires an external device which supports direct access. This sorting procedure sorts records 'in-place', that is, no additional

Table 4.14: Simulation results (number of passes) for external Quicksort.

n	$M = 5$	$M = 10$	$M = 20$
100	3.5272±0.0011	2.73194±0.00076	2.09869±0.00090
500	5.7057±0.0015	4.74526±0.00077	3.88463±0.00057
1000	6.6993±0.0021	5.69297±0.00095	4.77862±0.00059
5000	9.0555±0.0051	7.9773±0.0016	6.99252±0.00063
10000	10.0792±0.0071	8.9793±0.0026	7.97913±0.00090

files are required. External Quicksort seems to be an ideal sorting routine for direct access files.

This version of Quicksort will have an improved efficiency when sorting partially ordered files.

References:
[Monard, M.C., 80], [Cunto, W. *et al.*, to app.].

5 Selection Algorithms

5.1 Priority queues

We define **priority queues** as recursive data structures where an order relation is established between a node and its descendants. Without loss of generality this order relation will require that the keys in parent nodes be greater than or equal to keys in the descendant nodes. Consequently the root or head of the structure will hold the maximum element.

The algorithms that operate on priority queues need to perform two basic operations: add an element into the queue; extract and delete the maximum element of the queue. Additionally we may require other operations: construct a priority queue from a set of elements; delete and insert a new element in a single operation; inspect (without deleting) the maximum element and merge two queues into a single priority queue. Certainly some of these operations may be built using others. For each algorithm we will describe the most efficient or basic ones.

Typical calling sequence for these functions in Pascal

```
procedure insert(new : typekey; var pq : queue);
function extract(var pq : queue) : typekey;
function inspect(pq : queue) : typekey;
procedure delete(var pq : queue);
function merge(a, b : queue) : queue;
procedure delinsert(new : typekey; var pq : queue);
```

For the C implementation, the procedures which use **var** parameters are changed into functions which return the modified priority queue.

For some applications we may superimpose priority queue operations with the ability to search for any particular element; search for the successor (or predecessor) of a given element; delete an arbitrary element, and so on.

Searching structures which accept lexicographical ordering may be used as priority queues. For example, a binary search tree may be used as a priority queue. To add an element we use the normal insertion algorithm; the minimum is in the leftmost node of the tree; the maximum is in the rightmost node.

In all cases C_n^I will denote the number of comparisons required to insert an element into a priority queue of size n, C_n^E the number of comparisons to extract the maximum element and reconstruct the priority queue, and C_n^C the number of comparisons needed to construct a priority queue from n elements.

5.1.1 Sorted/unsorted lists

A **sorted list** is one of the simplest priority queues. The maximum element is the head of the list. Insertion is done after a sequential search finds the correct location. This structure may also be constructed using any list-sorting algorithm.

$$C_n^E = 0$$

$$E[C_n^I] = \frac{n(n+3)}{2(n+1)}$$

$$I_n = \frac{n(n+5)}{6}$$

where I_n is the average number of records inspected for all sequences of n operations which start and finish with an empty queue.

Sorted list insertion

```
list insert(new, pq)
list new, pq;

{struct rec r;
list p;
r.next = pq;
p = &r;
while (p->next != NULL && p->next->k > new->k)
  p = p->next;
```

```
new ->next = p ->next;
p ->next = new;
return(r.next);
};
```

Sorted list deletion

```
list delete(pq)
list pq;

{if (pq==NULL) Error /*** Delete from empty PQ ***/;
else return(pq ->next);
};
```

Sorted list inspection

```
typekey inspect(pq)
list pq;
{if (pq==NULL) Error /* inspect an empty PQ */;
else return(pq ->k);
};
```

A sorted list used as a priority queue is inefficient for insertions, because it requires $O(n)$ operations. However it may be a good choice when there are

(1) very few elements in the queue;

(2) special distributions which will produce insertions near the head of the list;

(3) no insertions at all (all elements are available and sorted before any extraction is done).

An **unsorted list**, at the other extreme, provides very easy addition of elements, but a costly extraction or deletion.

$$C_n^E = n$$

$$C_n^I = 0$$

Unsorted list insertion

```
list insert(new, pq)
list new, pq;

{new->next = pq;
return(new);}
```

Unsorted list deletion

```
list delete(pq)
list pq;

{struct rec r;
list p, max;
if (pq==NULL) Error /*** Deleting from empty PQ ***/;
else {r.next = pq;
      max = &r;
      for (p=pq; p->next != NULL; p=p->next)
            if (max->next->k < p->next->k) max = p;
      max->next = max->next->next;
      return(r.next);
      }
};
```

Unsorted list inspection

```
typekey inspect(pq)
list pq;

{list p;
typekey max;
if (pq==NULL) Error /*** Empty Queue ***/;
else {max = pq->k;
      for (p=pq->next; p!=NULL; p=p->next)
            if (max < p->k) max = p->k;
      return(max);
      }
};
```

An unsorted list may be a good choice when

(1) the elements are already placed in a list by some other criteria;

(2) there are very few deletions.

Merging sorted lists is an $O(n)$ process; merging unsorted lists is also an $O(n)$ process unless we have direct access to the tail of one of the lists.

References:
[Nevalainen, O. *et al.*, 79].

5.1.2 P-trees

P-trees or **priority trees** are binary trees with a particular ordering constraint which makes them suitable for priority queue implementations. This ordering can be best understood if we tilt the binary tree 45° clockwise and let the left pointers become horizontal pointers and the right pointers become vertical. For such a rotated tree the ordering is lexicographical.

We also impose the condition that the maximum and minimum elements of the tree both be on the leftmost branch, and so on recursively. This implies that any leftmost node does not have right descendants.

The top of the queue, the maximum in our examples, is kept at the leftmost node of the tree. The minimum is kept at the root. This requires some additional searching to retrieve the top of the queue. If we keep additional pointers and introduce pointers to the parent node in each node, the deletion and retrieval of the top element become direct operations. In any case, a deletion does not require any comparisons, only pointer manipulations.

Let L_n be the length of the left path in a queue with n elements. For each node inspected a key comparison is done. Then for a queue built from n random keys:

$$E[L_n] = 2H_n - 1$$

$$E[C_n^I] = \frac{H_{n+1}^2}{3} + \frac{10H_{n+1}}{9} - \frac{H_{n+1}^{(2)}}{3} - \frac{28}{27} \qquad (n \geq 2)$$

$$E[C_n^C] = 1 + \sum_{i=2}^{n-1} E[C_i^I] \qquad (n \geq 2)$$

where $H_n = \sum_{i=1}^{n} 1/i$ denotes harmonic numbers and $H_n^{(2)} = \sum_{i=1}^{n} 1/i^2$ denotes biharmonic numbers.

P-tree insertion

```
tree insert(new, pq)
tree new, pq;

{
tree p;
if (pq == NULL) return(new);
else if (pq->k >= new->k) {
        /*** Insert above subtree ***/
        new->left = pq;
        return(new);
        }
else {
    p = pq;
    while (p->left != NULL)
        if (p->left->k >= new->k) {
            /*** Insert in right subtree ***/
            p->right = insert(new, p->right);
            return(pq);
            }
        else p = p->left;
    /*** Insert at bottom left ***/
    p->left = new;
    };
return(pq);
};
```

P-tree deletion of maximum

```
tree delete(pq)
tree pq;
{
if (pq == NULL) Error /*** deletion on an empty queue ***/;
else if (pq->left == NULL) return(NULL);
else if (pq->left->left == NULL) {
    pq->left = pq->right;
    pq->right = NULL;
    }
else pq->left = delete(pq->left);
return(pq);
};
```

P-tree, retrieval of head of queue

```
typekey inspect(pq)
tree pq;
{
if (pq == NULL) Error /*** Inspecting an empty queue ***/;
while (pq->left != NULL) pq = pq->left;
return(pq->k);
};
```

With a relatively small change, P-trees allow the efficient extraction of the minimum as well as the maximum, so this structure is suitable for handling **double-ended** priority queues.

This priority queue is stable; equal keys will be retrieved first-in first-out.

Table 5.1 contains exact results (rounded to six digits). Simulation results are in excellent agreement with the theoretical ones.

Table 5.1: Exact results for P-trees.

n	$E[C_n^C]$	$E[L_n]$
5	7.66667	3.56667
10	27.1935	4.85794
50	347.372	7.99841
100	939.017	9.37476
500	8207.70	12.5856
1000	20001.3	13.9709
5000	147948.6	17.1890
10000	342569.2	18.5752

References:
[Jonassen, A.T. *et al.*, 75], [Nevalainen, O. *et al.*, 78].

5.1.3 Heaps

A **heap** is a perfect binary tree represented implicitly in an array. This binary tree has priority queue ordering: the key in the parent node is greater than or equal to any descendant key. The tree is represented in an array without the use of pointers. The root is located in position 1. The direct descendants of the node located in position i are those located in $2i$ and $2i+1$. The parent of node i is located at $\lfloor i/2 \rfloor$. The tree is 'perfect' in the sense that a tree with n

nodes fits into locations 1 to n. This forces a breadth-first, left-to-right filling of the binary tree.

For Williams' insertion algorithm, let C_n^I denote the number of comparisons and M_n the number of interchanges needed to insert the $n+1$th element, then

$$1 \leq C_n^I \leq \lfloor \log_2 n \rfloor$$

$$E[M_n] = E[C_n^I] - \frac{n-1}{n}$$

For an insertion into a random heap (all possible heaps being equally likely), when n is in the range $2^{k-1} - 1 \leq n < 2^k - 1$ we have:

$$E[C_{2^k-2}^I] \leq E[C_n^I] \leq E[C_{2^{k-1}-1}^I]$$

$$E[C_{2n}^I] < E[C_{2n-1}^I] \qquad (n > 1)$$

$$E[C_{2^k-2}^I] = 2 + O(k2^{-k})$$

$$E[C_{2^{k-1}-1}^I] = 2.60669... + O(k2^{-k})$$

A heap built by random insertions using Williams' insertion algorithm is not a random heap.

Williams' heap-insertion algorithm

```
procedure insert(new : ArrayEntry; var r : RecordArray);
var  i, j : integer;
     flag : boolean;

begin
n := n+1;
j := n;
flag := true;
while flag and (j>1) do begin
     i := j div 2;
     if r[i].k >= new.k then flag := false
          else begin r[j] := r[i]; j := i end
     end;
r[j] := new
end;
```

If all the elements are available at the same time, we can construct a heap more efficiently using Floyd's method. In this case

$$n - 1 \le C_n^C \le 2n - 2\nu(n) - \phi(n)$$

$$E[C_{2^k-1}^C] = (\alpha_1 + 2\alpha_2 - 2)2^k - 2k - 1 - \frac{6k+5}{9\,2^k} + O(k4^{-k})$$

$$E[C_n^C] = 1.88137...n + O(\log n)$$

where $\alpha_1 = \sum_{k\ge1} \frac{1}{2^k-1} = 1.60669...$ and $\alpha_2 = \sum_{k\ge1} \frac{1}{(2^k-1)^2} = 1.13733...$

$$0 \le M_n^C \le n - \nu(n)$$

$$E[M_{2^k-1}^C] = (\alpha_1 + \alpha_2 - 2)2^k - k - \frac{3k+4}{9\,2^k} + O(k4^{-k})$$

$$E[M_n^C] = 0.74403...n + O(\log n)$$

where $\nu(n)$ is the number of 1s in the binary representation of n and $\phi(n)$ is the number of trailing 0s of the binary representation of n.

Floyd's heap-construction algorithm

```
procedure siftup(var r : RecordArray; i,n : integer);

var  j : integer;
     tempr : ArrayEntry;
begin
     while 2*i<=n do begin
          j := 2*i;
          if j<n then
               if r[j].k < r[j+1].k then j := j+1;
          if r[i].k < r[j].k then begin
               tempr := r[j];
               r[j] := r[i];
               r[i] := tempr;
               i := j
               end
          else i := n+1
          end
end;
for i := (n div 2) downto 1 do siftup(r,i,n);
```

Worst-case lower and upper bounds:

$$C_n^I = \lfloor \log_2 (\lfloor \log_2 n \rfloor + 1) \rfloor + 1$$

$$C_n^E = \lfloor \log_2 n \rfloor + g(n) + O(1)$$

where $g(0) = 0$ and $g(n) = 1 + g(\lfloor \log_2 n \rfloor)$.

$$C_n^C = \frac{13}{8}n + O(\log n)$$

Average lower and upper bound:

$$1.36443...n + O(\log n) \le C_n^C \le 1.52128...n + o(n)$$

Extraction and reorganization:

$$C_n^E \le 2\lfloor \log_2 (n-1) \rfloor - \rho(n-1)$$

where $\rho(n)$ is 1 if n is a power of 2, 0 otherwise.

Heap extraction and reorganization

```
function extract(var r : RecordArray) : typekey;
begin
if n<1 then Error {*** extracting from an empty Heap ***}
else begin
     extract := r[1].k;
     r[1] := r[n];
     n := n-1;
     siftup(r, 1, n)
     end
end;
```

For a random delete-insert operation into a random heap we have:

$$2 \le C_n^E \le 2\lfloor \log_2 n \rfloor - \rho(n)$$

$$E[C_n^E] = \frac{2((n+1)k - \lfloor n/2 \rfloor - 2^k)}{n}$$

where $k = \lfloor \log_2 n \rfloor + 1$.

$$E[C_{2^k-1}^E] = \frac{(2k-3)2^k + 2}{2^k - 1}$$

Heap delete-insert algorithm

```
procedure delinsert(new : RecordEntry; var r : RecordArray);
begin
r[1] := new;
```

```
    siftup(r, 1, n)
    end;
```

The heap does not require any extra storage besides the elements themselves. These queues can be implemented just by using arrays and there are no requirements for recursion.

The insertion and extraction operations are guaranteed to be $O(\log n)$.

Whenever we can allocate vectors to store the records, the heap seems to be an ideal priority queue.

Merging two disjoint heaps is an $O(n)$ operation.

We can generalize the heap to any branch factor b other than two; in this case the parent of node i is located at $\lfloor (i-2)/b \rfloor + 1$ and the descendants are located at $\lceil b(i-1)+2 \rceil, ..., \lceil bi+1 \rceil$. This provides a tradeoff between insertion and extraction times: the larger b, the shorter the insertion time and longer the extraction time.

Table 5.2 gives figures for the number comparisons, C_n^C, required to build a heap by repetitive insertions, the number of comparisons required to insert the $n+1$th element, C_n^I and the number of comparisons required to extract all the elements from a heap constructed in this manner, C_n^D.

Table 5.2: Complexity of heaps created by insertions.

n	$E[C_n^C]$	$E[C_n^I]$	$E[C_n^D]$
5	5.133333	1.583333	5.8
10	13.95278	1.667027	25.54239
50	96.60725	1.983653	330.165±0.029
100	206.0169	2.135882	850.722±0.062
500	1103.952	2.116126	6501.21±0.26
1000	2237.752	2.253290	14989.06±0.53
5000	11348.8±3.2	2.330±0.015	98310.6±3.2
10000	22749.8±6.6	2.401±0.022	216592.0±6.2

References:

[Floyd, R.W., 64], [Williams, J.W.J., 64], [Knuth, D.E., 73], [Porter, T. *et al.*, 75], [Gonnet, G.H., 76], [Kahaner, D.K., 80], [Doberkat, E.E., 81], [Doberkat, E.E., 82], [Carlsson, S., 84], [Doberkat, E.E., 84], [Bollobas, B. *et al.*, 85], [Sack, J.R. *et al.*, 85], [Atkinson, M.D. *et al.*, 86], [Fredman, M.L. *et al.*, 86], [Gajewska, H. *et al.*, 86], [Gonnet, G.H. *et al.*, 86], [Sleator, D.D. *et al.*, 86], [Carlsson, S., 87], [Fredman, M.L. *et al.*, 87], [Fredman, M.L. *et al.*, 87], [Hasham, A. *et al.*, 87], [Stasko, J.T. *et al.*, 87], [Brassard, G. *et al.*, 88], [Draws, L. *et al.*, 88], [Driscoll, J.R. *et al.*, 88], [Frieze, A.M., 88], [Sedgewick, R., 88], [Carlsson, S. *et al.*, 89], [Manber, U., 89], [McDiarmid, C.J.H. *et al.*,

89], [Strothotte, T. *et al.*, 89], [Weiss, M.A. *et al.*, 89], [Cormen, T.H. *et al.*, 90], [Frederickson, G.N., 90], [Sack, J.R. *et al.*, 90].

5.1.4 Van Emde-Boas priority queues

Van Emde-Boas priority queues are queues which perform the operations insert, delete, extract maximum or minimum and find predecessor or successor in $O(\log \log N)$ operations. For these queues, N represents the size of the universe of keys and n the actual size of the subset of keys we include in the queue. It makes sense to use these queues when the keys are subsets of the integers 1 to N.

These queues are represented by one of various possible data structures. A queue is either

(1) empty, in which case it is represented by **nil**;

(2) a single element, in which case it is represented by the integer element itself;

(3) a boolean array of size N, if the universe is small ($N \leq m$);

(4) a structure composed of a queue of queues. The queue of queues is called the *top* part, and the element queues, which are arranged as an array, are called the *bottom* queues. Additionally we keep the maximum and minimum value occurring in the queue. The sizes of the top and bottom queues are as close to the square root of the cardinality of the universe as possible.

As a hyperrule, these priority queues have the definition:

$$\mathbf{vEB - N : [int, int, vEB - s(N), \{vEB - s(N)\}_1^{s(N)}]; \{bool\}_1^N; int; nil}$$

where $\mathrm{s(N)} = \lceil\sqrt{\mathrm{N}}\rceil$. The top queue is a queue on the indices of the bottom array. The index of every non-empty queue in the bottom is a key in the top queue.

Van Emde-Boas priority queue insertion

```
insert(new : integer; var pq);
case pq is nil:
        pq := NewSingleNode(new);
case pq is boolean array:
        turn on corresponding entry;
case pq is single element:
        expand entry to full node;
```

```
        seep into next case;
    case pq is full node:
        compute index based on "new"
        if bottom[index] <> nil then
                insert in bottom[index]
        else bottom[index] := NewSingleNode(new);
             insert index in top queue;
        adjust max and min if necessary;
    end;
```

Van Emde-Boas priority queue extraction

```
    extract(var pq) : integer;
    case pq is nil:
        Error;
    case pq is boolean array:
        Find last true entry;
        if only one entry remains then transform to SingleEntry;
    case pq is single element:
        return element;
        pq := nil;
    case pq is full node:
        return maximum;
        if bottom queue corresponding to maximum is single element
            then extract from top queue;
                max := max of bottom[max of top];
        else extract from bottom;
             max := max of bottom;
    end;
```

Let S_n^N be the storage utilized by a queue with n elements from a universe of size N. Then

$$S_n^N = O(\min(\sqrt{N} + nN^{1/4}, N))$$

The functions *extract minimum, test membership, find successor* and *find predecessor* can also be implemented in the same time and space.

References:
[van Emde-Boas, P. *et al.*, 77], [van Emde-Boas, P., 77].

5.1.5 Pagodas

The pagoda is an implementation of a priority queue in a binary tree. The binary tree is constrained to have priority queue ordering (parent larger than descendants). The structure of the pointers in the pagoda is peculiar; we have the following organization:

(1) the root pointers point to the leftmost and to the rightmost nodes;

(2) the right link of a right descendant points to its parent and its left link to its leftmost descendant;

(3) the left link of a left descendant points to its parent and its right link to its rightmost descendant.

The basic operation in a pagoda is merging two disjoint pagodas, which can be done very efficiently. An insertion is achieved by merging a single element with the main structure; an extraction is done by merging the two descendants of the root.

Merging pagodas is done bottom-up, merging the leftmost path of one with the rightmost path of the other.

Let C_{mn}^M be the number of comparisons needed to merge two pagodas of sizes m and n respectively. Then for pagodas built from random input we have

$$1 \leq C_{mn}^M \leq m+n-1$$

$$E[C_{mn}^M] = 2(H_n + H_m - H_{m+n})$$

$$1 \leq C_n^I \leq n$$

$$E[C_n^I] = 2 - \frac{2}{n+1}$$

$$0 \leq C_n^E \leq n-2$$

$$E[C_n^E] = 2(H_n - 2 + \frac{1}{n})$$

$$n-1 \leq C_n^C \leq 2n-3$$

$$E[C_n^C] = 2n - 2H_n$$

Merging two pagodas

```
function merge(a, b : tree) : tree;
var  bota, botb, r, temp : tree;

begin
if a=nil then merge := b
else if b=nil then merge := a
else begin
     {*** find bottom of a's rightmost edge ***}
     bota := a↑.right;  a↑.right := nil;
     {*** bottom of b's leftmost edge ***}
     botb := b↑.left;   b↑.left := nil;
     r := nil;
     {*** merging loop ***}
     while (bota<>nil) and (botb<>nil) do
          if bota↑.k < botb↑.k then begin
               temp := bota↑.right;
               if r=nil then  bota↑.right := bota
                    else  begin
                         bota↑.right := r↑.right;
                         r↑.right := bota
                         end;
               r := bota;
               bota := temp
               end
          else  begin
               temp := botb↑.left;
               if r=nil then  botb↑.left := botb
                    else  begin
                         botb↑.left := r↑.left;
                         r↑.left := botb
                         end;
               r := botb;
               botb := temp
               end;
     {*** one edge is exhausted, finish merge ***}
     if botb=nil then begin
          a↑.right := r↑.right;
          r↑.right := bota;
          merge := a
          end
     else begin
          b↑.left := r↑.left;
```

```
                r↑.left := botb;
                merge := b
                end
            end
        end;
```

Insertion in a pagoda

```
procedure insert(new : tree; var pq : tree);
begin
new↑.left := new; new↑.right := new;
pq := merge(pq, new)
end;
```

Deletion of head in a pagoda

```
procedure delete(var pq : tree);
var  le, ri : tree;

begin
if pq=nil then Error {*** deletion on empty queue ***}
else begin
     {*** find left descendant of root ***}
     if pq↑.left = pq then le := nil
          else begin
               le := pq↑.left;
               while le↑.left <> pq do le := le↑.left;
               le↑.left := pq↑.left
               end;
     {*** find right descendant of root ***}
     if pq↑.right = pq then ri := nil
          else begin
               ri := pq↑.right;
               while ri↑.right <> pq do ri := ri↑.right;
               ri↑.right := pq↑.right
               end;
     {*** merge descendants ***}
     pq := merge(le, ri)
     end
end;
```

Pagodas are remarkably efficient in their average behaviour with respect to the number of comparisons.

References:
[Francon, J. *et al.*, 78].

5.1.6 Binary trees used as priority queues

5.1.6.1 Leftist trees

A leftist tree is a binary tree with a priority queue ordering, which uses a count field at every node. This count field indicates the height (or distance) to the closest leaf. Leftist trees are arranged so that the subtree with the shortest path to a leaf is on the right descendant.

These trees are called leftist as their left branches are usually taller than their right ones.

An insertion can be done in the path to any leaf, so it is best to do it towards the rightmost leaf which is the closest to the root. A deletion is done through merging the two immediate descendants for the root. Leftist trees allow efficient, $O(\log n)$, merging of different trees.

Leftist tree insertion

```
procedure insert (new : tree; var pq : tree);

begin
if pq = nil then pq := new
else if pq↑.k > new↑.k then begin
            insert(new, pq↑.right);
            fixdist(pq)
            end
else begin
      new↑.left := pq;
      pq := new
      end
end;
```

Leftist tree deletion

```
function merge(a, b : tree) : tree;
begin
if a = nil then merge := b
else if b = nil then merge := a
else if a↑.k > b↑.k then begin
     a↑.right := merge(a↑.right, b);
     fixdist(a);
     merge := a
     end
else begin
     b↑.right := merge(a, b↑.right);
     fixdist(b);
     merge := b
     end
end;

procedure delete (var pq : tree);
begin
if pq = nil then Error {*** delete on an empty queue ***}
else pq := merge(pq↑.left, pq↑.right)
end;
```

Leftist tree distance

```
function distance(pq : tree) : integer;
begin
if pq=nil then distance := 0
     else distance := pq↑.dist
end;

procedure fixdist(pq : tree);
var temp : tree;
begin
if distance(pq↑.left) < distance(pq↑.right) then begin
     temp := pq↑.right;
     pq↑.right := pq↑.left;
     pq↑.left := temp
     end;
pq↑.dist := distance(pq↑.right) + 1
end;
```

The function *fixdist* recomputes the distance to the closest leaf by inspecting at the right descendant, if any.

All operations on the leftist trees require $O(\log n)$ time even in the worst case.

Table 5.3 summarizes simulation results on leftist trees. C_n^C indicates the number of comparisons required to build a leftist tree, *dist* indicates the distance from the root to the closest leaf and C_n^D the number of comparisons required to extract all the elements from the tree.

Table 5.3: Simulation results for leftist trees.

n	$E[C_n^C]$	$E[dist]$	$E[C_n^D]$
10	14.5955±0.0099	2.4314±0.0029	11.992±0.010
50	131.44±0.14	3.6807±0.0097	176.056±0.081
100	317.11±0.41	4.211±0.015	469.35±0.18
500	2233.6±4.7	5.497±0.041	3779.2±1.0
1000	5036±14	6.071±0.063	8817.3±2.2
5000	31845±155	7.45±0.16	58797±13
10000	69500±395	7.97±0.23	130312±22

5.1.6.2 Binary priority queues

We can construct a binary tree with a priority queue ordering instead of a lexicographical ordering. By doing this, most of the algorithms for binary trees can also be used for priority queues. There is a contradiction of goals however. While the best binary tree for searching is a tree as height balanced as possible, the best tree for a priority queue is one which is as thin or as tall as possible. With this in mind we can devise an algorithm to produce rather tall trees.

For simplicity of the algorithms we will impose the following conditions:

(1) the key in the node is larger than any other key in the descendant subtrees;

(2) if a subtree is non-null, then the left subtree is non-null;

(3) the key in the direct left descendant (if any) is larger than the key in the direct right descendant.

Binary priority queue insertion

```
procedure insert (new : tree; var pq : tree);

begin
if pq = nil then pq := new
else if pq↑.k <= new↑.k then begin
          new↑.left := pq;
          pq := new
          end
else if pq↑.left = nil then
          pq↑.left := new
else if pq↑.left↑.k <= new↑.k then
          insert(new, pq↑.left)
else      insert(new, pq↑.right)
end;
```

Binary priority queue deletion

```
procedure delete (var pq : tree);
var  temp : tree;
begin
if pq = nil then Error {*** deletion on an empty queue ***}
else if pq↑.right = nil then
     pq := pq↑.left
else begin
     {*** promote left descendant up ***}
     pq↑.k := pq↑.left↑.k;
     delete(pq↑.left);
     {*** rearrange according to constraints ***}
     if pq↑.left = nil then begin
          pq↑.left := pq↑.right; pq↑.right := nil  end;
     if pq↑.right <> nil then
          if pq↑.left↑.k < pq↑.right↑.k then begin
               {*** descendants in wrong order ***}
               temp := pq↑.right;
               pq↑.right := pq↑.left;
               pq↑.left := temp
               end
     end
end;
```

Table 5.4 summarizes the simulation results for binary priority queues. I_n indicates the number of iterations performed by the insertion procedure, C_n^C the number of comparisons to construct the queue and C_n^D the number of comparisons to extract all the elements from the queue.

Table 5.4: Simulation results for binary tree priority queues.

n	$E[I_n]$	$E[C_n^C]$	$E[C_n^D]$
10	18.3524±0.0079	23.384±0.016	7.1906±0.0071
50	148.56±0.13	232.65±0.26	139.80±0.12
100	353.51±0.40	578.59±0.80	396.03±0.40
500	2463.6±4.9	4287.1±9.8	3472.2±6.0
1000	5536±14	9793±28	8276±18
5000	34827±161	63258±322	56995±204
10000	75931±407	139071±814	127878±569

5.1.6.3 Binary search trees as priority queues

Binary search trees, in any of their variations, can be used as priority queues. The maximum is located at the rightmost node and the minimum is located at the leftmost node. The insertion algorithm is almost the same as for binary search trees, except that we are not concerned about duplicated keys. An extraction is done by deleting the rightmost node which is one of the easy cases of deletion.

The complexity measures for random insertions are the same as those for binary search trees (see Section 3.4.1).

Binary search tree insertion

```
procedure insert(new : tree; var t : tree);
begin
if t = nil then t := new
else if t↑.k < new↑.k then insert(new, t↑.right)
                 else      insert(new, t↑.left)
end;
```

Binary search tree, extraction of maximum

```
function extract(var pq : tree) : typekey;
begin
if pq=nil then Error {*** extraction from empty queue ***}
else if pq↑.right = nil then begin
      extract := pq↑.k;
      pq := pq↑.left
      end
else extract := extract(pq↑.right)
end;
```

Binary search trees used as queues behave as double-ended priority queues, since we can extract both the maximum and the minimum element. Binary search trees are not easy to merge as they require linear time in their total size.

This priority queue is stable; equal keys will be retrieved first-in first-out.

When used for insertions intermixed with extractions, this type of queue tends to degenerate into a skewed tree. For this reason it appears to be much safer to use any type of balanced binary tree.

References:
[Knuth, D.E., 73], [Aho, A.V. *et al.*, 74], [McCreight, E.M., 85], [Sleator, D.D. *et al.*, 85], [Atkinson, M.D. *et al.*, 86].

5.1.7 Binomial queues

Binomial queues use binary decomposition to represent sets of values with special structures of sizes 1, 2, 4, ... 2^k. A structure of size 2^k is called a B_k **tree** and has the following properties:

(1) the maximum element of the set is the root;

(2) the root has k descendants; one B_0, one B_1, ... , one B_{k-1} tree.

B_k trees are the natural structure that arises from a tournament between 2^k players.

Two B_k trees can be joined into a single B_{k+1} tree with one single comparison. Consequently a B_k tree can be constructed using $2^k - 1$ comparisons. This construction is optimal.

A binomial queue of size n is represented as a **forest** of B_k trees where there is at most one B_k tree for each k. This corresponds to the binary decomposition of n. For example, $n = 13 = 1101_2$ is represented by B_3, B_2, B_0

The maximum element of a binomial queue can be found by inspecting the head of all of its B_k trees which requires $\nu(n) - 1 \le \lfloor \log_2 n \rfloor$ comparisons (where $\nu(n)$ is the number of '1' digits in the binary representation of n).

Two binomial queues can be merged into a single queue by joining all equal-size B_k trees in a process which is identical to binary addition. Merging two queues with sizes m and n requires

$$C_{mn}^M = \nu(n) + \nu(m) - \nu(m+n)$$

An insertion of a single element into a queue with n elements is treated as a merge and hence requires

$$C_n^I = \nu(n) + 1 - \nu(n+1)$$

Constructing a binomial queue by repetitive insertions requires

$$C_n^C = n - \nu(n)$$

A deletion of an extraction is accomplished by removing the largest root of the B_k trees and merging all its descendants with the original queue. This operation requires

$$\nu(n) - 1 \le C_n^D \le 2\nu(n) + \lfloor \log_2 n \rfloor - \nu(n-1) - 1$$

Binomial queues can be implemented using binary trees. These implementations are simplified if we include the size of each B_k tree in the root node.

Binomial queues give excellent worst-case behaviour for insertions, constructions by insertions, deletions and merging of queues at the cost of two pointers per entry.

References:
[Brown, M.R., 77], [Brown, M.R., 78], [Vuillemin, J., 78], [Carlsson, S. *et al.*, 88], [Cormen, T.H. *et al.*, 90].

5.1.8 Summary

Table 5.5 shows an example of real relative total times for constructing a priority queue with 10007 elements by repetitive insertions and then extracting all its elements.

General references:
[Johnson, D.B., 75], [Pohl, I., 75], [Brown, M.R. *et al.*, 79], [Flajolet, P. *et al.*, 79], [Flajolet, P. *et al.*, 80], [Standish, T.A., 80], [Itai, A. *et al.*, 81], [Ajtai, M. *et al.*, 84], [Fischer, M.J. *et al.*, 84], [Mehlhorn, K., 84], [Mairson, H.G., 85], [Huang, S-H.S., 86], [Jones, D.W., 86], [Lentfert, P. *et al.*, 89], [Sundar, R., 89].

Table 5.5: Relative total times for priority queue algorithms.

Algorithm	*C*	*Pascal*
Sorted lists	55.1	52.9
Unsorted lists	240.2	146.7
P-trees	3.4	3.4
Heaps	1.0	1.0
Pagodas	1.5	1.6
Leftist trees	4.3	4.2
Binary priority queues	2.1	2.3
B.S.T as priority queues		1.7

5.2 Selection of kth element

The selection of the kth element is defined as searching for an element X in an unordered set such that $k-1$ elements from the set are less than or equal to X and the rest are greater than or equal to X.

Finding the first or last (minimum or maximum) is the most important special case and was treated in the first section of this chapter. Finding the median (or closest to the median) is another special case of selection.

Let $C_{k,n}$ denote the number of comparisons needed to find the kth element in a set containing n unordered elements. Let $C_{k,n}^{MM}$ denote the minimum maximum or minimum worst-case number of comparisons for the same problem. For the Floyd and Rivest algorithm we have:

$$E[C_{k,n}] \;=\; n + \min(k, n-k) + O(\sqrt{n})$$

For small k,

$$E[C_{k,n}] \;\leq\; n + O(k \;\ln\ln\; n)$$

For any selection algorithm we have the following average-case lower bound:

$$E[C_{k,n}] \;\geq\; n = \min(k, n-k) - O(1)$$

Table 5.6 summarizes the worst-case upper and lower bounds on the problem.

In the following algorithms, we assume that all the records are stored in an array. This array can be shuffled if necessary.

General references:
[Hoare, C.A.R., 61], [Blum, N. *et al.*, 73], [Knuth, D.E., 73], [Nozaki, A., 73], [Pratt, V. *et al.*, 73], [Aho, A.V. *et al.*, 74], [Noshita, K., 74], [Floyd, R.W. *et al.*, 75], [Fussenegger, F. *et al.*, 76], [Hyafil, L., 76], [Schonhage, A. *et al.*, 76],

Table 5.6: Upper and lower bounds for kth selection.

$C_{k,n}^{MM}$	
	Lower bounds
$k = 1$	$n - 1$
$k = 2$	$n - 2 + \lceil \log_2 n \rceil$
for any j	
$k = 3,\ n = 2^j + 1$	$n - 3 + 2\lceil \log_2(n-1) \rceil$
$k = 3,\ 3 \times 2^j < n \leq 4 \times 2^j$	$n - 4 + 2\lceil \log_2(n-1) \rceil$
$k = 3,\ 2 \times 2^j + 1 < n \leq 3 \times 2^j$	$n - 5 + 2\lceil \log_2(n-1) \rceil$
$2k - 1 \leq n < 3k$	$\lfloor \frac{3n+k-5}{2} \rfloor$
$3k \leq n$	$n + k - 3 + \sum_{j=0}^{k-2} \log_2 \lceil \frac{n-k+2}{k+j} \rceil$
$2k = n$	$2n + o(n)$
	Upper bounds
$k = 1$	$n - 1$
$k = 2$	$n - 2 + \lceil \log_2 n \rceil$
$k \geq 1$	$n - k + (k-1)\lceil \log_2 (n - k + 2) \rceil$
$2^s(2^{\lceil \log_2 k \rceil} + j) < n - k + 2$ and $n - k + 2 \leq 2^s(2^{\lceil \log_2 k \rceil} + j + 1)$ and $o\lfloor k/2 \rfloor > j\lceil \log_2 k \rceil$	$n - k + (k-1)\lceil \log_2(n-k+2) \rceil - \lfloor (k-1)/2 \rfloor + j\lceil \log_2 k \rceil$
$2k = n + 1$	$3n + O((n \log n)^{3/4})$
$5k \leq n$	$\approx n(1 + 2^{1-\lceil \log_2(n/5k) \rceil}) + 5k\lceil \log_2(n/5k) \rceil$

[Wirth, N., 76], [Yap, C.K., 76], [Reingold, E.M. *et al.*, 77], [Johnson, D.B. *et al.*, 78], [Reiser, A., 78], [Eberlein, P.J., 79], [Fussenegger, F. *et al.*, 79], [Galil, Z. *et al.*, 79], [Kronsjo, L., 79], [Allison, D.C.S. *et al.*, 80], [Frederickson, G.N. *et al.*, 80], [Munro, J.I. *et al.*, 80], [Dobkin, D. *et al.*, 81], [Kirkpatrick, D.G., 81], [Motoki, T., 82], [Yao, A.C-C. *et al.*, 82], [Cunto, W., 83], [Postmus, J.T. *et al.*, 83], [Devroye, L., 84], [Mehlhorn, K., 84], [Ramanan, P.V. *et al.*, 84], [Bent, S.W. *et al.*, 85], [Wirth, N., 86], [Baase, S., 88], [Brassard, G. *et al.*, 88], [Lai, T.W. *et al.*, 88], [Sedgewick, R., 88], [Cunto, W. *et al.*, 89], [Manber, U., 89], [Yao, A.C-C., 89], [Cormen, T.H. *et al.*, 90], [Frederickson, G.N., 90].

5.2.1 Selection by sorting

One of the simplest strategies for selection is to sort all the array and then directly select the desired element.

Selection by sorting

```
function select(i : integer; var r : RecordArray;
                lo, up : integer) : typekey;
begin
i := i+lo-1;
if (i<lo) or (i>up) then Error {*** selection out of bounds ***}
else begin
     sort(r, lo, up);
     select := r[i].k
     end
end;
```

This method is expensive for selecting a single element but should be preferred whenever several successive selections are performed.

5.2.2 Selection by tail recursion

This function uses a tail recursion technique. Each iteration starts by selecting a splitter element from the file. The file is then split into two subfiles: those elements with keys smaller than the selected one, and those elements with larger keys. In this way, the splitting element is placed in its proper final location between the two resulting subfiles. This procedure is repeated recursively on the subfile which contains the element to be selected.

For a randomly ordered file, the first selection of the kth element will require

$$E[C_{k,n}] = 2((n+1)H_n - (n+3-k)H_{n+1-k} - (n+2)H_k + n + 3)$$

$$E[C_{n/2,n}] \approx 3.38629...n$$

Selection by tail recursion

```
function select(s : integer; var r : RecordArray;
                lo, up : integer) : typekey;
```

```
var  i, j : integer;
     tempr : ArrayEntry;
begin
s := s+lo-1;
if (s<lo) or (s>up) then Error {*** selection out of bounds ***}
else begin
     while (up>=s) and (s>=lo) do begin
          i := lo;
          j := up;
          tempr := r[s];   r[s] := r[lo];   r[lo] := tempr;
          {*** split file in two ***}
          while i<j do begin
               while r[j].k > tempr.k do
                    j := j-1;
               r[i] := r[j];
               while (i<j) and (r[i].k<=tempr.k) do
                    i := i+1;
               r[j] := r[i]
               end;
          r[i] := tempr;
          {*** select subfile ***}
          if s<i then up := i-1
               else lo := i+1
          end;
     select := r[s].k
     end
end;
```

The above algorithm uses as a splitting element the one located at the selected position. For a random file, any location would provide an equivalently good splitter. However, if the procedure is applied more than once, any other element (for example, the first) may produce an almost worst-case behaviour.

As selections are done, the array is sorted into order. It is expected that later selections will cost less, although these will always use $O(n)$ comparisons.

Strategies which select, in place, a smaller sample to improve the splittings, cause an almost worst-case situation and should be avoided. Sampling, if done, should not alter the order of elements in the array.

Any of the distributive methods of sorting, for example, such as bucket sort (see Section 4.2.3) or top-down radix sort (see Section 4.2.4), can be modified to do selection. In all cases the strategy is the same: the sorting algorithms split the file into several subfiles and are applied recursively on to each subfile (divide and conquer). For selection, we do the same first step, but then we select only the subfile that will contain the desired element (by counting the sizes of the subfiles) and apply recursion only on one subfile (tail recursion).

5.2.3 Selection of the mode

The **mode of a set** is defined as the key value which occurs most frequently in a set. The number of times a key value is repeated is called the **multiplicity**. The mode is the key with largest multiplicity.

The selection of the mode is almost trivial when the set is ordered, as a single pass through the set is enough to determine it. The complexity of determining the mode is lower than the complexity of sorting; in fact if n_m is the multiplicity of the mode, then we have the upper and lower bound

$$C_n^{Mode} \approx n \log_2(n/n_m) - n \log_2(\log_2 n - P)$$

where $P = \sum_i n_i/n \log_2 n_i$ and n_i is the multiplicity of the ith different key value.

The following algorithm uses a divide-and-conquer technique to find the mode.

Determining the mode

```
function mode(S : SetOfKeys) : SetOfKeys;
var  A, A1, A2, A3 : SetOfKeys;
     Homog, Heter : set of SetOfKeys;
begin
Homog := [ ];
Heter := [S];
while LargestCardinality(Heter) > LargestCardinality(Homog) do
     begin
     A := LargestSet(Heter);
     med := median(A);
     split A into A1, A2, A3
               { with elements <med; =med; >med }
     Heter := (Heter - A) + A1 + A3;
     Homog := Homog + A2
     end;
LargestSet(Homog);
end;
```

This algorithm requires

$$C_n^{Mode} \approx kn \log_2(n/n_m)$$

comparisons, where kn is the number of comparisons required to find the median among n elements. For example, using the Schonhage, Paterson and Pippenger median algorithm (see Section 5.2), $k = 3$ and

$$C_n^{Mode} \approx 3n \log_2(n/n_m)$$

in the worst case.

Dobkin and Munro describe an optimal (within lower order terms) algorithm to find the mode. Their optimal algorithm is rather complicated, and mostly of theoretical interest.

References:

[Dobkin, D. *et al.*, 80].

Arithmetic Algorithms

6.1 Basic operations, multiplication/division

In this section we will discuss arithmetic algorithms to perform the basic operations. Given that addition and subtraction are relatively straightforward, we will concentrate on multiplication/division and other operations.

Our model of computation can be called **multiple-precision**, as we are interested in describing arithmetic operations in terms of operations in a much smaller domain. For example, some algorithms may implement decimal operations using ASCII characters as basic symbols, or we may implement extended precision using basic integer arithmetic, or integer arithmetic using bits, and so on. Without loss of generality we will call the basic unit of implementation a **digit**, and a logical collection of digits a **number**. Our complexity measures will be given in number of operations on digits as a function of the number of digits involved.

Let $M(n)$ denote the complexity of multiplying two n-digit numbers and let $Q_{f(x)}(n)$ denote the complexity of computing the function $f(x)$ with n-digit precision. ($Q_{\times}(n) = M(n)$). Then we have the following results:

$$Q_{+}(n) = Q_{-}(n) = Q_{kx}(n) = O(n)$$

for an integer constant k. The classical method of multiplication gives

$$M(n) = O(n^2)$$

By splitting the factors in two $(n/2)$-digit numbers and using

$$a = a_1 B^{n/2} + a_2, \qquad b = b_1 B^{n/2} + b_2$$

$$p_1 = a_1 b_1, \quad p_2 = a_2 b_2, \quad p_3 = (a_1 + a_2)(b_1 + b_2)$$

$$ab = p_1 B^n + (p_3 - p_2 - p_1) B^{n/2} + p_2$$

where B is the base of the numbering system, we obtain

$$M(n) = 3M(n/2) + O(n) = O(n^{1.58496...})$$

Similarly, by splitting the numbers in k (n/k)-digit components,

$$M(kn) = (2k-1)M(n) + O(n) = O(n^{\log_k (2k-1)})$$

By the application of a technique resembling the fast Fourier transform and modular arithmetic,

$$M(n) = O(n \log(n) \log(\log n))$$

Note that the complexity of multiplication is bounded above by the complexity of squaring and by the complexity of computing inverses. That is to say

$$Q_\times(n) \le 2Q_{x^2}(n) + O(n)$$

since

$$ab = \frac{(a+b)^2 - (a-b)^2}{4}$$

and

$$Q_{x^2}(n) \le 3Q_{1/x}(n) + O(n)$$

since

$$x^2 = \frac{1}{\frac{1}{x-1} - \frac{1}{x}} + x$$

For the next complexity results we will assume that we use an asymptotically fast multiplication algorithm, that is, one for which

$$M(n) = O(n(\log n)^k)$$

In such circumstances,

$$\sum_{k \ge 0} M(n\alpha^k) = \frac{M(n)}{1-\alpha}(1 + O(1/(\log n)))$$

Inverses ($x = 1/a$) can be computed using variable-precision steps with the second-order iterative formula:

$$x_{i+1} = x_i(2 - a x_i)$$

Each step requires two multiplications and one addition. Since this Newton-type iteration converges quadratically, the last iteration is done with n digits, the previous to the last with $\lceil n/2 \rceil$, the previous with $\lceil n/4 \rceil$, and so on.

$$\begin{aligned} Q_{1/x}(n) &= \sum_{i \ge 0} 2M(\lceil n/2^i \rceil) + O(n/2^i) \\ &\approx 3M(n) \end{aligned}$$

If we use a third-order method:

$$x_{i+1} = x_i(\epsilon_i^2 - \epsilon_i + 1) \qquad \epsilon_i = ax_i - 1$$

then $Q_{1/x}(n) \approx 3M(n)$ also. Consequently divisions can be computed in

$$Q_/(n) \approx 4M(n)$$

To evaluate $x = a^{-1/2}$ we can use the third-order iteration:

$$\epsilon_i = ax_i^2 - 1$$

$$x_{i+1} = x_i - x_i\epsilon_i \frac{4 - 3\epsilon_i}{8}$$

for which

$$Q_{1/\sqrt{x}}(n) \approx \frac{9M(n)}{2}$$

Consequently

$$Q_{\sqrt{x}}(n) \approx \frac{11M(n)}{2}$$

Derivatives can be computed from the formula

$$f'(x) = \frac{f(x+h) - f(x-h)}{2h} + O(h^3)$$

by making $h = O(f'(x)B^{-n/3})$. For this method

$$Q_{f'(x)}(n) = 2Q_{f(x)}(3n/2) + O(n)$$

The inverse of a function can be computed by using any iterative zero-finder with variable precision. By using the secant method:

$$x_{i+1} = x_i - f(x_i)\frac{x_i - x_{i-1}}{f(x_i) - f(x_{i-1})}$$

then

$$Q_{f^{-1}(x)}(n) \approx 15M(n) + Q_{f(x)}(n) + \sum_{i \geq 2} Q_{f(x)}(2n\rho^{-i})$$

where $\rho = (1 + \sqrt{5})/2$ is the golden ratio.

For the purpose of describing the algorithms we will use a common representation, based on arrays of digits. The digits may take values from 0 to $BASE - 1$ in their normalized form, although a digit may hold a maximum value $MAXD$. For example, for eight-bit characters on which we want to represent decimal numbers, $BASE = 10$ and $MAXD = 255$. The bound $MAXD$ may be any value including $BASE - 1$. For our algorithms we will assume that $MAXD \geq 2BASE^2$. With this assumption we do not have to use temporary variables for the handling of digits.

The data definition for our C algorithms is

```
typedef digit mp[ ];
```

mp[0] will be called the **header** and will be used to store control information about the number. Typical control information is sign, length, exponent (for floating-point implementations), and so on. We are not concerned about the organization of bits in the header, as long as we can store and retrieve its values. The lowest order digit is stored in *mp*[1]; the highest order digit is stored in *mp*[*length*(*mp*) – 1]. This organization, although not very common, is quite convenient.

The following procedure normalizes a multiple-precision number, adjusts its length, propagates carries and adjusts sign if needed.

Normalization of a multiple-precision number

```
normalize(a)
mp a;

{int cy, i, la;
la = length(a);
start:
cy = 0;
for (i=1; i<la; i++) {
        cy = (a[i] += cy) / BASE;
        a[i] -= cy*BASE;
        if (a[i]<0) {a[i] += BASE; cy--;}
        }
while (cy>0) {
        a[i++] = cy%BASE;
        cy /= BASE;}
if (cy<0)  {
        a[la-1] += cy*BASE;
        for (i=1; i<la; i++)  a[i] = -a[i];
        storesign(a, sign(a)==POS ? NEG : POS);
        goto start;
        }
while (a[i-1]==0 && i>2) i--;
storelength(a, i);
if (i==2 && a[1]==0) storesign(a, POS);
};
```

The following procedure computes a linear combination of two multiple-precision numbers. The integer coefficients should be in the range $-BASE$ to $BASE$. The result is computed, destructively, on the first argument.

Linear combination of two numbers

```
linear(a, ka, b, kb)
mp a, b;
int ka, kb;

/*** compute a*ka + b*kb --> a ***/
{int i, la, lb;
la = length(a);   lb = length(b);
for (i=1; i<la; i++) a[i] *= ka;
if (sign(a)!=sign(b)) kb = -kb;
if (lb>la) {
      storelength(a, lb);
      for (i=la; i<lb; i++) a[i] = 0;
      }
for (i=1; i<lb; i++) a[i] += kb*b[i];
normalize(a);
};
```

Multiple-precision multiplication

```
mulint(a, b, c)
mp a, b, c;
/*** multiply two integers.  a*b-->c ***/

{int i, j, la, lb;
/*** b and c may coincide ***/
la = length(a); lb = length(b);
for (i=0; i<la-2; i++) c[lb+i] = 0;
for (i=lb-1; i>0; i--) {
      for (j=2; j<la; j++)
            if ((c[i+j-1] += b[i]*a[j]) >
                  MAXD-(BASE-1)*(BASE-1)-MAXD/BASE) {
                  c[i+j-1] -= (MAXD/BASE)*BASE;
                  c[i+j] += MAXD/BASE;
                  }
      c[i] = b[i]*a[1];
      }
storelength(c, la+lb-2);
storesign(c, sign(a)==sign(b) ? POS : NEG);
normalize(c);
};
```

References:
[Knuth, D.E., 69], [Aho, A.V. *et al.*, 74], [Borodin, A. *et al.*, 75], [Floyd, R.W., 75], [Artzy, E. *et al.*, 76], [Brent, R.P., 76], [Brent, R.P., 76], [Collins, G.E. *et al.*, 77], [Dhawan, A.K. *et al.*, 77], [Knuth, D.E., 78], [Morris, R., 78], [Ja'Ja', J., 79], [Alt, H., 80], [Bruss, A.R. *et al.*, 80], [Head, A.K., 80], [Linnainmaa, S., 81], [Alt, H., 83], [Stockmeyer, L.J., 83], [Regener, E., 84], [Flajolet, P. *et al.*, 85], [Flajolet, P., 85], [Kaminski, M., 87], [Alt, H., 88], [Robertazzi, T.G. *et al.*, 88].

6.2 Other arithmetic functions

6.2.1 Binary powering

Binary powering is a tail recursion technique for powering a number. To compute a given power, we first compute the power to half the exponent and then square the result. If the exponent is odd, we additionally multiply the result by the base. Let $Q_{bp}(n)$ denote the number of multiplications required to compute the nth power of a number using binary powering. Then

$$Q_{bp}(n) = \lfloor \log_2 n \rfloor + \nu(n) - 1$$

where $\nu(n)$ is the number of 'one' digits in the binary representation of n.

Let $Q_{opt}(n)$ be the number of multiplications required by the optimal method of powering, that is, the method which minimizes the number of multiplications. Then

$$1 \le \frac{Q_{opt}(n)}{\lfloor \log_2 n \rfloor} \le 1 + \frac{1}{\log_2 \log_2 n} + O\left(\frac{\log_2 \log_2 \log_2 n}{(\log_2 \log_2 n)^2}\right)$$

and

$$Q_{opt}(n) \le Q_{bp}(n)$$

The first inequality is tight, but the latter is not. $n = 15$ is the smallest example for which they differ: we can compute x^{15} by computing x^2, x^3, x^6, x^{12} and x^{15} giving $Q_{opt}(15) = 5$ while $Q_{bp}(15) = 6$. Similarly, the smallest exponent for which the difference is 2 is 63, $Q_{opt}(63) = 8$ while $Q_{bp}(63) = 10$. (One of the optimal sequences of powers is 2,4,5,9,18,27,45,63.)

The problem of computing the optimal strategy for powering is related to the **addition chain** problem, which is how to construct an increasing sequence $a_1, a_2, \ldots, a_k$ for which every element is the sum of two previous elements and $a_1 = 1$ and $a_k = n$ for a minimal k.

Using the fact that $(a^x)^y = a^{xy}$, if the power is a composite number, then

$$Q_{opt}(pq) \le Q_{opt}(p) + Q_{opt}(q)$$

This inequality is not tight. For example, $Q_{opt}(33) = 6$ but $Q_{opt}(3) = 2$ and $Q_{opt}(11) = 5$.

It is always possible to do a squaring as the last step, which gives

$$Q_{opt}(2n) \leq Q_{opt}(n) + 1$$

but this bound is not tight either since $Q_{opt}(191) = 11$ and $Q_{opt}(382) = 11$.

For binary powering we can define an average value of the complexity, as if the bits of the power were randomly selected. For this definition

$$\overline{Q_{bp}}(n) = 2^{-k} \sum_{i=2^k}^{2^{k+1}-1} Q_{bp}(i) = \frac{3k}{2}$$

where $k = \lfloor \log_2 n \rfloor$.

When powering integers, as the powers grow in size, it is important to know the complexity of the multiplication method used. Let n denote the exponent and N the number of digits in the base number to be powered. If we use the classical algorithm, $M(N) = O(N^2)$ then

$$Q_{bp}(n) = \left(\frac{n^2}{3} + O(n)\right) M(N)$$

The iterative version of the powering algorithm runs in the order

$$Q_{iter}(n) = \frac{n(n-1)}{2} M(N)$$

If we use an asymptotically fast multiplication algorithm, $(M(N) = O(N(\log N)^k))$, then binary powering is definitely better than iterative powering:

$$Q_{bp}(n) \approx 2M(Nn)$$

as opposed to

$$Q_{iter}(n) \approx \frac{n^2}{2} M(N)$$

In the above cases it is assumed that the size of the result of powering an N-digit number to the nth power is an Nn-digit number. This may be too pessimistic sometimes.

Binary powering

```
function power(b : number; e : integer) : number;
begin
if e<0 then power := 1/power(b,-e)
```

```
else if e=0 then power := 1
else if e=1 then power := b
else if (e mod 2) = 0 then
            power := sqr(power(b, e div 2))
else        power := sqr(power(b, e div 2)) * b
end;
```

6.2.2 Arithmetic-geometric mean

The arithmetic-geometric mean (AG mean) constructs two sequences of numbers a_i and b_i from starting values a_0 and b_0 and the iteration formulas:

$$a_{i+1} = \frac{a_i + b_i}{2}$$

$$b_{i+1} = \sqrt{a_i b_i}$$

For $0 < a_0 \leq 1$ and $0 < b_0 \leq 1$ the sequences converge quadratically to their common limit denoted by $AG(a_0, b_0)$. Computing one step of the iteration requires one multiplication and one square root plus other $O(n)$ operations. Consequently the complexity of the AG computation is

$$Q_{AG}(n) \approx \frac{13}{2} M(n) \log_2 n$$

where n is the number of digits in the answer. The AG mean is related to the complete elliptic integrals as

$$\frac{\pi}{2AG(1, \cos\phi)} = \int_0^{\pi/2} \frac{d\theta}{\sqrt{1 - \sin^2\phi \sin^2\theta}}$$

The **Brent–Salamin** method for computing π which uses the AG mean and a Legendre's identity requires

$$Q_\pi(n) \approx \frac{15}{2} M(n) \log_2 n$$

Fast computation of π

```
function pi : number;
var a, b, t, x, tempa : number;

begin
a := 1;
```

```
b := sqrt(0.5);
t := 0.25;
x := 1;
while a-b>epsilon do begin
      tempa := a;
      a := (a+b) / 2;
      b := sqrt(tempa*b);
      t := t - x*sqr(a-tempa);
      x := 2*x
      end;
pi := sqr(a+b) / (4*t)
end;
```

Other classical methods for evaluating π are based on identities of the type

$$\pi = 16 \arctan(1/5) - 4 \arctan(1/239)$$

The function $\arctan(1/i)$ for integer i can be evaluated in time proportional to $O(n^2/\log i)$ using the Maclaurin expansion of $\arctan(x) = x - x^3/3 + x^5/5 - \ldots$.

6.2.3 Transcendental functions

Assuming that π and $\ln(BASE)$ are precomputed to the desired accuracy, we can compute $\ln(x)$ using the formula

$$\ln(x) = \frac{\pi}{2AG(1, 4/x)}(1 + O(x^{-2}))$$

If x is not large enough, we can simply scale it by multiplying by a suitable power of the $BASE$ (just a shift). For this method

$$Q_{\ln(x)}(n) \approx 13M(n) \log_2 n$$

Computation of natural logarithms by AG means

```
function ln (x : number) : number;
var   a, b, temp : number;
      shift, logbase : integer;

begin
logbase := crude_estimate_of_ln(x)/ln(BASE);
if 2*logbase<Digits then begin
      shift := Digits div 2 - logbase + 1;
```

```
            ln := ln(x * BASE**shift) - shift*LNBASE
            end
      else begin
            a := 1.0;   b := 4/x;
            while a-b>sqrteps do begin
                  temp := a;
                  a := (a+b) / 2;
                  b := sqrt(temp*b)
                  end;
            ln := Pi / (a+b)
            end
      end;
```

The above algorithm uses two pre-computed constants: *Pi* and *LNBASE*, with their obvious meanings. *LNBASE* can be computed with the above function by computing the logarithm of $BASE^{Digits}$ and then dividing the result by *Digits*. The global variable **Digits** indicates the precision of the computation, or how many significant digits in base *BASE* are kept. *epsilon* is a bound on the desired error and *sqrteps* is the square root of epsilon.

By computing inverses with a method of high order of convergence (in this case all the derivatives are easy to compute) we obtain

$$Q_{e^x}(n) \approx 13M(n) \log_2 n$$

By doing all the arithmetic operations with complex numbers, or by computing the arctan(x) as below, we can compute all the trigonometric functions and their inverses. For example,

$$\sin x = \frac{e^{ix} - e^{-ix}}{2i}$$

then

$$Q_{\sin x}(n) \approx Q_{\cos x}(n) \approx \cdots \approx Q_{\arcsin x}(n) \approx \cdots \approx 34M(n) \log_2 n$$

Computation of arctan(x) by AG means

```
      function arctan (x : number) : number;
      var   q, s, v, w : number;

      begin
      s := sqrteps;
      v := x / (1 + sqrt(1+x*x));
      q := 1;
      while 1-s > epsilon do begin
```

```
            q := 2*q / (1+s);
            w := 2*s*v / (1+v*v);
            w := w / (1 + sqrt(1-w*w));
            w := (v+w) / (1-v*w);
            v := w / (1 + sqrt(1+w*w));
            s := 2*sqrt(s) / (1+s)
            end;
      arctan := q * ln((1+v)/(1-v))
      end;
```

References:
[Knuth, D.E., 69], [Horowitz, E., 73], [Kedem, Z.M., 74], [Borodin, A. *et al.*, 75], [Winograd, S., 75], [Brent, R.P., 76], [Brent, R.P., 76], [Yao, A.C-C., 76], [Pippenger, N., 79], [Pippenger, N., 80], [Downey, P. *et al.*, 81], [Borwein, J.M. *et al.*, 84], [Brassard, G. *et al.*, 88], [Tang, P.T.P., 89].

6.3 Matrix multiplication

For any matrices

```
      a : array [1..m, 1..p] of basetype;
      b : array [1..p, 1..n] of basetype;
      c : array [1..m, 1..n] of basetype;
```

we define the matrix product $c = a \times b$ by

$$c_{ij} = \sum_{k=1}^{p} a_{ik} b_{kj}$$

The classical algorithm for matrix multiplication requires mnp multiplications and $mn(p-1)$ additions. Let $M_{\times}(n)$ be the number of multiplications used to multiply two $n \times n$ matrices. Then $M_{\times}(n) = n^3$ for the classical algorithm.

Classical algorithm

```
    for i:=1 to m do
          for j:=1 to n do begin
                c[i,j] := 0;
                for k:=1 to p do
                      c[i,j] := c[i,j] + a[i,k]*b[k,j]
                end;
```

Winograd's method of general matrix multiplication reduces the number of multiplications to about half with the formula:

$$c_{ij} = \sum_{k=1}^{p/2} (a_{i,2k} + b_{2k-1,j})(a_{i,2k-1} + b_{2k,j}) - d_i - e_j + a_{ip}b_{pn}\dagger$$

where

$$d_i = \sum_{k=1}^{p/2} a_{i,2k}a_{i,2k-1}$$

$$e_j = \sum_{k=1}^{p/2} b_{2k-1,j}b_{2k,j}$$

and the last term (†) is present only if p is odd. Winograd's matrix multiplication uses

$$M_\times(m,p,n) = mn\lceil p/2 \rceil + (m+n)\lfloor p/2 \rfloor$$

multiplications and

$$A_\times(m,p,n) = mn(p+2) + (mn+m+n)(\lfloor p/2 \rfloor - 1)$$

additions/subtractions.

6.3.1 Strassen's matrix multiplication

When $m = n = p = 2$, the product can be computed using 7 multiplications instead of 8 but using 15 additions instead of 4.

$$\begin{array}{llllll}
s_1 = & a_{21} + a_{22} & p_1 = & s_2 s_6 & s_8 = & s_6 - b_{21} \\
s_2 = & s_1 - a_{11} & p_2 = & a_{11} b_{11} & s_9 = & p_1 + p_2 \\
s_3 = & a_{11} - a_{21} & p_3 = & a_{12} b_{21} & s_{10} = & s_9 + p_4 \\
s_4 = & a_{12} - s_2 & p_4 = & s_3 s_7 & c_{11} = & p_2 + p_3 \\
s_5 = & b_{12} - b_{11} & p_5 = & s_1 s_5 & c_{12} = & s_9 + p_5 + p_6 \\
s_6 = & b_{22} - s_5 & p_6 = & s_4 b_{22} & c_{21} = & s_{10} - p_7 \\
s_7 = & b_{22} - b_{12} & p_7 = & a_{22} s_8 & c_{22} = & s_{10} + p_5
\end{array}$$

This can be applied not only to 2×2 matrices, but to any $n \times n$ matrix, partitioned into 4 $(n/2)\times(n/2)$ matrices (with proper 0 padding if necessary). The number of multiplications required by a recursive application of Strassen's algorithm to multiply two $2^k \times 2^k$ matrices is

$$M_\times(2^k) = 7^k$$

and in general

$$M_\times(n) = 7M_\times(\lceil n/2 \rceil) = O(n^{2.80735\ldots})$$

Let $A_\times(n)$ be the number of additions used to multiply two $n \times n$ matrices, then

$$A_\times(2^k) = 5(7^k - 4^k)$$

$$A_\times(n) = 7A_\times(\lceil n/2 \rceil) + 15\lceil n/2 \rceil^2 = O(n^{2.80735\ldots})$$

For the implementation of this algorithm we are interested in the total number of additions/ multiplications. Noticing that when n is odd, one of the recursive matrix multiplications can be done on $\lfloor n/2 \rfloor \times \lfloor n/2 \rfloor$ matrices and by shifting to the classical algorithm whenever it is more efficient, we obtain that the total number of operations is

$$\begin{aligned} M_\times(n) &= \min(M_\times(\lfloor n/2 \rfloor) + 6M_\times(\lceil n/2 \rceil) + 15\lceil n/2 \rceil^2,\ n^2(2n-1)) \\ &\approx 3.73177\ldots n^{2.80735\ldots} \end{aligned}$$

Even for this optimized implementation, n has to be larger than 1580 to save 50% or more of the operations with respect to the classical method.

6.3.2 Further asymptotic improvements

The following methods present asymptotic improvements to the number of operations necessary to do matrix multiplication. These improvements are only of theoretical interest, as their complexity for normal size problems is much worse than the classical algorithm. Furthermore, their numerical properties are unknown.

Pan devised a general multiplication scheme using trilinear forms which requires $n^3/3 + 6n^2 - 4n/3$ multiplications to multiply two $n \times n$ matrices. His method does not rely on product commutativity and can be composed in the same way as Strassen's. By selecting as a base 70×70 matrices we obtain

$$M_\times(n) = O(n^\omega)$$

where $\omega = \ln 143640 / \ln 70 = 2.79512\ldots$.

Bini *et al.* use an approximate (arbitrary precision approximating) method to multiply 12×12 matrices with 1000 multiplications and hence, extending their method gives

$$\omega = \frac{\ln 1000}{\ln 12} = 2.77988\ldots$$

Schonhage generalized the above method to obtain

$$\omega = 2.54799\ldots$$

where $16^{\omega/3} + 9^{\omega/3} = 17$. Pan further improved this with a construction that achieves

$$\omega = \frac{3 \ln 52}{\ln 110} = 2.52181...$$

Coppersmith and Winograd describe the construction of a faster matrix multiplication algorithm based on arithmetical progressions. Using the above idea they construct a method for which $\omega < 2.376...$.

A non-trivial lower bound in the number of additions and multiplications is

$$2n^2 + \frac{3}{46}n + O(1)$$

Matrix inversion, computation of determinants, solution of simultaneous equations and lower-upper triangular factoring can be done in terms of matrix multiplications and hence can be done in time proportional to matrix multiplication.

References:
[Winograd, S., 68], [Knuth, D.E., 69], [Strassen, V., 69], [Dobkin, D., 73], [Aho, A.V. *et al.*, 74], [Pratt, V., 74], [Savage, J.E., 74], [Borodin, A. *et al.*, 75], [Brockett, R.W. *et al.*, 76], [Cohen, J. *et al.*, 76], [Dobkin, D. *et al.*, 76], [Probert, R.L., 76], [Probert, R.L., 76], [Probert, R.L., 76], [Adleman, L. *et al.*, 78], [Pan, V.Y., 78], [Probert, R.L., 78], [Schachtel, G., 78], [Yuval, G., 78], [Bini, D. *et al.*, 79], [Ja'Ja', J., 79], [Kronsjo, L., 79], [Pan, V.Y., 79], [Ja'Ja', J., 80], [Lotti, G. *et al.*, 80], [Pan, V.Y., 80], [Santoro, N., 80], [Feig, E., 81], [Makarov, O.M., 81], [Pan, V.Y., 81], [Pan, V.Y., 81], [Schonhage, A., 81], [Coppersmith, D. *et al.*, 82], [Coppersmith, D., 82], [Hu, T.C. *et al.*, 82], [Hu, T.C., 82], [Romani, F., 82], [Schoor, A., 82], [Cohen, J., 83], [Feig, E., 83], [Ja'Ja', J., 83], [Pan, V.Y., 83], [Pan, V.Y., 84], [Santoro, N., 84], [Alekseyed, V.B., 85], [Ja'Ja', J. *et al.*, 85], [Strassen, V., 86], [Wilf, H., 86], [Alagar, V.S. *et al.*, 87], [Coppersmith, D. *et al.*, 87], [Atkinson, M.D. *et al.*, 88], [Baase, S., 88], [Manber, U., 89], [Cormen, T.H. *et al.*, 90].

6.4 Polynomial evaluation

The simplest method of evaluating a polynomial is with Horner's rule. Let

$$P(x) = a_0 + a_1 x + a_2 x^2 + ... + a_n x^n$$

then, $P(x)$ can be expressed as

$$P(x) = a_0 + x(a_1 + x(a_2 + x(\cdots x a_n)\cdots))$$

Let

> a : **array** $[0..n]$ **of** *basetype*;

be the array containing the coefficients of an nth degree polynomial. Then

Horner's polynomial evaluation

```
result := a[n];
for i:=n-1 downto 0 do
        result := result * x + a[i];
```

For evaluating a polynomial at a single point, Horner's rule is optimal with respect to the number of additions and multiplications. In general this method has very good numerical properties too.

If the same polynomial is evaluated at several points, then it is possible to do some set-up work to save time during the evaluation. This initial task is called **preconditioning**. It is normally assumed that the cost of preconditioning is not significant, that is, the polynomial will be evaluated so many times that the fixed initial cost can be discarded.

Table 6.1 shows upper bounds for evaluating polynomials over the real and over the complex numbers. The bounds are shown as pairs (m, a), where m indicates the number of multiplications and a indicates the number of additions.

Table 6.1: Upper bounds for polynomial evaluation with preconditioning.

Degree	*Reals*	*Complex*	*References*
4,6	$((n+2)/2,\ n+1)$	same	[Motzkin,65], [Knuth,62]
any	$(\lceil (n+3)/2 \rceil,\ n)$	same	[Eve,64], [Knuth,62]
even, $n \geq 8$	$((n+2)/2,$ $n+3$ or $n+1$† $)$	$((n+2)/2,\ n)$	[Pan,79], [Eve,74]
odd, $n \geq 11$	$((n+1)/2,\ n+4$†$)$	$((n+1)/2,\ n+2)$	[Pan,79]
9		$((n+1)/2,\ n+3)$	[Revah,75]

† indicates an additional shifting operation. Shifting the radix point can be considered as being of the same complexity as addition.

Table 6.2 shows lower bounds on the number of additions and multiplications for polynomial evaluation with preconditioning. Unless one of the values is missing, these bounds are simultaneous lower bounds; that is, no algorithm can perform polynomial evaluation with less than m multiplications and less than a additions.

References:

[Belaga, E.G., 58], [Knuth, D.E., 62], [Motzkin, T.S., 65], [Mesztenyi, C. *et al.*, 67], [Knuth, D.E., 69], [Moenk, R. *et al.*, 72], [Horowitz, E., 73], [Kung, H.T., 73], [Aho, A.V. *et al.*, 74], [Eve, J., 74], [Horowitz, E., 74], [Savage, J.E.,

Table 6.2: Lower bounds for polynomial evaluation with preconditioning.

Degree	$(\times, +)$	*References*
$n \geq 2$	$(\lceil (n+1)/2 \rceil, ...)$	[Motzkin,65]
odd, $n \geq 7$	$((n+3)/2, ...)$	[Motzkin,65], [Knuth,81], [Revah,75]
any	$(..., n)$	[Belaga,58]
4,6,8	$((n+2)/2, n+1)$	[Knuth,81], [Pan,79]
odd, $n \geq 11$	$((n+1)/2, n+2)$	[Knuth,62], [Revah,75]
odd, $n \geq 3$	$((n+3)/3, n)$	[Belaga,58], [Revah,75]

74], [Shaw, M. *et al.*, 74], [Strassen, V., 74], [Aho, A.V. *et al.*, 75], [Borodin, A. *et al.*, 75], [Hyafil, L. *et al.*, 75], [Lipton, R.J. *et al.*, 75], [Revah, L., 75], [Borodin, A. *et al.*, 76], [Chin, F.Y., 76], [Lipton, R.J. *et al.*, 76], [Schonhage, A., 77], [Shaw, M. *et al.*, 77], [Lipton, R.J., 78], [Pan, V.Y., 78], [van de Wiele, J.P., 78], [Kronsjo, L., 79], [Nozaki, A., 79], [Rivest, R.L. *et al.*, 79], [Brown, M.R. *et al.*, 80], [Dobkin, D. *et al.*, 80], [Heintz, J. *et al.*, 80], [Heintz, J. *et al.*, 80], [Mescheder, B., 80], [Schnorr, C.P. *et al.*, 80], [Pan, V.Y., 81], [Schnorr, C.P., 81], [Baase, S., 88], [Sedgewick, R., 88], [Hansen, E.R. *et al.*, 90].

7 Text Algorithms

Text searching is the process of finding a pattern within a string of characters. The answer may be (1) whether a match exists or not, (2) the place of (the first) match, (3) the total number of matches or (4) the total number of matches and where they occur.

We will divide the algorithms between those which search the text as given, those which require preprocessing of the text and other text algorithms. Text preprocessing is preferred for large static text databases (such as bibliographic databases, dictionaries or corpora), while smaller dynamic text (such as text editing or casual browsing) will benefit from direct text searching.

In this chapter, n will denote the length of the text to be searched, m will denote the length of the pattern being searched, k the number of errors allowed, and $|\Sigma| > 1$ the size of the underlying alphabet. A **random string** is a sequence of symbols chosen with uniform probability from the alphabet Σ. The average results are computed for searching random patterns over random strings.

7.1 Text searching without preprocessing

Direct text searching algorithms accept a pattern and a string of text, and will locate an exact match of the pattern in the given string. The pattern is itself a string. When successful the *search* function returns a pointer p to the matching text in C ($p[0]$, $p[1]$, ... is the first occurrence of the pattern in the text) or an offset i into the given text in Pascal ($text[i]$, $text[i+1]$, ... is the first match). When the pattern is not present in the text, *search* returns the null pointer in C and -1 in Pascal.

For each algorithm we will describe the most efficient or basic variations.

The typical calling sequence for these functions in C is:

```
char *search(pat, text)  char *pat, *text;
void preprocpat(pat, ....)  char *pat;
```

and in Pascal:

```
function search(pat : PATTERN; text : TEXT) : integer;
procedure preprocpat(pat : PATTERN; ...);
```

The Pascal compiler must support variable length strings to have the programs given here working.

These functions can be composed to search on external text files:

Composition to search external text files

```
int extsearch(pat, filedesc)
char *pat;
int filedesc;

{ int offs, i, m, nb, nr;
  char buff[BUFSIZ], *p;

  m = strlen(pat);
  if(m == 0) return(0);
  if(m >= BUFSIZ)
      return(-2);          /*** Buffer is too small ***/

  /*** Assume that the file is open and positioned ***/
  offs = 0;                /*** number of characters already read ***/
  nb = 0;                  /*** number of characters in buffer ***/
  while(TRUE) {
     if(nb >= m) {
          /*** try to match ***/
          p = search(pat,buff);
          if(p != NULL)
               return(p-buff+offs); /*** found ***/
          for(i=0; i < m; i++) buff[i] = buff[i+nb-m+1];
          offs += nb-m+1;
```

```
            nb = m-1;
            }
        /*** read more text ***/
        nr = read(filedesc, buff+nb, BUFSIZ-1-nb);
        if(nr <= 0) return(-1);  /*** not found ***/
        nb += nr;
        buff[nb] = EOS;
        }
}
```

Any preprocessing of the pattern should be done only once, at the beginning. Especially, if the buffer size is small. Also, the knowledge of the length of the buffer (text) should be used (for example, see Section 7.1.3).

Similarly, these functions can be adapted or composed to count the total number of matches. We use two special constants: $MAXPATLEN$ which is an upper bound on the size of the pattern, and $MAXCHAR$ which is the size of the alphabet (a power of 2).

Let A_n be the number of comparisons performed by an algorithm, then in the worst case we have the following lower and upper bounds

$$n - m + 1 \le A_n \le \frac{4}{3}n - \frac{1}{3}m$$

For infinitely many n's, $|\Sigma| > 2$, and odd $m \ge 3$ we have

$$A_n \ge n + \left\lfloor \frac{n}{2m} \right\rfloor$$

For random text

$$E[A_n] \ge \Theta\left(\frac{\lceil \log_{|\Sigma|} m \rceil}{m} n\right)$$

General references:
[Karp, R.M. *et al.*, 72], [Slisenko, A., 73], [Fischer, M.J. *et al.*, 74], [Sellers, P., 74], [Galil, Z., 76], [Rivest, R.L., 77], [Seiferas, J. *et al.*, 77], [Galil, Z. *et al.*, 78], [Yao, A.C-C., 79], [Aho, A.V., 80], [Galil, Z. *et al.*, 80], [Main, M. *et al.*, 80], [Sellers, P., 80], [Slisenko, A., 80], [Crochemore, M., 81], [Galil, Z. *et al.*, 81], [Galil, Z., 81], [Galil, Z. *et al.*, 83], [Galil, Z., 85], [Pinter, R., 85], [Li, M. *et al.*, 86], [Abrahamson, K., 87], [Baeza-Yates, R.A., 89], [Baeza-Yates, R.A., 89], [Vishkin, U., 90].

7.1.1 Brute force text searching

Brute force text searching scans the text from left to right and tries to match the pattern at every text position.

$$n \leq A_n \leq m(n-m+2)-1$$

$$E[A_n] = \frac{(n-m+1)|\Sigma|}{|\Sigma|-1}\left(1-\frac{1}{|\Sigma|^m}\right) < \frac{n|\Sigma|}{|\Sigma|-1}$$

Brute force text searching

```
function search(pat: PATTERN; text: TEXT): integer;

var i, j, m, n: integer;
    found: boolean;
begin
  m := length(pat);
  if m = 0 then search := 1
  else begin
       n := length(text);  search := 0;
       j := 1;  i := 1;  found := FALSE;
       while not found and (i <= n-m+1) do begin
            if pat = substr(text, i, m) then begin
                 search :=  i;  found := TRUE; end;
            i := i + 1;
            end;
       end;
end;
```

It is easy to force this algorithm into its $O(nm)$ worst-case by searching a pattern of all a's ended by a b in a text which is all a's. This function may inspect text characters more than once and may backtrack to inspect previous characters.

References:
[Barth, G., 84], [Wirth, N., 86], [Baase, S., 88], [Sedgewick, R., 88], [Baeza-Yates, R.A., 89], [Baeza-Yates, R.A., 89], [Manber, U., 89], [Cormen, T.H. *et al.*, 90].

7.1.2 Knuth–Morris–Pratt text searching

This algorithm scans the text from right to left. It uses knowledge of the previous characters compared to determine the next position of the pattern to use. A table of size m is computed preprocessing the pattern before the search. This table is used to decide which character of the pattern should be used. For this algorithm we have

$$n \le A_n \le 2n + O(m)$$

$$n \le E[A_n] \le \frac{|\Sigma| + 1}{|\Sigma|} n$$

Knuth–Morris–Pratt text searching

```
void preprocpat(pat, next)
char *pat;
int next[ ];

{ int i, j;
  i = 0;
  j = next[0] = -1;
  do { if(j==(-1) || pat[i]==pat[j]) {
          i++;
          j++;
          next[i] = (pat[j]==pat[i]) ? next[j] : j;
       }
       else j = next[j]; }
     while(pat[i] != EOS);
}

char *search(pat, text)
char *pat, *text;

{ int next[MAXPATLEN], j;

  if(*pat == EOS) return(text);
  preprocpat(pat, next);

  for(j=0; *text != EOS;) {
     if(j==(-1) || pat[j] == *text) {
         text++; j++;
         if(pat[j] == EOS) return(text-j);
     }
     else j = next[j];
  }
  return(NULL);
}
```

This function may inspect some characters more than once, but will never backtrack to inspect previous characters. It is an **on-line algorithm**, that is, characters are inspected (may be more than once) strictly left to right.

References:
[Aho, A.V. *et al.*, 74], [Knuth, D.E. *et al.*, 77], [Barth, G., 81], [Salton, G. *et al.*, 83], [Barth, G., 84], [Meyer, B., 85], [Takaoka, T., 86], [Wirth, N., 86], [Baase, S., 88], [Brassard, G. *et al.*, 88], [Sedgewick, R., 88], [Baeza-Yates, R.A., 89], [Baeza-Yates, R.A., 89], [Manber, U., 89], [Cormen, T.H. *et al.*, 90].

7.1.3 Boyer–Moore text searching

This algorithm performs the comparisons with the pattern from the right to the left. After a mismatching position is found, it computes a shift, that is, an amount by which the pattern is moved to the right, before a new matching attempt is tried. This shift is computed using two heuristics, one based in the table used in the Knuth–Morris–Pratt algorithm (see Section 7.1.2), and the second based on matching the next character of the pattern that matches the character of the text that caused the mismatch. Both heuristic tables are built before the search using $O(m + |\Sigma|)$ comparisons and extra space.

Boyer–Moore preprocessing

```
void preprocpat(pat, skip, d)
char *pat;
int skip[ ], d[ ];

{    int j, k, m, t, t1, q, q1;
     int f[MAXPATLEN];              /*** auxiliary table ***/

     m = strlen(pat);
     for(k=0; k<MAXCHAR; k++) skip[k] = m;
     for(k=1; k<=m; k++) {
          d[k-1] = (m << 1) - k;
          skip[pat[k-1]] = m-k;
     }
     t = m + 1;
     for(j=m; j > 0; j--) {
          f[j-1] = t;
          while(t <= m && pat[j-1] != pat[t-1])
          {
               d[t-1] = min(d[t-1], m-j);
               t = f[t-1];
```

```
            }
            t--;
        }
        q = t;  t = m + 1 - q;  q1 = 1;  t1 = 0;
        for(j=1; j<=t; j++) {
            f[j-1] = t1;
            while(t1 >= 1 && pat[j-1] != pat[t1-1])
                t1 = f[t1-1];
            t1++;
        }
        while(q < m)
        {
            for(k=q1; k<=q; k++) d[k-1] = min(d[k-1], m+q-k);
            q1 = q + 1;  q = q + t - f[t-1];  t = f[t-1];
        }
}
```

There are several versions of this algorithm. The one presented here is the one given in Knuth–Morris–Pratt's paper. The running time is $O(n + rm)$ where r is the number of occurrences found. For any version of this algorithm we have

$$A_n \geq \frac{n}{m}$$

Table 7.1 shows the best known upper bound for different variations of the Boyer–Moore algorithm when there are no occurrences of the pattern in the text.

Table 7.1: Worst-case of Boyer–Moore type algorithms.

A_n	*References*
$3n$	[Boyer *et al.*, 77], [Knuth *et al.*, 77]
$14n$	[Galil, 79]
$2n$	[Apostolico *et al.*, 86]
$3n/2$	[Colussi *et al.*, 90]
$4n/3$	[Colussi *et al.*, 90]

For several variations of this algorithm

$$E[A_n] = O\left(\frac{\log m}{m} n\right)$$

which is optimal. For large patterns, the maximum shift will also depend on the alphabet size.

The idea of this algorithm can be extended to a Boyer–Moore automaton, a finite state machine that compares the pattern from right to left in the text. By keeping all the comparison information this automaton never inspects a character twice, and always shifts the pattern as much as possible. However, there are patterns such that the associated automaton needs $O(m^3)$ states (for any alphabet size bigger than 1). It is not known if this bound is tight (a trivial upper bound is $2^m - 1$).

Boyer–Moore text searching

```
char *search(pat, text, n)
char *pat, *text;
int n;

{ int j, k, m, skip[MAXCHAR], d[MAXPATLEN];

  m = strlen(pat);
  if(m == 0) return(text);
  preprocpat(pat, skip, d);

  for(k=m-1; k<n; k += max(skip[text[k] &(MAXCHAR-1)],d[j])) {
        for(j=m-1; j >= 0 && text[k] == pat[j]; j--) k--;
        if(j ==(-1)) return(text+k+1);
        }
  return(NULL);
}
```

This function may inspect text characters more than once and may backtrack to inspect previous characters. We receive the length of the text as a paremeter, such that we do not need to compute it. Otherwise, we lose the good average performance of this algorithm. This function works even if the text contains a character code that is not in the alphabet. If we can ensure that the text only has valid characters, the anding with $MAXCHAR - 1$ can be eliminated.

In practice, it is enough to use only the heuristic which always matches the character in the text corresponding to the mth character of the pattern. This version is called the Boyer–Moore–Horspool algorithm. For large m,

$$E[A_n] \geq \frac{n}{|\Sigma| - 1}$$

Boyer–Moore–Horspool text searching

```
char *search(pat, text, n)
char *pat, *text;
int n;

{ int i, j, k, m, skip[MAXCHAR];

  m = strlen(pat);
  if(m==0) return(text);
  for(k=0; k<MAXCHAR; k++) skip[k] = m;
  for(k=0; k<m-1; k++) skip[pat[k]] = m-k-1;

  for(k=m-1; k < n; k += skip[text[k] & (MAXCHAR-1)]) {
      for(j=m-1, i=k; j>=0 && text[i] == pat[j]; j--) i--;
      if(j == (-1)) return(text+i+1);
      }
  return(NULL);
}
```

This algorithm may require $O(nm)$ comparisons in the worst-case, but this happens with very low probability or for pathological cases. Recently it has been suggested that the first character in the text after the actual position of the pattern should be used. In practice, this is equivalent to having a pattern one character longer.

References:
[Boyer, R. *et al.*, 77], [Galil, Z., 79], [Bailey, T.A. *et al.*, 80], [Guibas, L.J. *et al.*, 80], [Horspool, R.N.S., 80], [Rytter, W., 80], [Salton, G. *et al.*, 83], [Moller-Nielsen, P. *et al.*, 84], [Apostolico, A. *et al.*, 86], [Wirth, N., 86], [Baase, S., 88], [Brassard, G. *et al.*, 88], [Schaback, R., 88], [Sedgewick, R., 88], [Baeza-Yates, R.A., 89], [Baeza-Yates, R.A., 89], [Baeza-Yates, R.A., 89], [Manber, U., 89], [Baeza-Yates, R.A. *et al.*, 90], [Cormen, T.H. *et al.*, 90].

7.1.4 Searching sets of strings

A natural extension of the Knuth–Morris–Pratt algorithm, without being as general as a deterministic finite automaton (DFA), is to define a **pattern matching machine** (PMM). Pattern matching machines search for any of several strings simultaneously. A pattern matching machine consists of a current state, a **transition table** ('go to' table) as in a finite automaton, a **failure function** to economize transitions and an **output function** to determine, upon reaching an accepting state, which string actually matched.

While searching, if the character read is one of the go to transitions, we change state accordingly, and we read the next character. Otherwise, we use the failure transition, and we compare the current character again in the new state. Let m be the total number of characters in the strings being searched. The size of the transition table is $O(m)$, independent of the alphabet size. The number of character inspections is independent of m,

$$n \leq A_n \leq 2n$$

Pattern matching machine

```
state := 1;
for i := 1 to n do begin
    while trans(state, text[i]) = FAIL do
        state := failure(state);
    state := trans(state, text[i]);
    if output(state) <> {} then
        {*** a match was found ***};
    end;
```

The advantage of the PMM over a DFA is that the transition table is smaller at the cost of sometimes inspecting characters more than once. This function will never backtrack to inspect previous characters. It is an on-line algorithm.

The construction and optimizations of the table are beyond the scope of this handbook. More efficient automata are fully described in Section 7.1.6.

There also exist pattern matching machines based on the Boyer–Moore algorithm (Section 7.1.3). In this case, the search is done from right to left in the set of strings. If a mismatch is found, the set of strings is shifted to the right.

References:
[Aho, A.V. *et al.*, 74], [Aho, A.V. *et al.*, 75], [Commentz-Walter, B., 79], [Bailey, T.A. *et al.*, 80], [Meyer, B., 85], [Sedgewick, R., 88], [Baeza-Yates, R.A. *et al.*, 90].

7.1.5 Karp–Rabin text searching

This algorithm searches a string by computing a signature or a hashing value, of each m characters of the string to be searched. A **signature** is an integer value computed from a string, which is useful for quickly detecting inequality. This algorithm achieves its efficiency by computing the signature for position i from the signature in position $i-1$.

The number of characters inspected is

$$A_n = 2n$$

Karp–Rabin text searching

```
function search(pat: PATTERN; text: TEXT): integer;

const B = 131;
var hpat, htext, Bm, j, m, n: integer;
   found: boolean;
begin
  found := FALSE; search := 0;
  m := length(pat);
  if m=0 then begin
      search := 1; found := TRUE; end;

  Bm := 1;
  hpat := 0; htext := 0;
  n := length(text);
  if n >= m then                    {*** preprocessing ***}
      for j := 1 to m do begin
           Bm := Bm*B;
           hpat := hpat*B + ord(pat[j]);
           htext := htext*B + ord(text[j]);
           end;

  j := m;                           {*** search ***}
  while not found do begin
      if (hpat = htext) and (pat = substr(text,j-m+1,m)) then
           begin search := j-m+1; found := TRUE; end;
      if j < n then begin
           j := j+1;
           htext := htext*B - ord(text[j-m])*Bm + ord(text[j]);
           end
      else found := TRUE;
      end;
end;
```

The above implementation avoids the computation of the *mod* function at every step, instead it uses the implicit modular arithmetic given by the hardware. The value of B is selected such that $B^k \bmod 2^r$ has maximal cycle (cycle of length 2^{r-2}) for r in the range 8 to 64. $B = 131$ has this property.

References:
[Harrison, M.C., 71], [Karp, R.M. *et al.*, 87], [Sedgewick, R., 88], [Baeza-Yates, R.A., 89], [Cormen, T.H. *et al.*, 90], [Gonnet, G.H. *et al.*, 90].

7.1.6 Searching text with automata

Any regular language can be recognized by a DFA, hence it is interesting to construct and search with such automata. We will use the following definition of an automaton:

Automata definition

```
typedef struct automrec {
    short d;        /*** size of the alphabet (0, ..., d-1) ***/
    short st;       /*** number of states (0, ..., st-1)    ***/
    short **nextst; /*** transition function: nextst[st][ch] ***/
    short *final;   /*** state i is final if final[i] != 0   ***/
    } *automata;

automata stringautom(str) char *str;
automata starautom(aut) automata aut;
automata unionautom(aut1, aut2) automata aut1, aut2;
automata concatautom(aut1, aut2) automata aut1, aut2;
```

In addition to the above definition, when automata are used for string matching, we will encode final states in the transition table as the complement of the state number. This allows a single quick check in a crucial part of the search loop. For an accepting state, *final* will encode the length of the match, whenever this is possible.

With this definition, the searching function is:

Deterministic-finite-automata text searching

```
char *search(pat, text)
char *pat, *text;

{ short st, **states;
  automata a;

  if(pat[0] == EOS) return(text);
  a = stringautom(pat);
```

```
    states = a ->nextst;
    for(st=0; st < a ->st; st++) states[st][EOS] = -1;
    st = 0;
    while((st = states[st][*text++ & (MAXCHAR-1)]) >= 0);
    if(*(text-1) == EOS)
        return(NULL);
    else return(text - a ->final[-st]);
}
```

This function will inspect each character once, and will never backtrack to inspect previous characters. This function works even if the text contains a character code that is not in the alphabet. If we can ensure that the text only has valid characters, the anding with $MAXCHAR-1$ can be eliminated. It is an on-line algorithm. The automata is modified to produce a false acceptance upon recognition of the end-of-string (EOS) character.

Regular expressions can be built from strings, concatenation, union, Kleene's closure or star (∗) and complement. We will therefore give functions to perform the above operations, and consequently any regular expression can be built using them.

To generate an automaton which recognizes a string we use the *stringautom* function.

Build an automaton which recognizes a string

```
automata stringautom(pat)
char *pat;

{ short back, i, st;
  char ch;
  automata a;

  a = (automata)malloc(sizeof(struct automrec));
  a ->d = MAXCHAR;
  a ->st = strlen(pat)+1;
  a ->nextst = (short **)calloc(a ->st, sizeof(short *));
  a ->final = (short *)calloc(a ->st, sizeof(short));

  for(st=0; st < a ->st; st++) {
      a ->nextst[st] = (short *)calloc(MAXCHAR, sizeof(short));
      if(st < a ->st-2) a ->nextst[st][pat[st]] = st+1;
      }
  a ->nextst[a ->st-2][pat[a ->st-2]] = 1-a ->st;
  /* set final state (with the match length) */
```

```
    a->final[a->st-1] = a->st-1;

    /* Set backwards transitions */
    for(st=1; st < a->st; st++)
        for(back=st-1; back >= 0; back--) {
            ch = pat[back];
            if(a->nextst[st][ch] == 0)
                for(i=1; i<=st; i++)
                    if((st==i || strncmp(pat,pat+i,st-i)==0)
                        && ch == pat[st-i]) {
                        a->nextst[st][ch] = st-i+1;
                        break;
                        }
            }

    return(a);
  }
```

The next function produces the union of two automata.

Build the union of two automata

```
  short mergestates();

  automata unionautom(aut1, aut2)
  automata aut1, aut2;

  { short *st1, *st2, ts;
    automata a;

    if(aut1->d != aut2->d)
        return(NULL); /*** different alphabets ***/
    a = (automata)malloc(sizeof(struct automrec));
    a->d = aut1->d;
    a->st = 0;
    ts = aut1->st + aut2->st;
    a->nextst = (short**) malloc(ts * sizeof(short*));
    a->final = (short*) malloc(ts * sizeof(short));
    st1 = (short*) calloc(ts, sizeof(short));
    st2 = (short*) calloc(ts, sizeof(short));
    mergestates(0, 0, aut1, aut2, a, st1, st2);
    free(st1);   free(st2);
```

```
    return(a);
}

short mergestates(s1, s2, aut1, aut2, newaut, st1, st2)
short s1, s2, *st1, *st2;
automata aut1, aut2, newaut;

{ short as1, as2, i, j;

  /*** find if state is already stored ***/
  for(i=0; i < newaut->st; i++)
      if(st1[i]==s1 && st2[i]==s2)
          return(s1<0 || s2<0 ? -i : i);

  /*** create new state ***/
  st1[i] = s1;    st2[i] = s2;
  newaut->st++;
  as1 = s1 < 0 ? -s1 : s1;    as2 = s2 < 0 ? -s2 : s2;
  newaut->nextst[i] = (short*) malloc(newaut->d * sizeof(short));
  for(j=0; j<newaut->d; j++)
      newaut->nextst[i][j] =
          mergestates(aut1->nextst[as1][j], aut2->nextst[as2][j],
                      aut1, aut2, newaut, st1, st2);
  if(s1 < 0) {
      newaut->final[i] =
          (s2<0) ? max(aut1->final[-s1], aut2->final[-s2])
                 : aut1->final[-s1];
      return(-i);
      }
  else if(s2 < 0) {
      newaut->final[i] = aut2->final[-s2];
      return(-i);
      }
  return(i);
}
```

References:
[Thompson, K., 68], [Aho, A.V. *et al.*, 74], [Hopcroft, J.E. *et al.*, 79], [Salton, G. *et al.*, 83], [Sedgewick, R., 88], [Myers, E. *et al.*, 89], [Cormen, T.H. *et al.*, 90].

7.1.7 Shift-or text searching

This algorithm uses a word of m bits, one bit for every character in the pattern, to represent the state of the search. The ith bit is a zero if the first i characters of the pattern have matched the last i character of the text, otherwise it is a one. A match is detected when the mth bit is a zero. We have

$$A_n = n \left\lceil \frac{m}{w} \right\rceil$$

where w is the word size.

To update the current state after a new character is read, we perform a bit shift of the state and a logical or with a precomputed table indexed on the new character. This table depends on the pattern and the alphabet. The following program uses the variable *bits* to keep track of the state of the search, and the table $mask[MAXCHAR]$ to update the state after reading a new character. The value of $mask[x]$ ($x \in \Sigma$) is such that it has a zero bit in the ith position if $pat[i] = x$, otherwise it is a one bit. For example, if x does not appear in the pattern, $mask[x]$ is a sequence of 1s.

Shift-or text searching

```
char *search(pat, text)
char *pat, *text;

{ int B, bits, i, m, mask[MAXCHAR];

  if(pat[0]==EOS) return(text);
  B = 1;
  for(m=0; m<MAXCHAR; m++) mask[m] = ~0;
  for(m=0; B != 0 && pat[m] != EOS; m++) {
       mask[pat[m]] &= ~B;
       B <<= 1;
       }

  B = 1<<(m-1);
  for(bits= ~0; *text != EOS; text++) {
       bits = bits<<1 | mask[*text & (MAXCHAR-1)];
       if((bits&B) == 0) {
            for(i=0; pat[m+i] != EOS && pat[m+i]==text[i+1]; i++);
            if(pat[m+i]==EOS) return(text-m+1);
            }
       }
  return(NULL);
}
```

This function will inspect each character once, and will never backtrack to inspect previous characters. This function works even if the text contains a character code that is not in the alphabet. If we can ensure that the text only has valid characters, the anding with $MAXCHAR - 1$ can be eliminated. It is an on-line algorithm.

This algorithm extends to classes of characters, by modifying the preprocessing of the table *mask*, such that every position in the pattern can be a class of characters, a complement of a class or a 'don't care' symbol. Similarly, we may allow 'don't care' symbols in the text, by defining a special symbol x such that $mask[x] = 0$. This is the fastest algorithm to solve this generalization of string searching. There exist algorithms with better asymptotic complexity to solve this problem, but these are not practical.

References:
[Abrahamson, K., 87], [Baeza-Yates, R.A. *et al.*, 89], [Baeza-Yates, R.A., 89], [Kosaraju, S.R., 89].

Table 7.2: Algorithms for string matching with mismatches.

Worst-case A_n	$E[A_n]$	*Extra space*	*Reference*
$kn + km \log m$	$kn + km \log m$	km	[Landau *et al.*, 85]
$kn + m \log m$	$kn + m \log m$	m	[Galil *et al.*, 85]
$m(n + m - 2k)$	$\alpha(m)(k+1)n$ $(\alpha(m) < 1)$	m	[Baeza-Yates, 89]
$n \log m + m^{k+1}\|\Sigma\|$	$n \log m + m^{k+1}\|\Sigma\|$	$m^{k+1}\|\Sigma\|$	[Baeza-Yates, 89]
$m \log k(n + \|\Sigma\|)/w$	$m \log k(n + \|\Sigma\|)/w$	$\|\Sigma\| m \log k/w$	[Baeza-Yates *et al.*, 89]
$n \log m + rm$	$n \log m + rm$	m	[Grossi *et al.*, 89]
$mn + k\|\Sigma\|$	$kn(k/\|\Sigma\| + 1/(m-k)) + k\|\Sigma\|$	$k\|\Sigma\|$	[Tarhio *et al.*, 90]

7.1.8 String similarity searching

There are two main models of string similarity. The simplest one just counts characters which are unequal. That is, the distance, or the editing cost between two strings of the same length m, is defined as the number of corresponding characters that mismatch (this is also called **Hamming distance**). The problem of string searching with k mismatches consists in finding the first substring of length m in the text, such that the Hamming distance between the pattern and the substring is at most k. When $k = 0$ the problem reduces to simple string searching.

Table 7.2 shows the worst-case and expected-case complexity of algorithms that solve this problem, where w denotes the computer word size and r the number of occurrences found.

The brute force algorithm for this problem is presented below. We have

$$(k+1)n \leq A_n \leq mn$$

$$E[A_n] \leq \frac{(k+1)|\Sigma|}{|\Sigma|-1} n \leq 2(k+1)n$$

Brute force text searching with k mismatches

```
char *search(k, pat, text)
int k;
char *pat, *text;

{ int j, m, count;

  m = strlen(pat);
  if(m <= k) return(text);

  for(; *text != EOS; text++) {
     for(count=j=0; j < m && count <= k; j++)
         if(pat[j] != text[j]) count++;
     if(count <= k) return(text);
     }

  return(NULL);
}
```

The second model is more general and considers that characters could be inserted, deleted, or replaced to produce the matching. Let A_D be the cost of deleting a character from the pattern, A_I the cost of inserting a character, and $A_{x,y}$ the cost of replacing symbol x for symbol y. We define the distance, $d(a,b)$, between two strings a and b as the minimal cost of transforming a into b.

Let $T_{i,j}$ be the minimal distance between the first i characters of the pattern and the substring of the text ending at j such that

$$T_{i,j} = \min_q(d(p_{1\ldots i}, t_{q\ldots j}))$$

Clearly $T_{0,j} = 0$ (no errors because $q = j$) and $T_{i,0} = iA_D$ (i deletions).

The problem of string searching with errors of cost k or less consists in finding all substrings of the text such that $T_{m,j} \leq k$. The table $T_{m,j}$ can be computed using dynamic programming with the following formula:

$$T_{i,j} = \min(T_{i-1,j-1} + A_{pat_i,text_j}, T_{i,j-1} + A_I, T_{i-1,j} + A_D)$$

with the initial conditions indicated above. The starting position(s) of each occurrence must be computed by backtracking each of the $T_{m,j}$.

The most commonly used cost values are $A_D = A_I = 1$, and $A_{x,y} = 1$ if $x \neq y$ or 0 otherwise (this is called **Levenshtein distance**). In this case, the searching problem is called approximate string matching with k errors.

The following function shows the dynamic programming algorithm for the Levenshtein distance. Instead of storing the complete T matrix of size $n \times m$, the function uses just one column of it, needing only $O(m)$ extra space. The total number of operations is $O(nm)$.

String matching with k errors

```
char *search(k, pat, text, n)  /*** at most k errors ***/
int k, n;
char *pat, *text;

{ int T[MAXPATLEN+1];
  int i, j, m, tj, tj1;

  m = strlen(pat);
  if(m <= k) return(text + n);
  T[0] = 0;                          /*** initial values ***/
  for(j=1; j<=m; j++) T[j] = j;

  for(i=1; i<=n; i++) {              /*** search ***/
      tj1 = 0;
      for(j=1; j<=m; j++) {
          tj = T[j];
          if(text[n-i] != pat[m-j]) tj1++;
          if(tj+1 < tj1) tj1 = tj+1;
          if(T[j-1]+1 < tj1) tj1 = T[j-1]+1;
          T[j] = tj1;
          tj1 = tj;
          }
      if(T[m] <= k) return(text+n-i);
      }
  return(NULL);
}
```

Table 7.3 shows the worst-case and expected time complexity of several algorithms for solving the Levenshtein case (see also Section 7.3.1), where $Q \leq \min(3^m, 2^k|\Sigma|^k m^{k+1})$.

Table 7.3: Algorithms for string matching with errors.

Worst-case A_n	$E[A_n]$	*Extra space*	*Reference*
mn	mn	m^2	Dynamic prog.
mn	kn	m^2	[Ukkonen, 85]
$n \log m + mQ \log Q$	$n \log m + mQ \log Q$	mQ	[Ukkonen, 85]
$k^2 n + m \log m$	$k^2 n + m \log m$	km	[Landau *et al.*, 88]
$kn + m^2$	$kn + m^2$	km	[Galil *et al.*, 89]
$kn + m^2 + \|\Sigma\|$	$kn + m^2 + \|\Sigma\|$	$m^2 + \|\Sigma\|$	[Ukkonen *et al.*, 90]
$mn + (m+k)\|\Sigma\|$	$kn(k/(\|\Sigma\| + 2k^2) + 1/m) + (m+k)\|\Sigma\|$	$m\|\Sigma\|$	[Tarhio *et al.*, 90]
$nk + m$	$2(k+1)n \log_b m/(m-k)$ $(k \leq m/(\log_b m + O(1)))$	m	[Chang *et al.*, 90]

References:
[Levenshtein, V., 65], [Levenshtein, V., 66], [Sellers, P., 74], [Wagner, R.E. *et al.*, 74], [Wagner, R.E., 75], [Wong, C.K. *et al.*, 76], [Hall, P.A.V. *et al.*, 80], [Bradford, J., 83], [Johnson, J.H., 83], [Sankoff, D. *et al.*, 83], [Ukkonen, E., 83], [Landau, G.M. *et al.*, 85], [Ukkonen, E., 85], [Ukkonen, E., 85], [Galil, Z. *et al.*, 86], [Landau, G.M. *et al.*, 86], [Landau, G.M. *et al.*, 86], [Landau, G.M., 86], [Krithivasan, K. *et al.*, 87], [Baase, S., 88], [Ehrenfeucht, A. *et al.*, 88], [Baeza-Yates, R.A. *et al.*, 89], [Baeza-Yates, R.A., 89], [Galil, Z. *et al.*, 89], [Grossi, R. *et al.*, 89], [Manber, U., 89], [Eppstein, D. *et al.*, 90], [Tarhio, J. *et al.*, 90], [Ukkonen, E. *et al.*, 90].

7.1.9 Summary of direct text searching

Table 7.4 shows relative total times of direct text searching algorithms written in C. These values were generated from searching the patterns 'to be or not to be' and 'data' in the whole text of *The Oxford English Dictionary* (2nd Edition), about 570 million characters in length. The timings consider the preprocessing and search time, and the reading of the file.

7.2 Searching preprocessed text

Large, static, text files may require faster searching methods than the ones described in the previous section, which are all basically linear in the length of the text.

In this section we will describe algorithms which require preprocessing of the text, most often building an index or some other auxiliary structure, to speed up later searches.

Table 7.4: Direct searching over *The Oxford English Dictionary*.

Algorithm	*'to be or not to be'*	*'data'*
Brute force	1.23	1.74
Knuth–Morris–Pratt	2.16	2.93
Boyer–Moore	1.33	1.16
Boyer–Moore–Horspool	1.00	1.00
Karp–Rabin	2.64	3.69
Automaton	1.19	1.67
Shift-or	1.41	2.10
Brute force ($k = 1$)	2.81	4.03
Dynamic programming ($k = 1$)	7.52	36.90

Usually there are some restrictions imposed on the indices and consequently on the later searches. Examples of these restrictions are: a **control dictionary** is a collection of words which will be indexed. Words in the text which are not in the control dictionary will not be indexed, and hence are not searchable. **Stop words** are very common words (such as articles or prepositions) which for reasons of volume or precision of recall will not be included in the index, and hence are not searchable. An **index point** is the beginning of a word or a piece of text which is placed into the index and is searchable. Usually such points are preceded by space, punctuation marks or some standard prefixes. In large text databases, not all character sequences are indexed, just those which are likely to be interesting for searching.

The most important complexity measures for preprocessed text files are: the extra space used by the index or auxiliary structures S_n, the time required to build such an index T_n and the time required to search for a particular query, A_n. As usual, n will indicate the size of the text database, either characters or number of index points.

General references:
[Gonnet, G.H., 83], [Larson, P., 83], [Faloutsos, C., 85], [Galil, Z., 85].

7.2.1 Inverted files

Inversion is a composition (as described in Section 2.2.2.1) of two searching algorithms, where we first search for an attribute name, which returns an index and on this index we search for an attribute value. The result of a search on an inverted file is a set of records (or pointers to records).

In text databases the records to be searched are variable-length portions of text, possibly subdivided in fields. For example, in a bibliographic database each work is a record and fields can be title, abstract, authors, and so on. Every word in any of the fields, is considered an index point.

The result of searching a term in an inverted index is a set of record numbers. All these sets are typically stored sequentially together in an external file. The set can be identified by its first and last position in the external file.

Let n be the total number of words indexed. The complexity of building the index is that of sorting n records, each one of length $\lceil \log_2 nfk \rceil$ bits where k is the size of the control dictionary and f is the number of fields in any record.

$$S_n = n \lceil \log_2 nk \rceil \text{ bits}$$

$$T_n = O(n \log n \lceil \log_2 nfk \rceil)$$

$$A_n = O(\log_2 k)$$

The data structures defining an inverted index are:

$$\textbf{ControlDict} : \{[\textbf{word}]\}_1^{\mathbf{k}}.$$

$$\textbf{FieldIndex} : \{\textbf{FieldName}, \{\textbf{first}, \textbf{last}\}_1^{\mathbf{k}}\}_1^{\mathbf{f}}.$$

word : **string**. **FieldName** : **string**.

first : **int**. **last** : **int**.

Building inverted files can be done following these steps:

(1) Assume that the control dictionary can be kept in main memory. Assign a sequential number to each word, call this the **word number** (an integer between 1 and k).

(2) Scan the text database and for each word, if in the control dictionary, output to a temporary file the record number, field number, and its word number.

(3) Sort the temporary file by field number, word number, and record number.

(4) For each field, compact the sorted file to distinct record numbers alone. During this compaction, build the inverted list from the end points of each word. This compacted file becomes the main index for that field.

(5) For certain applications the multiplicity of occurrences is also interesting. The multiplicities can be easily recorded during the compaction phase.

For a single term search, the location of the answer and the size of the answer are immediately known. Further operations on the answers, intersections, unions, and so on, will require time proportional to the size of the sets.

The operations of union, intersection and set difference can be made over the set of pointers directly (all these sets will be in sorted order) without any need for reading the text.

References:
[Knuth, D.E., 73], [Grimson, J.B. *et al.*, 74], [Stanfel, L., 76], [McDonell, K.J., 77], [Nicklas, B.M. *et al.*, 77], [Jakobsson, M., 80], [Salton, G. *et al.*, 83], [Sankoff, D. *et al.*, 83], [Waterman, M.S., 84], [Blumer, A. *et al.*, 87], [Rao, V.N.S. *et al.*, 88], [Coulbourn, C.J. *et al.*, 89].

7.2.2 Trees used for text searching

A **semi-infinite string** (or **sistring**) is a substring of the text database, defined by a starting position and continuing to the right as far as necessary. (The database may be viewed as having an infinite number of null characters at its right end.) Sistrings are compared lexicographically, character by character. For any database, no two sistrings in different positions compare equal. Since a sistring is defined by an offset and the text in the database, then assuming that the text is available, each sistring can be represented by an integer. An index of the text database will be any search structure based on the sistrings of all the index points.

Any search structure which allows for range searches can be used to search on the set of all sistrings. In particular, most algorithms based on trees are good candidates. Note that hashing algorithms are not suitable, as these neither allow range searching, nor an easy way of computing a hashing value for a semi-infinite string.

The most suitable trees to store this information are digital trees (Section 3.4.4), in particular Patricia trees. A Patricia tree built on all the sistrings of a text database is called a PAT tree. The PAT structure has two advantages: (1) the search is done over the tree alone scanning bits of the string to be searched, but it does not need to compare the text during the search; (2) the whole set of sistrings answering a query is contained in a single subtree and hence the searching time is independent of the size of the answer. For a Patricia tree we have

$$S_n = n\ ExtNodes\ +\ (n-1) IntNodes$$

$$T_n\ =\ O(n \log n)$$

Prefix searching Every subtree of the PAT tree contains all the sistrings with a given prefix, by construction. Hence prefix searching in a PAT tree consists of searching the prefix in the tree up to the point where we exhaust the prefix or up to the point where we reach an external node. At this point we need to verify whether we could have skipped bits. This is done with a single comparison of any of the sistrings in the subtree (considering an external node as a subtree of size one). If this comparison is successful then all the sistrings in the subtree (which share the common prefix) are the answer, otherwise there are no sistrings in the answer. We have

$$E[A_n] \leq \min(m, \log_2 n + 1 + O(n^{-1}))$$

where m is the bit length of the prefix.

The search ends when the prefix is exhausted or when we reach an external node and at that point all the answer is available (regardless of its size) in a single subtree. By keeping the size of each subtree in each internal node we can trivially find the size of any matched subtree (knowing the size of the answer is very appealing for information retrieval purposes.)

Range searching Searching for all the strings within a certain range of values (lexicographical range) can be done equally efficiently. More precisely, range searching is defined as searching for all strings which lexicographically compare between two given strings. For example the range 'abc' .. 'acc' will contain strings like 'abracadabra', 'acacia', 'aboriginal' but not 'abacus' or 'acrimonious'.

To do range searching on a PAT tree we search each of the defining intervals and then collect all the subtrees between (and including) them. Only $O(height)$ subtrees will be in the answer even in the worst-case (the worst-case is $2\ height - 1$) and hence only $O(\log n)$ time is necessary in total on the average.

Longest repetition searching The longest repetition of a text is defined as the match between two different positions of a text where this match is the longest (the most number of characters) in the entire text. For a given text the longest repetition will be given by the tallest internal node in the PAT tree. Hence, the tallest internal node gives a pair of sistrings which match for the most number of characters. In this case tallest means considering not only the shape of the tree but also the skipped bits. For a given text the longest repetition can be found while building the tree and it is a constant, that is, it will not change unless we change the tree (that is, the text).

It is also possible to search for the longest repetition not just for the entire tree/text, but for a subtree. This means searching for the longest repetition among all the strings which share a common prefix. This can be done in $O(height)$ time by keeping one bit of information at each internal node, which will indicate on which side we have the tallest subtree. By keeping such a bit we can find one of the longest repetitions starting with an arbitrary prefix in $O(\log n)$ time. If we want to search for all of the longest repetitions we need two bits per internal node (to indicate equal heights as well) and the search becomes logarithmic in height and linear in the number of matches.

'Most significant' or 'most frequent' searching This type of search has great practical interest, but is slightly difficult to describe. By 'most significant' or 'most frequent' we mean the most frequently occurring strings within the text database. For example, finding the 'most frequent' trigram is finding a sequence of three letters which appears the greatest number of times within our text.

In terms of the PAT tree, and for the example of the trigrams, the number of occurrences of a trigram is given by the size of the subtree at distance three characters from the root. So finding the most frequent trigram is equivalent to finding the largest subtree at distance three characters from the root. This can be achieved by a simple traversal of the PAT tree which is at most $O(n/average\ size\ of\ the\ answer)$ but usually much faster.

Searching for trigrams (or n-grams) is simpler than searching, for example, for the 'most common' word. A word could be defined as any sequence of characters delimited by a blank space. This type of search will also require a traversal, but in this case the traversal is only done in a subtree (the subtree of all sistrings starting with a space) and does not have a constant depth; it traverses the tree at the place where the second blank appears.

We may also apply this algorithm over any arbitrary subtree. This is equivalent to finding the most frequently occurring trigram, word, ... that follows some given prefix.

In all cases, finding the most frequent string with a certain property requires a subtree selection and then a tree traversal which is at most $O(n/k)$ but typically is much smaller. Here k is the average size of each group of strings of the given property. Techniques similar to alpha-beta pruning can be used to improve this search.

References:
[Fredkin, E., 60], [Morrison, D.R., 68], [Weiner, P., 73], [Aho, A.V. *et al.*, 74], [McDonell, K.J., 77], [Nicklas, B.M. *et al.*, 77], [Majster, M. *et al.*, 80], [Comer, D. *et al.*, 82], [Orenstein, J.A., 82], [Gonnet, G.H., 83], [Salton, G. *et al.*, 83], [Apostolico, A. *et al.*, 85], [Apostolico, A., 85], [Chen, M.T. *et al.*, 85], [Merrett, T.H. *et al.*, 85], [Kemp, M. *et al.*, 87], [Gonnet, G.H., 88], [Baeza-Yates, R.A., 89].

7.2.3 Searching text with automata

In this section we present an algorithm which can search for arbitrary regular expressions in an indexed text of size n in time sublinear in n on the average. For this we simulate a DFA on a binary trie built from all the sistrings of a text (searching an arbitrary regular expression in $O(n)$ is done in Section 7.1.6).

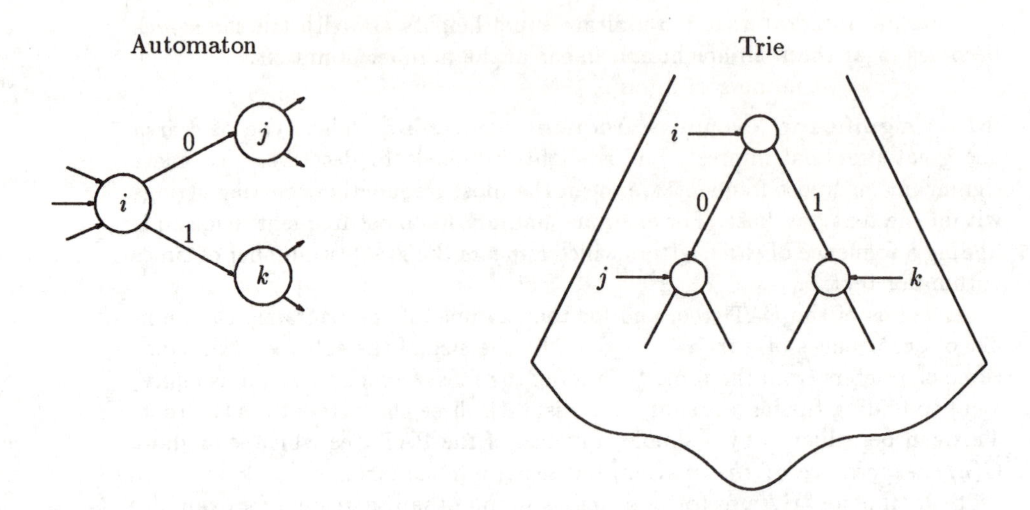

Figure 7.1: Simulating the automaton on a binary digital tree.

The main steps of the algorithm are:

(1) Convert the query regular expression into a partial DFA (a partial DFA will not represent transitions which can never reach an accepting state).

(2) Eliminate outgoing transitions from final states, eliminate all dead states, and minimize the DFA. This may require exponential space/time with respect to the query size but is independent of the size of the text.

(3) Convert the character DFA into a binary DFA using the binary encoding of the input alphabet; each state will then have at most two outgoing transitions, one labelled 0 and one labelled 1.

(4) Simulate the binary DFA on the binary trie from all sistrings of text using the same binary encoding as in step (2). That is, associate the root of the tree with the initial state, and, for any internal node associated with state i, associate its left descendant with state j if $i \rightarrow j$ for a bit 0 on the DFA, and associate its right descendant with state k if $i \rightarrow k$ for a 1 (see Figure 7.1).

(5) For every node of the index associated with a final state, accept the whole subtree and halt the search in that subtree. (For this reason, we do not need outgoing transitions in final states.)

(6) On reaching an external node, run the remainder of the automaton on the single string determined by this external node.

A depth-first traversal to associate automaton states with trie nodes ensures $O(\log n)$ space for the simulation in the case of random text.

The expected number of internal nodes visited is

$$E[N_n] = O(\log^{m-1}(n)\; n^{\alpha})$$

where $\alpha = \log_2 |\lambda|$, and λ is the largest eigenvalue of the incidence matrix of the DFA with multiplicity m. For any binary DFA $|\lambda| < 2$ and hence $\alpha < 1$. The expected number of external nodes visited is proportional to N_n, and the expected number of comparisons needed in every external node is $O(1)$. Therefore, the total searching time is given by $O(N_n)$.

References:
[Gonnet, G.H., 88], [Baeza-Yates, R.A. *et al.*, 89], [Baeza-Yates, R.A., 89], [Baeza-Yates, R.A. *et al.*, 90].

7.2.4 Suffix arrays and PAT arrays

A PAT array is a compact representation of a PAT tree (Section 7.2.2), because it stores only the external nodes of the tree. Thus, we need only one pointer per indexing point. The definition for PAT arrays is

$$\{[\mathbf{string}]\}_{\mathbf{0}}^{\mathbf{N-1}} .$$

Building a PAT array is similar to sorting variable-length records, thus

$$T_n = O(n \log n)$$

Any Patricia tree operation can be simulated in a PAT array within a factor of $O(\log n)$ time (by doing a binary search on the next bit to determine the left and right subtrees). However, it turns out that it is not necessary to simulate the PAT tree for prefix and range searching and we obtain algorithms which are $O(\log n)$ instead of $O(\log^2 n)$ for these operations. Actually prefix searching and range searching become very similar operations. Both can be implemented by doing an indirect binary search over the array with the results of the comparisons being less than, equal (or included in the case of range searching) and greater than. In this way the searching takes at most

$$A_n \le m(2\log_2 n - 1) \qquad \text{(character comparisons)}$$
$$A_n \le 4\log_2 n \qquad \text{(disk accesses)}$$

where m is the length of given prefix (query).

Prefix searching in a PAT array

```
int search(pat, index, n)
char *pat, *index[ ];
int n;                    /* size of the PAT array */

{ int m, left, right, low, high, i;

  m = strlen(pat);
  /* search left end */
  if(strncmp(pat, index[0], m) != 1) left = 0;
  else if(strncmp(pat, index[n-1], m) == 1) left = n;
  else { /* binary search */
      for(low=0, high=n; high-low > 1;) {
          i = (high+low)/2;
          if(strncmp(pat, index[i], m) != 1) high = i;
          else     low = i;
      }
      left = high;
  }
  /* search right end */
  if(strncmp(pat, index[0], m) == -1) right = -1;
  else if(strncmp(pat, index[n-1], m) != -1) right = n-1;
  else { /* binary search */
      for(low=0, high=n; high-low > 1;) {
          i = (high+low)/2;
          if(strncmp(pat, index[i], m) != -1) low = i;
          else     high = i;
      }
      right = low;
  }
  return(right-left+1);
}
```

PAT arrays are also called **suffix arrays**. With additional information about the longest common prefixes of adjacent index points in the array, it is possible to speed up a prefix search to

$$A_n = m + \lceil \log_2 n \rceil$$

Searching for two strings s_1 and s_2 ($|s_1| \leq m$) such that s_2 is at most k characters after s_1 can be done in time $O(n^{1/4})$ using a PAT array and extra information of size $O((k+m)n)$. If we are interested only in the number of occurrences, the query time is reduced to $O(\log n)$. This kind of search is called **proximity searching**.

References:
[Gonnet, G.H., 86], [Manber, U. *et al.*, 90], [Manber, U. *et al.*, to app.].

7.2.5 DAWG

The Directed Acyclic Word Graph (DAWG) is a deterministic finite automaton that recognizes all possible substrings of a text. All states in the DAWG are accepting (final) states. Transitions which are not defined are assumed to go to a non-accepting dead state.

For any text of size $n > 2$ we have

$$n + 1 \leq \ states \ \leq 2n - 1$$

$$n \leq \ transitions \ \leq 3n - 4$$

$$E[states] = \frac{n}{\ln(|\Sigma|)}(|\Sigma| \ln |\Sigma| - (|\Sigma| - 1) \ln(|\Sigma| - 1)) \ + \ nP(n)$$

$$\begin{aligned} E[transitions] \quad = \quad & \frac{n}{\ln(|\Sigma|)} \left\{ \frac{|\Sigma|^2 - |\Sigma| + 1}{|\Sigma|} \ln \left(\frac{|\Sigma|^2 - |\Sigma| + 1}{|\Sigma|} \right) \right. \\ & \left. -(|\Sigma| - 1) \ln(|\Sigma| - 1) \right\} \ + \ n(1 + P(n)) \end{aligned}$$

where $P(n)$ is an oscillating function with an exponentially increasing period, small amplitude, and averages to zero.

Building DAWGs for a fixed finite alphabet Σ requires

$$S_n = O(n)$$

$$T_n = O(n)$$

To search a substring in the DAWG we simply run the string through the DFA as in the search function of Section 7.1.6. If the DAWG is implemented as DFAs like in Section 7.1.6, the running time is

$$A_n = m$$

transitions for a string of length m.

Figure 7.2 shows the DAWG for the string *sciences*.

A similar DFA can be defined for all possible subsequences in a text: the Directed Acyclic Subsequence Graph (DASG). The DASG has at most $O(n \log_2 n)$ states and transitions.

References:
[Blumer, A. *et al.*, 85], [Crochemore, M., 85], [Blumer, A. *et al.*, 87], [Baeza-Yates, R.A., to app.].

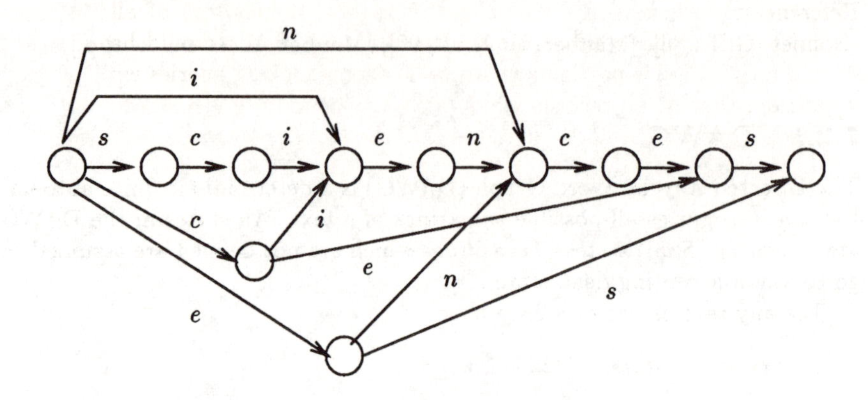

Figure 7.2: DAWG for *sciences*.

7.2.6 Hashing methods for text searching

The main idea of hashing methods (signature encoding) is to build a signature from the set of all words associated with each record (usually a document) in the text. A **signature file** is a file with all the signatures of the records. The signature of a word is (usually) a fixed-length bit sequence, that has a prespecified number of bits set to 1.

A signature file is basically a probabilistic membership tester. Using the signature file we can know if a word is not in the text. A positive answer does not necessarily mean that the word is in the record. The probability of error can be made arbitrarily small by adequately choosing the length of the signature, the number of bits set to 1, etc.

Independently of the signature method used, a search works as follows. We scan the signature file sequentially searching for the signature value of the given query. Qualifying records are either checked (to verify that they are part of the answer) or they are accepted as is (in this case there may be a small number of incorrect answers, or **false drops**). The size of signature files can be set to be around 10% to 20% of the text size. Although the search is linear, it is much faster than the algorithms presented in Section 7.1 for most queries. A_n, T_n and S_n are all $O(n)$.

The simplest signature record is to hash every word in a document to a fixed-length signature, and then to concatenate all the signatures. To improve space and retrieve performance stop words are usually ignored. Let B be the size of each signature. Then

$$S_n = \frac{n\ B}{\text{average word size}} \text{ bits .}$$

A different signature technique is based on superimposed coding. The signature for the record is the superimposition (logical or) of all the word signatures. For this method the signatures of the words should have fewer 1 bits. This method is particularly attractive for searching queries with an 'and' condition, that is, all records which have two or more given words. An 'and' search is done by searching the 'or' of all the word signatures of the query.

In this method we divide each document into sets of words of size W (logical blocks), and we hash every distinct word from each block in bit patterns of length B. The signature of a block is obtained by superimposing those bit patterns. Finally, the document signature is the concatenation of all block signatures. In this case, the optimal number of bits set to 1 (that is, to minimize false drops) is

$$\frac{B \ln 2}{W}$$

for single word queries. We have

$$S_n = \frac{Bn}{W \times \text{average word size}} \text{ bits} .$$

These techniques can be extended to handle subword searches, and other boolean operations. Other variations include compression techniques.

References:
[Harrison, M.C., 71], [Bookstein, A., 73], [Knuth, D.E., 73], [Rivest, R.L., 74], [Rivest, R.L., 76], [Burkhard, W.A., 79], [Cowan, R. *et al.*, 79], [Comer, D. *et al.*, 82], [Tharp, A.L. *et al.*, 82], [Larson, P., 83], [Ramamohanarao, K. *et al.*, 83], [Sacks-Davis, R. *et al.*, 83], [Salton, G. *et al.*, 83], [Faloutsos, C. *et al.*, 84], [Faloutsos, C. *et al.*, 87], [Karp, R.M. *et al.*, 87], [Sacks-Davis, R. *et al.*, 87], [Faloutsos, C., 88].

7.2.7 P-strings

Text is sometimes used to describe highly structured information, such as, dictionaries, scientific papers, and books. Searching such a text requires not only string searching, but also consideration of the structure of the text. Large structured texts are often called **text-dominated databases**. A text-dominated database is best described by a schema expressed as a grammar.

Just as numeric data is structured in a business database, string data must be structured in a text-dominated database. Rather than taking the form of tables, hierarchies, or networks, grammar-based data takes the form of **parsed strings**, or **p-strings**.

A p-string is the main data structure of a text-dominated database and it is formed from a text string and its parse tree (or derivation tree, see [Hopcroft *et al.* 79, pages 82–87]). Notice that we do not require to have a parseable string (with the schema grammar) but instead we keep both the string and its parsing tree together.

Since p-strings represent database instances, they are subject to alteration via operations in a data manipulation language. It follows that as a result of data manipulation, the text in a p-string may not be unambiguously parseable by the associated grammar; thus it is necessary to implement p-strings containing both the text and the parse tree.

A p-string is an abstract data type with three logical components: text, an implicit grammar, and parse structure.

Example grammar:

```
author   :=  surname ',' ( ' ' initial | ' '  name) + ;
surname  :=  char + ;
initial  :=  char '.' ;
name     :=  char + ;
```

For the string Doe, John E. we have the p-string shown in Figure 7.3.

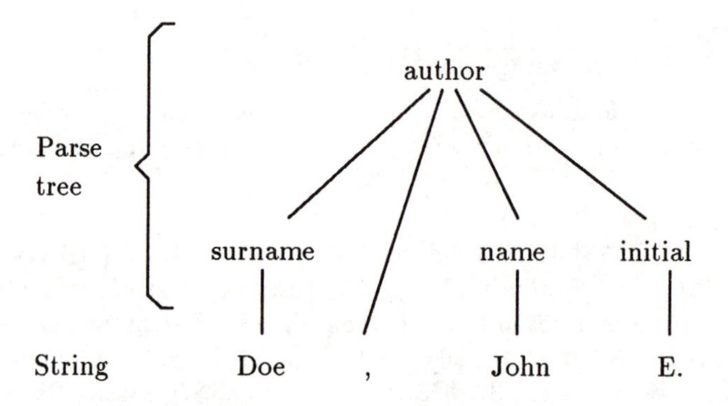

Figure 7.3: P-string example.

Data conversion between strings and p-strings is fundamental to text processing. The operator *string* returns the complete text of the p-string passed as its argument. Conversely the operator *parsed by* takes a string and a non-terminal symbol and creates an instance associated with the string and having a parse tree constructed according to the schema and rooted by the nonterminal. Thus, for example *string*('Jones' *parsed by* `surname`) yields 'Jones'.

Other operators allow us to manipulate, construct and split p-strings as required.

The operator *in* takes a non-terminal symbol and a p-string and returns the p-string whose root is the non-terminal that is first encountered when the argument parse tree is traversed by a pre-order search. For example, `surname` *in* `E` (or equivalently `surname` *in* `author` *in* `E`) thus returns the first p-string with root labelled `surname` in `E`.

The *every ... in* operator retrieves a vector of p-strings representing each subtree whose root is the non-terminal, in order of encounter in a pre-order traversal.

The above operators allow structured search within the text database. String searching algorithms can be composed with the above. For example,

search('Doe' , *string*(*every* **surname** *in* **author** *in* **E**));

References:
[Gonnet, G.H. *et al.*, 87], [Smith, J. *et al.*, 87].

7.3 Other text searching problems

Most of the problems in this section are on general sequences of symbols (for example, genetic sequences) or extensions of text searching to other dimensions (for example, 2-dimensional text searching). The problems with genetic sequences are heavily biased towards approximate matching, while the interest in 2-dimensional searching comes from picture searching where every pixel (or small groups of pixels) can be considered a character.

General references:
[Maier, D., 78], [Tzoreff, T. *et al.*, 88], [Myers, E. *et al.*, 89], [Amir, A. *et al.*, 90], [Manber, U. *et al.*, to app.].

7.3.1 Searching longest common subsequences

A subsequence of a string s is any sequence of characters from s obtained by deleting 0 or more characters from s. The longest common subsequence (LCS) of two strings s_1 and s_2 is the longest string that is both a subsequence of s_1 and s_2. This problem can be solved by using dynamic programming (as in Section 7.1.8). The matching of two strings by their longest common subsequence is a subset of the **alignment problem** of the genetic/biochemical community.

Consider two strings of the same length n. Let r be the number of matching points (that is, all pairs (i,j) such that $s_1[i] = s_2[j]$), and ℓ the length of the longest common subsequence. For every matching point (i,j), we say that its rank is k if the LCS of $s_1[1..i]$ and $s_2[1..j]$ has length k. The matching point (i,j) is k-dominant if it has rank k and for any other matching point (i',j') with the same rank either $i' > i$ and $j' \leq j$ or $i' \leq i$ and $j' > j$. Let d be the total number of dominant points (all possible ranks). We have

$$0 \leq \ell \leq d \leq r \leq n^2$$

and $\ell \leq n$.

To compute the LCS of two strings it is enough to determine all dominant points. Table 7.5 shows the time and space complexities of several algorithms that find the length of the LCS (in general, more time and space is needed to find one LCS).

Table 7.5: Complexity of algorithms for finding the length of a LCS.

Worst-case time	*Space*	*References*
n^2	n^2 or n	[Hirschberg, 75]
$n\ell + n \log n$	$n\ell$	[Hirschberg, 77]
$(n+1-\ell)\ell \log n$	$(n+1-\ell)^2 + n$	[Hirschberg, 77]
$(r+n) \log n$	$(r+n)$	[Hunt *et al.*, 77]
$(n-\ell)n$	n^2	[Nakatsu *et al.*, 82]]
$(r+n) \log n$	$(r+n)$	[Mukhopadhay, 80]
$\ell n + d(1 + \log(\ell n/d))$	d	[Hsu *et al.*, 84], [Apostolico, 87]
$(n-\ell)n$	$(n-\ell)n$ or n	[Myers, 86]
$n \log n + (n-\ell)^2$	n	[Myers, 86]
$n \log n + d \log(n^2/d)$	$d+n$	[Apostolico, 86], [Apostolico *et al.*, 87]
$n(n-\ell)$	n	[Kumar *et al.*, 87]

The dynamic programming algorithm can be extended to find the longest common subsequence of a set of strings, also called the multiple alignment problem. The algorithm for this case has complexity $O(n^L)$ for L strings of length n.

A related problem is to find the shortest common supersequence (SCS) of a set of strings. That is, the shortest string such that every string in the set is a subsequence of it.

References:
[Hirschberg, D.S., 75], [Aho, A.V. *et al.*, 76], [Hirschberg, D.S., 77], [Hunt, J. *et al.*, 77], [Hirschberg, D.S., 78], [Maier, D., 78], [Dromey, R.G., 79], [Mukhopadhay, A., 80], [Nakatsu, N. *et al.*, 82], [Hsu, W.J. *et al.*, 84], [Hsu, W.J. *et al.*, 84], [Apostolico, A., 86], [Crochemore, M., 86], [Myers, E., 86], [Apostolico, A. *et al.*, 87], [Apostolico, A., 87], [Kumar, S.K. *et al.*, 87], [Cormen, T.H. *et al.*, 90], [Eppstein, D. *et al.*, 90], [Baeza-Yates, R.A., to app.], [Myers, E., to app.].

7.3.2 Two-dimensional searching

The problem consists in finding a 2-dimensional pattern in a 2-dimensional text. **Two-dimensional** text will be defined as a rectangle $n_1 \times n_2$ consisting in n_1 lines, each one n_2 characters long. For example, finding a small bit pattern in a bit-mapped screen. To simplify the formulas we use $n_1 = n_2 = n$.

Note that now the size of the text is n^2 instead of n. For this problem, the brute force algorithm may require $O(n^2m^2)$ time, to search for a pattern of size $m \times m$ in a text of size $n \times n$.

Table 7.6 shows the time and space required by 2-dimensional pattern matching algorithms. Some of these algorithms can be extended to allow scaling of the pattern or approximate matching. However, there are no efficient algorithms that allow arbitrary rotations of the pattern.

Table 7.6: Comparison of two-dimensional pattern matching algorithms.

Worst-case A_n	$E[A_n]$	*Extra space*	*References*
m^2n^2	$\vert\Sigma\vert n^2/(\vert\Sigma\vert - 1)$	1	Brute force
$n^2 + m^2$	$n^2 + m^2$	$n + m^2$	[Bird, 77], [Baker, 78]
$K(n^2 + m^2)$	$K(n^2 + m^2)$ $(K >> 1)$	m^2	[Karp *et al.*, 87]
m^2	$\min(m^2, \log n)$	n^2	[Gonnet, 88]
$n^2 + m^2$	$n^2 + m^2$	n^2	[Zhu *et al.*, 89]
$n^2 + m^2$	$n^2 \log m/m + m^2$	n^2	[Zhu *et al.*, 89]
mn^2	$n^2/m + m^2$	m^2	[Baeza-Yates *et al.*, 90]
$n^2 + m^3 + \vert\Sigma\vert$	$\alpha(m)n^2/m + m^3 + \vert\Sigma\vert$ $(\alpha(m) < 1)$	$m^2 + \vert\Sigma\vert$	[Baeza-Yates *et al.*, 90]

References:
[Bird, R., 77], [Baker, T., 78], [Davis, L.S. *et al.*, 80], [Karp, R.M. *et al.*, 87], [Krithivasan, K. *et al.*, 87], [Gonnet, G.H., 88], [Zhu, R.F. *et al.*, 89], [Baeza-Yates, R.A. *et al.*, 90].

Linear time algorithms The algorithms by Bird and Baker require

$$n^2 \leq A_n \leq 4n^2 .$$

These algorithms decompose the pattern in a set of unique row-pattern strings, and search them in every row of the pattern using the pattern matching machine (see Section 7.1.4). The output of this machine is the index of the string (if any) which was matched. This index is used to search by column for the sequence of strings that compose the pattern. The vertical search is done with the Knuth–Morris–Pratt algorithm (see Section 7.1.2). For example, if the pattern is composed of the row-strings $(p_1, p_2, p_2, p_3, p_1)$, we search in every column for an output sequence $R = (1, 2, 2, 3, 1)$ (see Figure 7.4). By performing the multiple string searching left to right, top to bottom, and the n KMP searches in parallel, top to bottom, only $O(n)$ extra space (for the KMP states) is needed.

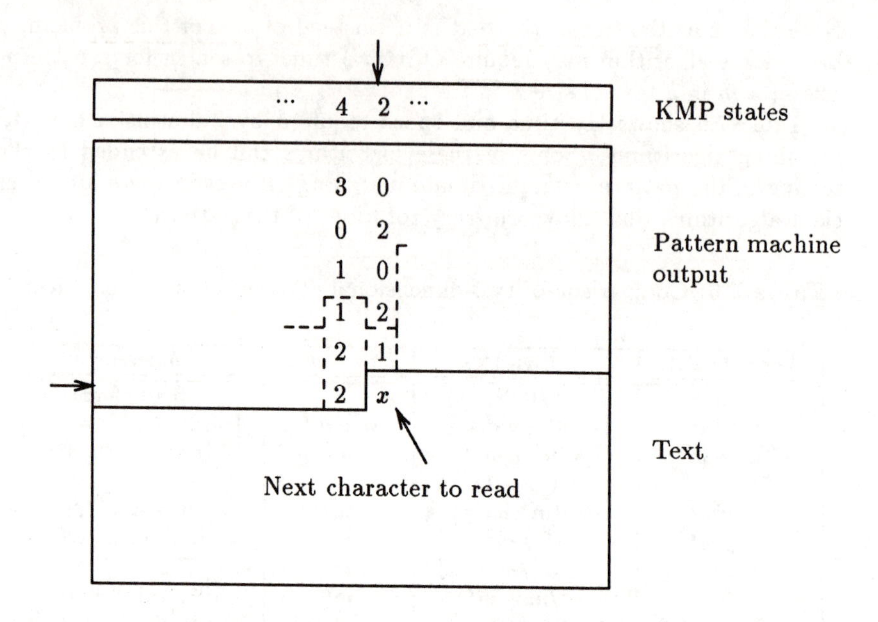

Figure 7.4: Linear time 2-dimensional searching.

Fast algorithm on average An algorithm using

$$\frac{n^2}{m^2} \leq A_n = O(n^2 + m^3)$$

is obtained by searching the patterns only in rows $m, 2m, ..., \lfloor \frac{n}{m} \rfloor m$ of the text using any multiple-string searching algorithm (see Section 7.1.4). If a row-string is found, the algorithm checks above/below that row for the rest of the pattern (see Figure 7.5). On average we have

$$E[A_n] = f(m) \frac{n^2}{m}$$

with $f(m) < 1$.

This algorithm can be improved to avoid repeating comparisons in the checking phase if we have overlapped occurrences. It can also be extended to non-rectangular pattern shapes, or higher dimensions.

Algorithm with preprocessing of the text In this section we will describe how PAT trees can be used to search in two dimensions, in particular search for subpictures ($m \times m$ text squares) inside a bigger picture (an $n_1 \times n_2$ text rectangle), or among many bigger pictures.

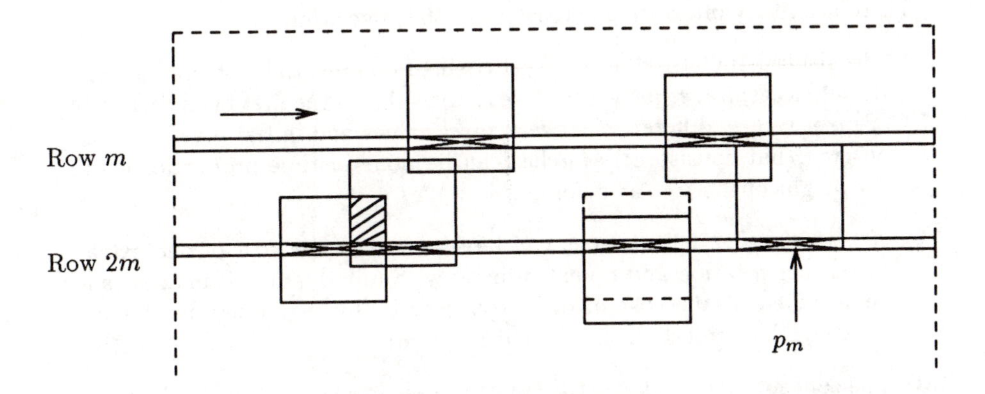

Figure 7.5: Faster 2-dimensional pattern matching.

Let a collection of disjoint pictures be an **album**. The size of an album is n, the total number of pixels of all its pictures. We will solve the problem of exact matching of a given subpicture into any of the pictures of an album in $O(\log n)$ time. To do this we will preprocess the album using at most $O(n)$ storage and $O(n \log n)$ time.

The crux of the algorithm is in devising the equivalent of semi-infinite strings for each of the pixels. The sistrings had the right context (linear to the right) for text, but for pictures, the context is two dimensional. Hence the equivalent of the sistring for a pixel is a **semi-infinite spiral** centred at the pixel.

The comparing sequence for a semi-infinite spiral, **sispiral** is:

		...	17
7	6	5	16
8	1	4	15
9	2	3	14
10	11	12	13

where the integers indicate the ordinal position of the comparison for the pixel marked as 1 (the sispiral comparing sequence).

The main data structure for subpicture searching is a PAT tree (see Section 7.2.2 for the complexity measures) built on sispirals for each pixel. As with sistrings, every time that we step outside the picture we should use a 'null' character which is not used inside any of the pictures.

To search a square in the album, we just locate its centre, that is, a pixel that will develop a spiral which covers the square, and search the sispiral starting at this pixel in the PAT tree. The searching time is independent of the number of matches found.

There are many interesting extensions of this algorithm:

(1) The sispiral PAT tree can be relativized to its grey scale, by computing the whole sispiral sequence relative to the value of the first pixel (instead of pixel values, difference between pixel values and initial pixel). Then off-grey (but consistent) searches can be done in time proportional to the height of the tree as before.

(2) 90°, 180° and 270° rotations can be searched at the cost of one extra search per rotation and no extra memory. Similarly, mirror images can be searched at the cost of one extra search (by searching the mirror image of the sispiral on the searched square).

(3) The concept of longest repetition in this case means the largest identical square that repeats in the album.

APPENDIX I

Distributions Derived from Empirical Observation

In this appendix we will describe some probability distributions arising from empirical situations. The distributions described here may be used with other well-known distributions to test algorithms under various conditions. Some of these distributions are related directly to data processing.

I.1 Zipf's law

Zipf observed that the frequency of word usage (in written English) follows a simple pattern. When word frequencies are listed in decreasing order, we have the relation

$$f_1 \approx i\, f_i$$

where f_i denotes the frequency of the ith most frequent word. Zipf observed that the population of cities in the USA also follows this relation closely. From this observation we can easily define a Zipfian probability distribution as

$$p_i = \frac{1}{iH_n} \quad 1 \leq i \leq n$$

The first moments and variance of this distribution are

$$\mu_1' = \frac{n}{H_n}$$

$$\mu_2' = \frac{n(n+1)}{2H_n}$$

$$\sigma^2 = \frac{n}{H_n}\left(\frac{n+1}{2} - \frac{n}{H_n}\right)$$

This distribution can be generalized in the two following ways.

I.1.1 First generalization of a Zipfian distribution

In this case the probabilities are defined by

$$p_i = \frac{1}{a(i+b)} \qquad (1 \le i \le n,\ b > -1)$$

here $a = \psi(n+b+1) - \psi(b+1)$. The first moments and variance are

$$\mu_1' = \frac{n}{a} - b$$

$$\mu_2' = \frac{n(n+1) - 2nb + 2ab^2}{2a}$$

$$\sigma^2 = \frac{n}{2a}(n+1+2b-2n/a)$$

Choosing b to be an integer allows us to represent truncated Zipfian distributions. Giving b a small non-integer value may provide a better fit for the first few frequencies.

I.1.2 Second generalization of a Zipfian distribution

This generalization introduces a parameter θ so that we may define

$$p_i = \frac{1}{i^\theta H_n^{(\theta)}} \qquad 0 \le \theta < 1$$

Zipf found that some word frequencies matched this distribution closely for values of θ other than 1. In this case the first moments and variance are

$$\mu_1' = \frac{H_n^{(\theta-1)}}{H_n^{(\theta)}} = \frac{n(1-\theta)}{2-\theta} + O(n^\theta)$$

$$\mu_2' = \frac{H_n^{(\theta-2)}}{H_n^{(\theta)}} = \frac{n^2(1-\theta)}{3-\theta} + O(n^{1+\theta})$$

$$\sigma^2 = \frac{H_n^{(\theta-2)}H_n^{(\theta)} - (H_n^{(\theta-1)})^2}{(H_n^{(\theta)})^2} = \frac{n^2(1-\theta)}{(3-\theta)(2-\theta)^2} + O(n^{1+\theta})$$

References:
[Zipf, G.K., 49], [Johnson, N.L. *et al.*, 69], [Knuth, D.E., 73].

I.2 Bradford's law

Bradford's law was first observed in experiments dealing with the number of references made to a selection of books in search of information. This principle can be described in the following way. Assume that we have a collection of n books which treat a given topic, and that these books are placed on a shelf in decreasing order according to the number of times each book is referenced. Thus the most referenced book is first and so on. We then divide these books into k contiguous groups such that each group receives the same number of references. Bradford's law now states that the number of books in each successive division follows the ratio $1 : m : m^2 : ... : m^{k-1}$ for some constant m.

To translate this description into mathematical terms, we let r_i be the expected value of the number of references to the ith most referenced book on our shelf. Thus we have $r_1 \geq r_2 \geq ... \geq r_n$. Let $R(j)$ be the partial sum of the expected values of these references:

$$R(j) = \sum_{i=1}^{j} r_i$$

and so

$$R(n) = T$$

where T is the total expected number of references. To divide the n books into k divisions satisfying the given ratio, the number of books in each division must be $\frac{n(m-1)}{m^k-1}, \frac{nm(m-1)}{m^k-1}, ... , \frac{nm^{k-1}(m-1)}{m^k-1}$. Since each division receives the same number of references, this number must be T/k. Consequently the total expected number of references to the first division will be

$$\sum_{i=1}^{\frac{n(m-1)}{m^k-1}} r_i = R\left(\frac{n(m-1)}{m^k-1}\right) = \frac{T}{k}$$

In general, for the first j divisions we have the equation

$$R\left(\frac{(m^j-1)n}{m^k-1}\right) = \frac{jT}{k} \tag{I.1}$$

Now the quantities k and m are related to one another, since for any valid k, Bradford's law predicts the existence of a unique m. Examination of $R(x)$ for different values of k and m shows that in order for the law to be consistent, the quantity $m^k - 1 = b$ must be constant. This constant b defines the shape of the distribution. From equation I.1 we can solve for $R(x)$ and obtain

$$R(x) = \frac{T}{k} \log_m \left(\frac{bx}{n} + 1\right)$$

Let p_i be the probability that a random reference refers to the ith book. From the above discussion we have

$$p_i = \frac{R(i) - R(i-1)}{T} = \frac{1}{k} \log_m \left(\frac{bi + n}{b(i-1) + n}\right)$$

Since $m^k - 1 = b$, we have $k \ln m = \ln(b+1)$; this allows us to simplify the given probability to

$$p_i = \log_{b+1} \left(\frac{bi + n}{b(i-1) + n}\right)$$

The first moment of the above distribution is

$$\mu_1' = \sum_{i=1}^{n} i p_i = n \log_{b+1} \left(\frac{n(b+1)}{b}\right) - \log_{b+1} \left(\frac{\Gamma(n(b+1)/b)}{\Gamma(n/b)}\right) \quad \text{(I.2)}$$

$$= n \left(\frac{1}{\ln(b+1)} - \frac{1}{b}\right) + \frac{1}{2} + \frac{b^2}{12n(b+1)\ln(b+1)} + O(n^{-3})$$

The second moment is given by

$$\mu_2' = \frac{n^2}{b^2} - \frac{n}{b} + \frac{1}{3} + \frac{1}{\ln(b+1)} \left(\frac{n^2(b-2)}{2b} + n - \frac{b}{6(b+1)} + \frac{b^2}{12(b+1)n}\right) + O(n^{-2}) \quad \text{(I.3)}$$

The variance is

$$\sigma^2 = \frac{n^2}{\ln(b+1)} \left(\frac{b+2}{2b} - \frac{1}{\ln(b+1)}\right) + O(1) \quad \text{(I.4)}$$

This distribution behaves very much like the generalized harmonic (or the first generalization of Zipf's distribution). When the parameter $b \to 0$ Bradford's distribution coincides with the discrete rectangular distribution.

Although the process of accessing information from books is rarely automated, there is a significant number of automatic processes in which the accessing of information is similar to the situation of referencing books. In these cases Bradford's law may provide a good model of the access probabilities.

References:

[Pope, A., 75].

I.3 Lotka's law

Lotka observed that the number of papers in a given journal written by the same author closely followed an inverse square distribution. In other words, if we were to choose an author at random from the list of contributors to the journal, the probability that he or she had contributed exactly i papers would be proportional to i^{-2}. Later it was observed that for some journals an inverse cube law fit the data more precisely. We will generalize these two laws in the following way. Let n be the total number of authors who published at least one paper in a given journal. The probability that a randomly chosen author contributed exactly i papers will be given by

$$p_i = \frac{1}{\zeta(\theta) i^{\theta}}$$

The first moment of this distribution corresponds to the average number of papers published by each author; it is given by

$$\mu_1' = \sum_{i=1}^{\infty} i p_i = \frac{\zeta(\theta - 1)}{\zeta(\theta)}$$

We immediately conclude that this law will be consistent only for $\theta > 2$, as has been noted by several other authors; otherwise this first moment will be unbounded, a situation which does not correspond with reality. Note that $n\mu_1'$ denotes the expected number of papers published in a journal which has n contributors.

For $\theta \leq 3$, the variance of the distribution under discussion diverges. For $\theta > 3$, the variance is given by

$$\sigma^2 = \frac{\zeta(\theta - 2)}{\zeta(\theta)} - \left(\frac{\zeta(\theta - 1)}{\zeta(\theta)}\right)^2$$

The median number of papers by the most prolix author can be approximated by

$$median \approx \left(\frac{n}{\ln(2)\zeta(\theta)(\theta - 1)}\right)^{1/(\theta-1)}$$

References:
[Lotka, A.J., 26], [Murphy, L.J., 73], [Radhakrishnan, T. *et al.*, 79].

I.4 80%–20% rule

The 80%–20% rule was proposed as a probabilistic model to explain certain data-processing phenomena. In computing folklore it is usually given as: 80%

of the transactions are on the most active 20% of the records, and so on recursively. Mathematically, let $p_1 \geq p_2 \geq p_3 \geq \ldots \geq p_n$ be the independent probabilities of performing a transaction on each of the n records. Let $R(j)$ be the cumulative distribution of the p_i's, that is,

$$\sum_{i=1}^{j} p_i = R(j) \qquad R(n) = 1$$

The 80%–20% rule is expressed in terms of the function $R(j)$ by

$$R(n \times 20\%) = 80\%$$

This rule may be applied recursively by requiring that the relation hold for any contiguous subset of p_is that includes p_1. This requirement yields the necessary condition:

$$R(0.2j) = 0.8R(j)$$

More generally we may consider an $\alpha\% - (1-\alpha)\%$ rule given by

$$R((1-\alpha)j) = \alpha R(j), \qquad \frac{1}{2} \leq \alpha \leq 1 \tag{I.5}$$

The above functional equation defines infinitely many probability distributions for each choice of α. One simple solution that is valid for all real j is

$$R(i) = \frac{i^\theta}{n^\theta}$$

where $\theta = \frac{\ln(\alpha)}{\ln(1-\alpha)}$. Thus $0 \leq \theta \leq 1$. This formula for $R(i)$ implies

$$p_i = \frac{i^\theta - (i-1)^\theta}{n^\theta} \tag{I.6}$$

Note that this probability distribution also possesses the required monotone behaviour, that is, $p_i \geq p_{i+1}$.

The parameter θ gives shape to the distribution. When $\theta = 1$ ($\alpha = \frac{1}{2}$) the distribution coincides with the discrete rectangular distribution. The moments and variance of the distribution described by equation I.6 are

$$\mu_1' = \sum_{i=1}^{n} i p_i = \frac{\theta n}{\theta+1} + \frac{1}{2} - \frac{\zeta(-\theta)}{n^\theta} - \frac{\theta}{12n} + O(n^{-3})$$

$$\mu_2' = \sum_{i=1}^{n} i^2 p_i = \frac{\theta n^2}{\theta+2} + \frac{\theta n}{\theta+1} + \frac{2-\theta}{6} - \frac{2\zeta(-\theta-1)+\zeta(-\theta)}{n^\theta} + O(n^{-1})$$

$$\mu_3' = \sum_{i=1}^{n} i^3 p_i = \frac{\theta n^3}{\theta+3} + \frac{3\theta n^2}{2(\theta+2)} + \frac{\theta(3-\theta)n}{4(\theta+1)} + O(1)$$

$$\mu_k' = \sum_{i=1}^{n} i^k p_i = \frac{\theta n^k}{\theta+k} + \frac{\theta k n^{k-1}}{2(\theta+k-1)} + \frac{\theta(k-\theta)kn^{k-2}}{12(\theta+k-2)} + O(n^{k-3}) + O(n^{-\theta})$$

$$\sigma^2 = \frac{\theta n^2}{(\theta+1)^2(\theta+2)} + O(n^{1-\theta})$$

For large n, the tail of the distribution coincides asymptotically with $p_i \approx i^{\theta-1}$. For the 80%–20% rule, $\theta = 0.138646...$; consequently the distribution which arises from this rule behaves very similarly to the second generalization of Zipf's distribution.

References:
[Heising, W.P., 63], [Knuth, D.E., 73].

APPENDIX II

Asymptotic Expansions

This appendix contains a collection of asymptotic expansions of functions or expressions commonly used in the analysis of algorithms. The criterion used for the length of the expansion, that is order, is rather artificial and depends upon computability and number of terms in the numerator, and is at most 7.

It is assumed that the expansions are for $n \to \infty$ unless otherwise specified. It is also assumed that a, b, c and z are all $O(1)$ when $n \to \infty$.

In the following, $\zeta(z)$ is the classical Riemann zeta function, defined by

$$\zeta(z) = \sum_{n=1}^{\infty} n^{-z}$$

$\Gamma(z)$ denotes the gamma function, defined by

$$\begin{aligned}\Gamma(z+1) &= \int_0^{\infty} t^z e^{-t} dt \\ &= z\Gamma(z)\end{aligned}$$

$\psi(z)$ denotes the psi function, defined by

$$\psi(z+1) = \frac{\Gamma'(z+1)}{\Gamma(z+1)} = \psi(z) + \frac{1}{z}$$

and γ will denote Euler's constant,

$$\gamma = \lim_{n\to\infty} H_n - \ln(n) = 0.5772156649...$$

II.1 Asymptotic expansions of sums

$$\sum_{k=1}^{n-1}[k(n-k)]^{-1/2} = \pi + n^{-1/2}\left(2\zeta(1/2) + \frac{\zeta(-1/2)}{n} + \frac{3\zeta(-3/2)}{4n^2}\right. \tag{II.1}$$

$$\left. + \frac{5\zeta(-5/2)}{8n^3} + \frac{35\zeta(-7/2)}{64n^4} + \frac{63\zeta(-9/2)}{128n^5} + \frac{231\zeta(-11/2)}{512n^6} + \cdots\right)$$

$$\sum_{k=1}^{n-1}[k(n-k)]^{1/2} = \frac{n^2\pi}{8} + \sqrt{n}\left(2\zeta(-1/2) - \frac{\zeta(-3/2)}{n} - \frac{\zeta(-5/2)}{4n^2}\right. \tag{II.2}$$

$$\left. - \frac{\zeta(-7/2)}{8n^3} - \frac{5\zeta(-9/2)}{64n^4} - \frac{7\zeta(-11/2)}{128n^5} - \frac{21\zeta(-13/2)}{512n^6} - \cdots\right)$$

$$\sum_{k=1}^{n-1}[k(n-k)]^{-s} = (n/2)^{1-2s}\sqrt{\pi}\frac{\Gamma(1-s)}{\Gamma(3/2-s)} + 2n^{-s}\left(\zeta(s) + \frac{s\zeta(s-1)}{n}\right. \tag{II.3}$$

$$\left. + \frac{s(s+1)\zeta(s-2)}{2n^2} + \ldots + \frac{\Gamma(s+i)\zeta(s-i)}{\Gamma(s)i!n^i} + \ldots\right) \quad [s \neq 2, 3, 4, \ldots]$$

$$\sum_{k=1}^{n} k^k = n^n\left(1 + \frac{1}{en} + \frac{e+2}{2e^2n^2} + \frac{7e^2+48e+24}{24e^3n^3}\right. \tag{II.4}$$

$$+ \frac{9e^3+160e^2+216e+48}{48e^4n^4} + \frac{743e^4+30720e^3+84240e^2+46080e+5760}{5760e^5n^5}$$

$$\left. + \frac{1075e^5+97792e^4+486000e^3+491520e^2+144000e+11520}{11520e^6n^6} + \cdots\right)$$

$$\sum_{k=1}^{n} k^n = n^n\left(\frac{e}{e-1} - \frac{e(e+1)}{2(e-1)^3n} + \frac{e(e+5)(11e^2+2e-1)}{24(e-1)^5n^2} + \ldots\right) \tag{II.5}$$

$$\sum_{k=1}^{n} k^{-s} = \zeta(s) + n^{-s}\left(\frac{n}{(1-s)} + \frac{1}{2} - \frac{s}{12n} + \frac{\Gamma(s+3)}{720\Gamma(s)n^3}\right. \tag{II.6}$$

$$\left. - \frac{\Gamma(s+5)}{30240\Gamma(s)n^5} - \frac{\Gamma(s+7)}{1209600\Gamma(s)n^7}\cdots\right) \quad [s \neq 1]$$

$$\sum_{k=1}^{n} \frac{z^k}{k} = -\ln(1-z) + \frac{(z-1)^{-1} z^{n+1}}{(n+1)} + \ldots + \frac{z^{n+i}(i-1)!n!}{(z-1)^i (n+i)!} + \ldots \qquad \text{(II.7)}$$

$$= -\ln(1-z) + \frac{z^{n+1}}{(z-1)n}\left(1 + \frac{1}{(z-1)n} + \frac{z+1}{(z-1)^2 n^2} + \frac{z^2+4z+1}{(z-1)^3 n^3} + \frac{(z+1)(z^2+10z+1)}{(z-1)^4 n^4} + \frac{z^4+26z^3+66z^2+26z+1}{(z-1)^5 n^5} + \ldots\right) \qquad [0 \le z < 1]$$

$$\sum_{k=1}^{n} \frac{2^k}{k} = \frac{2^n}{n} \sum_{i=0}^{n} \frac{1}{\binom{n}{i}}$$

$$\sum_{k=1}^{n} \frac{(1-z/n)^k}{k} = \ln(n/z) - E_1(z) \qquad \text{(II.8)}$$

$$+e^{-z}\left(\frac{z+1}{2n} - \frac{3z^3+z^2+2z+2}{24n^2} + \frac{z-3}{48n^3} z^4 - \frac{15z^7 - 135z^6 + 230z^5 - 2z^4 - 8z^3 - 24z^2 - 48z - 48}{5760n^4} + \cdots\right) \qquad [z > 0]$$

where $E_1(z) = \int_z^\infty \frac{e^{-t}}{t} dt$ is the exponential integral.

$$\sum_{k=1}^{\infty} z^{k^2+ak} = \frac{z^{-a^2/4}}{2}\sqrt{-\frac{\pi}{\ln z}} - \frac{a+1}{2} + a(a^2-1)\left(\frac{\ln z}{12}\right. \qquad \text{(II.9)}$$

$$-\frac{(a^2+1)\ln^2 z}{120} + \frac{(3a^4+3a^2+10)\ln^3 z}{5040} - \frac{(a^2+3)(a^4-2a^2+7)\ln^4 z}{30240}$$

$$\left. + \frac{(a^8+a^6+a^4-21a^2+210)\ln^5 z}{665280} + \cdots\right) + o(\ln^k z) \quad \text{(for any } k\text{)}$$

$$\sum_{k=1}^{\infty} \frac{1}{z^k - 1} = -\log_z(z-1) + \frac{\gamma}{\ln z} + \frac{3}{4} + \frac{5\ln z}{144} - \frac{31\ln^3 z}{86400} + O(\ln^5 z) \quad \text{(II.10)}$$

$$\sum_{k=1}^{\infty} \frac{z^k}{a+k} = z^{-a}\left(\psi(1+a) - \gamma - \ln(1-z) - a(z-1)\right. \qquad \text{(II.11)}$$

$$- \frac{a(a-1)(z-1)^2}{4} - \ldots - \left. \frac{a^{\underline{i}}(z-1)^i}{i\, i!} \right) \qquad [0 < z < 1]$$

where $a^{\underline{i}} = a(a-1)(a-2)\cdots(a-i+1)$ denotes the descending factorial.

$$\sum_{k=1}^{\infty} \frac{z^k}{k^2} = I(z) = \frac{\pi^2}{6} - \ln(1-z)\ln(z) - I(1-z) \tag{II.12}$$

$$= \frac{\pi^2}{6} - \ln(1-z)\ln z - (1-z) - \frac{(1-z)^2}{4} - \frac{(1-z)^3}{9} - \cdots$$

$$\sum_{k=1}^{n} \binom{n}{k} (-1)^k k^a = -(\ln n)^{-a} \sum_{m \ge 0} \frac{(-1)^m \Gamma^{(m)}(1)}{m!\Gamma(1-m-a)} \frac{1}{\ln^m n} \tag{II.13}$$

$$= -\frac{1}{(-a)!} \sum_{m=0}^{-a} \binom{-a}{m} (-1)^m \Gamma^{(m)}(1) \ln^{-a-m} n$$

$$+ O\left(\frac{\ln^{-a} n}{n}\right) \qquad [a = -1, -2, -3, \ldots]$$

II.2 Gamma-type expansions

$$\sum_{k=1}^{n} \frac{1}{k} = \psi(n+1) + \gamma = H_n = \gamma + \ln n + \frac{1}{2n} - \frac{1}{12n^2} \tag{II.14}$$

$$+ \frac{1}{120n^4} - \frac{1}{252n^6} + \frac{1}{240n^8} \cdots$$

$$\sum_{k=1}^{n} \frac{1}{k^2} = H_n^{(2)} = \frac{\pi^2}{6} - \frac{1}{n} + \frac{1}{2n^2} - \frac{1}{6n^3} + \frac{1}{30n^5} - \frac{1}{42n^7} \cdots \tag{II.15}$$

$$\sum_{k=1}^{n} \ln k = \ln \Gamma(n+1) = (n+1/2)\ln n - n + \frac{\ln 2\pi}{2} \tag{II.16}$$

$$+ \frac{1}{12n} - \frac{1}{360n^3} + \frac{1}{1260n^5} - \frac{1}{1680n^7} \cdots$$

$$n! = \Gamma(n+1) = \left(\frac{n}{\mathrm{e}}\right)^n \sqrt{2\pi n}\left(1 + \frac{1}{12n} + \frac{1}{288n^2} - \frac{139}{51840n^3} - \frac{571}{2488320n^4} + \frac{163879}{209018880n^5} + \cdots\right) \tag{II.17}$$

$$= n^n\sqrt{2\pi(n+1/6)}\mathrm{e}^{-n}\left(1 + \frac{1}{144n^2} + O(n^{-3})\right)$$

II.3 Exponential-type expansions

$$(1+z/n)^n = \mathrm{e}^z\left(1 - \frac{z^2}{2n} + \frac{3z+8}{24n^2}z^3 - \frac{(z+2)(z+6)}{48n^3}z^4 + \frac{15z^3+240z^2+1040z+1152}{5760n^4}z^5 - \frac{(z+4)(z^3+68z^2+408z+480)}{11520n^5}z^6 + \frac{63z^5+2520z^4+35280z^3+211456z^2+526176z+414720}{2903040n^6}z^7 - \cdots\right) \tag{II.18}$$

$$(1+1/n)^n = \mathrm{e}\left(1 - \frac{1}{2n} + \frac{11}{24n^2} - \frac{7}{16n^3} + \frac{2447}{5760n^4} - \frac{959}{2304n^5} + \frac{238043}{580608n^6} - \cdots\right) \tag{II.19}$$

$$(1-1/n)^n = e^{-1}\left(1 - \frac{1}{2n} - \frac{5}{24n^2} - \frac{5}{48n^3} - \frac{337}{5760n^4} - \frac{137}{3840n^5} - \frac{67177}{2903040n^6} - \cdots\right) \tag{II.20}$$

$$\left(1+\frac{b}{n^2}\right)^n = 1 + \frac{b}{n} + \frac{b^2}{2n^2} + \frac{b-3}{6n^3}b^2 + \frac{b-12}{24n^4}b^3 + \frac{b^2-30b+40}{120n^5}b^3 + \frac{b^2-60b+330}{720n^6}b^4 + \cdots \tag{II.21}$$

$$\left(1+\frac{c}{n^3}\right)^n = 1 + \frac{c}{n^2} + \frac{c^2}{2n^4} - \frac{c^2}{2n^5} + \frac{c^3}{6n^6} - \frac{c^3}{2n^7}\cdots \tag{II.22}$$

II.4 Asymptotic expansions of sums and definite integrals containing e^{-x^2}

$$\int_1^{\infty} e^{-x^2/n}dx = \frac{\sqrt{n\pi}}{2} - 1 + \frac{1}{3n} - \frac{1}{10n^2} + \frac{1}{42n^3} - ... + \frac{(-1)^{i-1}}{i!(2i+1)n^i} \qquad \text{(II.23)}$$

$$\int_1^{\infty} \frac{e^{-x^2/n}}{x}dx = -\frac{\gamma}{2} + \frac{\ln n}{2} + \frac{1}{2n} - \frac{1}{8n^2} + \frac{1}{36n^3} - ... + \frac{(-1)^{i-1}}{i!\,2i\,n^i} \qquad \text{(II.24)}$$

$$\int_1^{\infty} \frac{e^{-x^2/n}}{x^s}dx = \frac{e^{-1/n}}{s-1} - \frac{2}{(s-1)n}\int_1^{\infty}\frac{e^{-x^2/n}}{x^{s-2}}dx \quad (s > 1) \qquad \text{(II.25)}$$

$$\int_0^{\infty} e^{-x^2/n}\ln(1+x)dx = \sqrt{n\pi}\frac{\ln(n/4)-\gamma}{4} - \frac{\gamma}{2} + 1 + \frac{\ln n}{2} + \frac{\sqrt{\pi/n}}{2} - \frac{\ln n + 5/3 - \gamma}{6n} - \frac{\sqrt{\pi/n^3}}{6} \qquad \text{(II.26)}$$

$$\int_0^{\infty} e^{-x^2/n}\ln(1+x)x\,dx = \frac{n}{2}\int_0^{\infty}\frac{e^{-x^2/n}}{1+x}dx \quad \text{[see II.29]} \qquad \text{(II.27)}$$

$$\int_0^{\infty} e^{-x^2/n}dx = \frac{\sqrt{\pi n}}{2} \qquad \text{(II.28)}$$

$$\int_0^{\infty} \frac{e^{-x^2/n}}{1+x}dx = \frac{\ln n - \gamma}{2} + \sqrt{\pi/n} - \frac{\ln n + 1 - \gamma}{2n} - \frac{2\sqrt{\pi/n^3}}{3} + \frac{\ln n + 3/2 - \gamma}{4n^2} + \frac{4\sqrt{\pi/n^5}}{15} - ... \qquad \text{(II.29)}$$

$$\int_0^{\infty} \frac{e^{-x^2/n}}{[1+x]^s}dx = T(s) = \frac{1}{s-1} + \frac{2[T(s-2)-T(s-1)]}{n(1-s)} \quad [s > 1] \qquad \text{(II.30)}$$

$$\sum_{k=1}^{\infty}\frac{e^{-k^t/n}}{k^s} = \frac{n^{\frac{1-s}{t}}\Gamma(\frac{1-s}{t})}{t} + \zeta(s) - \frac{\zeta(s-t)}{n} + \frac{\zeta(s-2t)}{2n^2} - ... \qquad \text{(II.31)}$$

$$[s - mt \neq 1 \text{ for } m = 0, 1, 2, \ldots]$$

$$= \frac{(-n)^{-m}}{tm!}(\ln n + \psi(m+1)) + \zeta(s) - \frac{\zeta(s-t)}{n}$$

$$+\frac{\zeta(s-2t)}{2n^2}-\frac{\zeta(s-3t)}{6n^3}+\dots \quad [s-mt=1,\ (m=0,1,2,\dots) \text{ and } \zeta(s-mt) \text{ interpreted as } \gamma]$$

$$\begin{aligned}\sum_{k=1}^{\infty}\frac{e^{-k^t/n}}{k^s}\ln k &= \frac{n^{\frac{1-s}{t}}\Gamma(\frac{1-s}{t})(\ln(n)+\Psi(\frac{1-s}{t}))}{t^2}-\zeta'(s)+\frac{\zeta'(s-t)}{n} \\ &\quad -\frac{\zeta'(s-2t)}{2n^2}+\dots \qquad [s-mt\neq 1 \text{ for } m=0,1,2,\cdots] \\ &= \frac{(-n)^{-m}\left(\frac{\pi^2}{3}-\Psi'(m+1)+(\ln(n)+\Psi(m+1))^2\right)}{2t^2m!} \\ &\quad -\zeta'(s)+\frac{\zeta'(s-t)}{n}-\frac{\zeta'(s-2t)}{2n^2}+\dots. \\ &\quad [s-mt=1 \text{ and } \zeta'(s-mt) \text{ interpreted as } \gamma_1, \\ &\quad \text{where } \gamma_1=-\lim_{x=1}\zeta'(x)+\frac{1}{(x-1)^2}]\end{aligned} \tag{II.32}$$

II.5 Doubly exponential forms

In the functions below, it is assumed that $P(x)$ is some periodic function with period 1.

$$\begin{aligned}\sum_{k\geq 0}\left(1-\frac{z}{n}\right)^{2^k} &= -\log_2\left(\log_2\frac{n-z}{n}\right)+\frac{1}{2}+\frac{\gamma}{\ln 2} \\ &\quad +P\left(\log_2\left(\log_2\frac{n-z}{n}\right)\right)+\frac{z}{n}+\frac{z^2}{3n^2}+\frac{4z^3}{21n^3}+\frac{41z^4}{315n^4}+\frac{136z^5}{1395n^5}+\dots \\ &\quad |\,P(x)\,|\leq 0.00000316\dots\end{aligned} \tag{II.33}$$

$$\begin{aligned}\sum_{k\geq 0}(1-\frac{z}{n})^{\beta^k} &= -\log_\beta(\log_\beta\frac{n-z}{n})+\frac{1}{2}+\frac{\gamma}{\ln\beta} \\ &\quad +P\left(\log_\beta\left(\log_\beta\frac{n-z}{n}\right)\right)+\frac{z}{(\beta-1)n}+\frac{\beta z^2}{2(\beta^2-1)n^2} \\ &\quad +\frac{\beta(2+\beta+2\beta^2)z^3}{6(\beta+1)(\beta^3-1)n^3}+\frac{\beta(6+\beta+13\beta^2+\beta^3+6\beta^4)z^4}{24(\beta^4-1)(\beta^2+\beta+1)n^4}+\dots\end{aligned} \tag{II.34}$$

$$\sum_{k\geq 0}(1-e^{-ne^{-k}}) = \ln n + \gamma + \frac{1}{2} + P(\ln n) + e^{-n} + e^{-ne} + e^{-ne^2} + \ldots \quad (II.35)$$

$$|P(x)| \leq 0.0001035$$

$$\sum_{k\geq 0}(1-n^{-\beta^k}) = -\frac{\ln(\ln n)+\gamma}{\ln \beta} + P(\log_\beta(\ln n)) + n^{-\beta^{-1}} \quad (II.36)$$

$$+n^{-\beta^{-2}} + \ldots \qquad [\beta < 1]$$

$$\sum_{k\geq 0}(2^k e^{-n2^{-k}} - 2^k + n) = n\log_2 n + 1 + \left(\frac{\gamma-1}{\ln 2} + \frac{1}{2}\right)n \quad (II.37)$$

$$+nP(\log_2 n) - \frac{e^{-2n}}{2} - \frac{e^{-4n}}{4} - \ldots$$

$$|P(x)| \leq 0.000000173$$

$$\sum_{k\geq 0} 1 - \left(1-\frac{1}{a^k}\right)^n = 1 - \sum k \geq 1 \frac{\binom{n}{k}(-1)^k}{a^k - 1} \quad (II.38)$$

$$= \log_a n + \frac{\gamma}{\ln a} + \frac{1}{2} + \frac{k_1}{n} + \frac{k_2}{n^2} + \cdots + P(\log_a n)$$

where

$$k_1 = \frac{1}{2}\sum_{k=-\infty}^{k=\infty}\frac{e^{-a^{-k}}}{a^{2k}} \quad \text{and} \quad k_2 = \frac{a+1}{24a}\sum_{k=-\infty}^{k=\infty}\frac{(8a^k-3)e^{-a^{-k}}}{a^{4k}}$$

II.6 Roots of polynomials

For the following polynomials we find an asymptotic expression of the real root closest to 1. We use the transcendental function $w(x)$ defined by

$$w(x)e^{w(x)} = x$$

It is known that $w(x) = \ln x - \ln(\ln x) + o(1)$ when $x \to \infty$.

$$ax^n + bx^{n-1} + f(n) = 0 \quad (II.39)$$

$$x = 1+\frac{y}{n}+\frac{y}{n^2}\left(\frac{b}{a+b}+\frac{y}{2}\right)+\frac{y}{n^3}\left(\frac{b^2+b(b+a/2)y}{(a+b)^2}+\frac{y^2}{6}\right)+O(y^4n^{-4})$$

$$\text{where} \quad y = \ln\left(-\frac{f(n)}{a+b}\right)$$

$$x^n - x^{n-1} + f(n) = 0 \qquad \text{(II.40)}$$

$$x = 1+\frac{y}{n}+\frac{(y+2)y^2}{2(y+1)n^2}+\frac{(4y^3+23y^2+40y+24)y^3}{24(y+1)^3n^3}+O(y^4n^{-4})$$

$$\text{where} \quad y = w(-nf(n))$$

$$(a+n)x^n + (b-n)x^{n-1} + f(n) = 0 \qquad \text{(II.41)}$$

$$\begin{aligned} x &= 1+\frac{y-a-b}{n} \\ &\quad +\frac{2a(a+b)+((a+b)^2-4a-2b)y+2(1-a-b)y^2+y^3}{2(y+1)n^2}+O(y^3n^{-3}) \end{aligned}$$

$$\text{where} \quad y = w(-\mathrm{e}^{a+b}f(n))$$

$$(a+n)x^n + (b-cn)x^{n-1} + f(n) = 0 \qquad [c \neq 1] \qquad \text{(II.42)}$$

$$\begin{aligned} x &= 1+\frac{y}{n}+\frac{y}{n^2}\left(\frac{c}{c-1}+\frac{y}{2}\right) \\ &\quad +\frac{y}{n^3}\left(\frac{b+ac+c^2+c(c-1/2)y}{(c-1)^2}+\frac{y^2}{6}\right)+O(y^4n^{-4}) \end{aligned}$$

$$\text{where} \quad y = \ln\left(\frac{f(n)}{(c-1)n-b-a}\right)$$

II.7 Sums containing descending factorials

For the following formulas we will denote

$$g(x) = \sum_{i\geq 0} f(x^i)$$

or alternatively

$$g(x) = \sum_{i \geq 0} a_i x^i$$

the sum being convergent in some region around 0, and

$$g_k(x) = \sum_{i \geq 0} i^{\underline{k}} f(x^i)$$

Descending factorials are denoted by $i^{\underline{k}} = i(i-1)(i-2)...(i-k+1)$.
In all cases, $\alpha = n/m$.

$$\sum_{i \geq 0} f\left(\frac{n^{\underline{i}}}{m^i}\right) = g(\alpha) + \frac{g'(\alpha) - g_1'(\alpha)}{2m} \tag{II.43}$$

$$+ \frac{3\alpha(g_2'(\alpha) - g_1'(\alpha) + g'(\alpha)) - g_2'(\alpha) - g_1'(\alpha) + g'(\alpha)}{24\,mn}$$

$$+ O(n^{-3})$$

$$\sum_{i \geq 0} a_i \frac{n^{\underline{i}}}{m^i} = g(\alpha) - \frac{\alpha g'(\alpha)}{2m} + \frac{\alpha}{24m^2}(3\,\alpha\, g^{iv}(\alpha) + 8g''(\alpha)) \tag{II.44}$$

$$-\frac{\alpha}{48m^3}(\alpha^2\, g^{vi}(\alpha) + 8\alpha\, g^{v}(\alpha) + 12g^{iv}(\alpha)) + O(m^{-4})$$

$$\sum_{i \geq 0} f\left(\frac{n^{\underline{i}}}{m^{\underline{i}}}\right) = g(\alpha) + \frac{(g'(\alpha) - g_1'(\alpha))(1-\alpha)}{2m} \tag{II.45}$$

$$+ \frac{(1-\alpha)[3\alpha(1-\alpha)(g_2' - g_1' + g') - (7\alpha+1)g_2' + (5\alpha-1)(g_1' - g')]}{24mn} + O(n^{-3})$$

$$\sum_{i \geq 0} a_i \frac{n^{\underline{i}}}{m^{\underline{i}}} = g(\alpha) - \frac{\alpha(1-\alpha)g'(\alpha)}{2m} \tag{II.46}$$

$$+ \frac{\alpha(1-\alpha)}{24m^2}[3\alpha(1-\alpha)g^{iv} + 8(1-2\alpha)g'' - 12g']$$

$$+ \frac{\alpha(1-\alpha)}{48m^3}[-\alpha^2(1-\alpha)^2 g^{vi} - 8\alpha(1-\alpha)(1-2\alpha)g^{v}$$

$$-12(1-6\alpha+6\alpha^2)g^{iv} + 48(1-2\alpha)g'' - 24g'] + O(m^{-4})$$

$$\sum_n f(n^{\frac{k}{}}) = (n - \frac{k-1}{2} + \frac{1-k^2}{24n} - \frac{(k-1)^2(k+1)}{48n^2} - \frac{(k-1)(k+1)(73k^2-240k+143)}{5760n^3} + \cdots)T_{-1}(n^{\frac{k}{}})$$

$$-\frac{f(n^{\frac{k}{}})}{2} + k(\frac{1}{12n} + \frac{k-1}{24n^2} + \frac{19(k-1)(k-2)}{720n^3} + \cdots)n^{\frac{k}{}} f'(n^{\frac{k}{}})$$

$$+\left(\frac{1-k^2}{24n} - \frac{(k-1)^2(k+1)}{48n^2} - \frac{(k-1)^2(k+1)(7k-17)}{576n^3} + \cdots\right) T_1(n^{\frac{k}{}})$$

$$-\frac{k^2(k-1)(n^{\frac{k}{}})^2 f'(n^{\frac{k}{}})}{240n^3} - \frac{k^3(n^{\frac{k}{}})^3 f''(n^{\frac{k}{}})}{720n^3}$$

$$-\frac{(k-3)(k-1)(k+1)(k+3)}{640n^3} T_3(n^{\frac{k}{}}) + \cdots \tag{II.47}$$

where

$$T_i(x) = \frac{x^{i/k}}{k} \int f(x)\, x^{-i/k-1}\, dx$$

II.8 Summation formulas

Euler–Maclaurin summation formula

$$\sum_{k=1}^{n-1} f(k) = \int_1^n f(x)dx + \left(\sum_{i=1}^{\infty} \frac{B_i f^{(i-1)}(x)}{i!}\right)_{x=1}^{x=n} \tag{II.48}$$

where B_i are the Bernoulli numbers $B_0 = 1$, $B_1 = -1/2$, $B_2 = 1/6$, $B_4 = -1/30$, $B_6 = 1/42$, $B_8 = -1/30$, ...

$$= \int_1^n f(x)dx + \left(-\frac{f(x)}{2} + f'\frac{(x)}{12} - f''\frac{(x)}{720} + \frac{f^{(5)}(x)}{30240} - \frac{f^{(7)}(x)}{1209600} + \cdots\right)_{x=1}^{x=n}$$

If we write

$$\sum_{k=1}^{n-1} f(k) = \int_a^n f(x)dx + C(f,a) + \sum_{i=1}^{\infty} \frac{B_i f^{(i-1)}(n)}{i!}$$

then, if $f(x) = \sum_i a_i x^i + \sum_i b_i \ln x \, x^i + \sum_i c_i \ln^2(x) x^i + \cdots$ (i varying over the reals),

$$\begin{aligned} C(f,1) &= \sum_i a_i \phi(i) + \sum_i b_i \phi'(i) + \sum_i c_i \phi'(i) \cdots \qquad \text{(II.49)} \\ &= \left(\frac{\pi^2}{6} - 1\right) a_{-2} + \gamma a_{-1} + \frac{a_0}{2} + \frac{5a_1}{12} + \frac{a_2}{3} + \frac{31a_3}{120} \\ &\quad + \frac{a_4}{5} + \frac{41a_5}{252} + \cdots + \frac{b_0(\ln(2\pi) - 2)}{2} + \ldots \end{aligned}$$

where $\phi(i) = \zeta(-i) + \frac{1}{i+1}$, $\phi(-1) = \gamma$; $\phi'(-1) = -\gamma_1$, and if $a_{-1} = 0$,

$$\begin{aligned} C(f,0) &= \sum_i a_i \zeta(-i) - \sum_i b_i \zeta'(-i) + \cdots \qquad \text{(II.50)} \\ &= -\frac{a_0}{2} - \frac{a_1}{12} + \frac{a_3}{120} - \frac{a_5}{252} + \cdots + \frac{b_0 \ln 2\pi}{2} \cdots \end{aligned}$$

General references:
[de Bruijn, N.G., 70], [Abramowitz, M. *et al.*, 72], [Knuth, D.E., 73], [Knuth, D.E., 73], [Bender, E.A., 74], [Gonnet, G.H., 78], [Greene, D.H. *et al.*, 82], [Graham, R.L. *et al.*, 88].

APPENDIX III

References

III.1 Textbooks

The following are fine textbooks recommended for further information on their topics.

1. Aho, A.V., Hopcroft, J.E. and Ullman, J.D.: *The Design and Analysis of Computer Algorithms;* Addison-Wesley, Reading, Mass, (1974). (2.1, 2.2, 3.2.1, 3.3, 3.4.1, 3.4.1.3, 3.4.2.1, 4.1.3, 4.1.5, 4.2.1, 4.2.4, 4.2.6, 5.1.6, 5.2, 5.2.2, 6.1, 6.3, 6.4, 7.1.2, 7.1.4, 7.1.6, 7.2.2).

2. Aho, A.V., Hopcroft, J.E. and Ullman, J.D.: *Data Structures and Algorithms;* Addison-Wesley, Reading, Mass, (1983). (3.3, 3.4.1, 3.4.2, 4.1, 4.2).

3. Baase, S.: *Computer Algorithms: Introduction to Design and Analysis;* Addison-Wesley, Reading, Mass, (1988). (3.2.1, 3.4.1.7, 4.1.2, 4.1.3, 4.1.4, 4.1.5, 4.2.1, 4.2.4, 4.4, 5.2, 6.3, 6.4, 7.1.1, 7.1.2, 7.1.3, 7.1.8).

4. Borodin, A. and Munro, J.I.: *The Computational Complexity of Algebraic and Numeric Problems;* American Elsevier, New York, NY, (1975). (6.1, 6.2, 6.3, 6.4).

5. Brassard, G. and Bratley, P.: *Algorithmics - Theory and Practice;* Prentice-Hall, Englewood Cliffs, NJ, (1988). (3.2.1, 3.3.1, 3.4.1.7, 4.1.3, 4.2.1, 5.1.3, 5.2, 6.2, 7.1.2, 7.1.3).

6. Cormen, T.H., Leiserson, C.E. and Rivest, R.L.: *Introduction to Algorithms;* MIT Press, Cambridge, Mass., (1990). (3.3, 3.4.1, 3.4.1.8, 3.4.1.9, 3.4.2, 3.4.2.4, 4.1.3, 4.1.5, 4.2.3, 4.2.4, 5.1.3, 5.1.7, 5.2, 6.3, 7.1.1, 7.1.2, 7.1.3, 7.1.5, 7.1.6, 7.3.1).

7. de Bruijn, N.G.: *Asymptotic Methods in Analysis;* North-Holland, Amsterdam, (1970). (II).

8. Flores, I.: *Computer Sorting;* Prentice-Hall, Englewood Cliffs, NJ, (1969). (4.1, 4.2, 4.4).

9. Gotlieb, C.C. and Gotlieb, L.R.: *Data Types and Structures;* Prentice-Hall, Englewood Cliffs, NJ, (1978). (2.1, 3.1.1, 3.2.1, 3.2.2, 3.3, 3.4.1, 3.4.2, 3.4.3, 3.4.4, 4.1.2, 4.1.3, 4.2).
10. Greene, D.H. and Knuth, D.E.: *Mathematics for the Analysis of Algorithms;* Birkhauser, Boston, Mass, (1982). (3.3.2, 3.3.12, II).
11. Graham, R.L., Knuth, D.E. and Patashnik, O.: *Concrete Mathematics: A Foundation for Computer Science;* Addison-Wesley, Reading, Mass, (1988). (3.3.10, II).
12. Hopcroft, J.E. and Ullman, J.D.: *Introduction to Automata Theory, Languages, and Computation;* Addison-Wesley, Reading, Mass, (1979). (7.1.6).
13. Horowitz, E. and Sahni, S.: *Fundamentals of Data Structures;* Computer Science Press, Potomac, Maryland, (1976). (3.2, 3.3, 3.4.1, 3.4.2, 3.4.4, 4.1.2, 4.1.3, 4.1.5, 4.2.1, 4.4.2, 4.4.4, 4.3.1).
14. Hu, T.C.: *Combinatorial Algorithms;* Addison-Wesley, Reading, Mass, (1982). (3.4.1.7, 6.3).
15. Jensen, K. and Wirth, N.: *Pascal User Manual and Report;* Springer-Verlag, Berlin, (1974). (1).
16. Johnson, N.L. and Kotz, S.: *Discrete Distributions;* Houghton Mifflin, Boston, Mass, (1969). (I.1).
17. Kernighan, B.W. and Ritchie, D.M.: *The C Programming Language;* Prentice-Hall, Englewood Cliffs NJ, (1978). (1).
18. Knuth, D.E.: *The Art of Computer Programming, vol. I: Fundamental Algorithms;* Addison-Wesley, Reading, Mass, (1973). (3.4.1.2, II).
19. Knuth, D.E.: *The Art of Computer Programming, vol. II: Seminumerical Algorithms;* Addison-Wesley, Reading, Mass, (1969). (6.1, 6.2, 6.3, 6.4).
20. Knuth, D.E.: *The Art of Computer Programming, vol. III: Sorting and Searching;* Addison-Wesley, Reading, Mass, (1973). (3.1.1, 3.1.2, 3.1.4, 3.2.1, 3.3, 3.3.2, 3.3.4, 3.3.5, 3.3.6, 3.3.8.1, 3.3.11, 3.3.12, 3.3.1, 3.4.1, 3.4.1.1, 3.4.1.6, 3.4.1.7, 3.4.1.3, 3.4.1.4, 3.4.1.9, 3.4.2, 3.4.4, 3.4.4.5, 4.1.1, 4.1.2, 4.1.3, 4.1.4, 4.1.5, 4.2.1, 4.2.3, 4.2.4, 4.3.1, 4.3.2, 4.3.3, 4.4.1, 4.4.2, 4.4.3, 4.4.4, 4.4.5, 5.1.3, 5.1.6, 5.2.2, 5.2, 7.2.1, 7.2.6, I.1, I.4, II).
21. Kronsjo, L.: *Algorithms: their complexity and efficiency;* John Wiley, Chichester, England, (1979). (3.1.1, 3.2.1, 3.3, 3.4.1, 4.1, 4.4, 5.2, 6.3, 6.4).
22. Lorin, H.: *Sorting and Sort Systems;* Addison-Wesley, Reading, Mass, (1975). (4.1, 4.4).
23. Manber, U.: *Introduction to Algorithms: A Creative Approach;* Addison-Wesley, Reading, Mass, (1989). (3.2.1, 3.2.2, 3.3, 3.4.1, 3.4.1.3, 4.1.3, 4.1.5, 4.2.1, 4.2.3, 4.2.4, 5.1.3, 5.2, 6.3, 7.1.1, 7.1.2, 7.1.3, 7.1.8).
24. Mehlhorn, K.: *Data Structures and Algorithms, vol. I: Sorting and Searching;* Springer-Verlag, Berlin, (1984). (3.1, 3.2, 3.3, 3.4.1, 3.4.2, 3.4.4, 4.1, 4.2, 4.3, 4.4, 5.1, 5.2).
25. Mehlhorn, K.: *Data Structures and Algorithms, vol. III: Multidimensional Searching and Computational Geometry;* Springer-Verlag, Berlin, (1984). (3.5, 3.6).
26. Reingold, E.M. and Hansen, W.J.: *Data Structures;* Little, Brown, Boston, Mass, (1983). (3.3, 3.4.1, 4.1, 4.2, 4.4).
27. Reingold, E.M., Nievergelt, J. and Deo, N.: *Combinatorial Algorithms: Theory and Practice;* Prentice-Hall, Englewood Cliffs NJ, (1977). (3.1.1, 3.2.1, 3.3, 3.4.1.1, 3.4.1.3, 3.4.1.4, 3.4.1.7, 3.4.2, 3.4.4, 4.1.1, 4.1.2, 4.1.3, 4.1.5, 4.2.4, 4.3, 5.2).

28. Salton, G. and McGill, M.J.: *Introduction to Modern Information Retrieval;* McGraw-Hill, New York NY, (1983). (7.1.2, 7.1.3, 7.1.6, 7.2.1, 7.2.2, 7.2.6).

29. Sankoff, D. and Kruskal, J.B.: *Time Warps, String Edits and Macromolecules;* Addison-Wesley, Reading, Mass, (1983). (7.1.8, 7.2.1).

30. Sedgewick, R.: *Algorithms;* Addison-Wesley, Reading, Mass, (1988). (3.1.1, 3.2.1, 3.3.4, 3.3.5, 3.3.11, 3.3.13, 3.3.1, 3.4.1, 3.4.1.7, 3.4.2, 3.4.2.4, 3.4.4, 3.4.4.5, 3.6, 4.1.1, 4.1.2, 4.1.3, 4.1.4, 4.1.5, 4.2.4, 4.3, 4.4, 5.1.3, 5.2, 6.4, 7.1.1, 7.1.2, 7.1.3, 7.1.4, 7.1.5, 7.1.6).

31. Standish, T.A.: *Data Structure Techniques;* Addison-Wesley, Reading, Mass, (1980). (3.1, 3.3, 3.4.1, 3.4.2, 4.1.3, 4.1.5, 5.1).

32. Salzberg, B.: *File Structures: An Analytic Approach;* Prentice-Hall, (1988). (3.3.13, 3.3.14, 3.4.2).

33. Wilf, H.: *Algorithms and Complexity;* Prentice-Hall, Englewood Cliffs, NJ, (1986). (4.1.3, 6.3).

34. Wirth, N.: *Algorithms + Data Structures = Programs;* Prentice-Hall, Englewood Cliffs, NJ, (1976). (2.1, 2.2, 3.1, 3.3.6, 3.4.1, 3.4.2, 4.1, 4.2, 5.2).

35. Wirth, N.: *Algorithms and Data Structures;* Prentice-Hall, Englewood Cliffs, NJ, (1986). (2.1, 2.2, 3.1, 3.3.6, 3.3.1, 3.4.1, 3.4.2, 4.1, 4.2, 5.2, 7.1.1, 7.1.2, 7.1.3).

36. Zipf, G.K.: *Human Behaviour and the Principle of Least Effort;* Addison-Wesley, Cambridge, Mass, (1949). (I.1).

III.2 Papers

The following are research papers that contain some in-depth information on the topics covered in the indicated sections of the handbook. Technical reports and unpublished manuscripts are not included in this list.

1. Abrahamson, K.: Generalized String Matching; SIAM J on Computing, 16:1039-1051, (1987). (7.1, 7.1.7).

2. Abramowitz, M. and Stegun, I.: *Handbook of Mathematical Functions;* Dover, New York, (1972). (II).

3. Ackerman, A.F.: Quadratic Search for Hash Tables of Size p^n; C.ACM, 17(3):164, (Mar 1974). (3.3.6).

4. Adel'son-Vel'skii, G.M. and Landis, E.M.: An Algorithm for the organization of information; Dokladi Akademia Nauk SSSR, 146(2):263-266, (1962). (3.4.1.3).

5. Adleman, L., Booth, K.S., Preparata, F.P. and Ruzzo, W.L.: Improved Time and Space Bounds for Boolean Matrix Multiplication; Acta Informatica, 11(1):61-70, (1978). (6.3).

6. Aggarwal, A. and Vitter, J.S.: The Input/Output Complexity of Sorting and Related Problems; C.ACM, 31(9):1116-1127, (Sep 1988). (4.4).

7. Aho, A.V. and Corasick, M.: Efficient String Matching: An Aid to Bibliographic Search; C.ACM, 18(6):333-340, (June 1975). (7.1.4).

8. Aho, A.V., Hirschberg, D.S. and Ullman, J.D.: Bounds on the Complexity of the Longest Common Subsequence Problem; J.ACM, 23:1-12, (1976). (7.3.1).

9. Aho, A.V. and Lee, D.T.: Storing a Sparse Dynamic Table; Proceedings FOCS, Toronto, Canada, 27:55-60, (Oct 1986). (3.3.16).
10. Aho, A.V., Steiglitz, K. and Ullman, J.D.: Evaluating Polynomials at Fixed Points; SIAM J on Computing, 4(4):533-539, (Dec 1975). (6.4).
11. Aho, A.V.: Pattern Matching in Strings; Formal Language Theory: Perspectives and Open Problems, Academic Press, London, :325-347, (1980). (7.1).
12. Ajtai, M., Fredman, M.L. and Komlos, J.: Hash Functions for Priority Queries; Information and Control, 63(3):217-225, (Dec 1984). (3.3.1, 5.1).
13. Ajtai, M., Komlos, J. and Szemeredi, E.: There is no Fast Single Hashing Algorithm; Inf. Proc. Letters, 7(6):270-273, (Oct 1978). (3.3.2).
14. Akdag, H.: Performance of an Algorithm Constructing a Nearly Optimal Binary Tree; Acta Informatica, 20(2):121-132, (1983). (3.4.1.7).
15. Akl, S.G. and Meijer, H.: On the Average-Case Complexity of Bucketing Algorithms; J of Algorithms, 3(1):9-13, (Mar 1982). (4.2.3).
16. Akl, S.G. and Meijer, H.: Recent Advances in Hybrid Sorting Algorithms; Utilitas Mathematica, 21C:325-343, (May 1982). (4.2.5).
17. Alagar, V.S., Bui, T.D. and Thanh, M.: Efficient Algorithms for Merging; BIT, 23(4):410-428, (1983). (4.3.2).
18. Alagar, V.S. and Probst, D.K.: A Fast, Low-Space Algorithm for Multiplying Dense Multivariate Polynomials; ACM TOMS, 13(1):35-57, (Mar 1987). (6.3).
19. Aldous, D., Flannery, B. and Palacios, J.L.: Two Applications of Urn Processes: The Fringe Analysis of Search Trees and the Simulation of Quasi-Stationary Distributions of Markov Chains; Probability in the Eng. and Inf. Sciences, 2:293-307, (1988). (3.4.2, 3.4.2.1).
20. Aldous, D.: Hashing with Linear Probing, Under Non-Uniform Probabilities; Probability in the Eng. and Inf. Sciences, 2:1-14, (1988). (3.3.4).
21. Alekseyed, V.B.: On the Complexity of Some Algorithms of Matrix Multiplication; J of Algorithms, 6(1):71-85, (Mar 1985). (6.3).
22. Allen, B. and Munro, J.I.: Self-Organizing Search Trees; J.ACM, 25(4):526-535, (Oct 1978). (3.4.1.6, 3.1).
23. Allen, B.: On the Costs of Optimal and Near-Optimal Binary Search Trees; Acta Informatica, 18(3):255-263, (1982). (3.4.1.6, 3.4.1.7).
24. Allison, D.C.S. and Noga, M.T.: Selection by Distributive Partitioning; Inf. Proc. Letters, 11(1):7-8, (Aug 1980). (5.2).
25. Allison, D.C.S. and Noga, M.T.: Usort: An Efficient Hybrid of Distributive Partitioning Sorting; BIT, 22(2):135-139, (1982). (4.2.5).
26. Alt, H., Mehlhorn, K. and Munro, J.I.: Partial Match Retrieval in Implicit Data Structures; Inf. Proc. Letters, 19(2):61-65, (Aug 1984). (3.6.2).
27. Alt, H.: Comparing the Combinatorial Complexities of Arithmetic Functions; J.ACM, 35(2):447-460, (Apr 1988). (6.1).
28. Alt, H.: Functions Equivalent to Integer Multiplication; Proceedings ICALP, Lecture Notes in Computer Science 85, Springer-Verlag, Noordwijkerhovt, Holland, 7:30-37, (1980). (6.1).
29. Alt, H.: Multiplication is the Easiest Nontrivial Arithmetic Function; Proceedings FOCS, Tucson AZ, 24:320-322, (Nov 1983). (6.1).
30. Amble, O. and Knuth, D.E.: Ordered Hash Tables; Computer Journal, 17(2):135-142, (May 1974). (3.3.7).

31. Amir, A., Landau, G.M. and Vishkin, U.: Efficient Pattern Matching with Scaling; Proceedings SODA, San Francisco CA, 1:344-357, (Jan 1990). (7.3).

32. Anderson, H.D. and Berra, P.B.: Minimum Cost Selection of Secondary Indexes for Formatted Files; ACM TODS, 2(1):68-90, (1977). (3.4.3).

33. Anderson, M.R. and Anderson, M.G.: Comments on Perfect Hashing Functions: A Single Probe Retrieving Method for Static Sets; C.ACM, 22(2):104-105, (Feb 1979). (3.3.16).

34. Andersson, A. and Carlsson, S.: Construction of a Tree from Its Traversals in Optimal Time and Space; Inf. Proc. Letters, 34(1):21-25, (1983). (3.4.1).

35. Andersson, A. and Lai, T.W.: Fast Updating of Well Balanced Trees; Proceedings Scandinavian Workshop in Algorithmic Theory, SWAT'90, Lecture Notes in Computer Science 447, Springer-Verlag, Bergen, Norway, 2:111-121, (July 1990). (3.4.1).

36. Andersson, A.: Improving Partial Rebuilding by Using Simple Balance Criteria; Proceedings Workshop in Algorithms and Data Structures, Lecture Notes in Computer Science 382, Springer-Verlag, Ottawa, Canada, 1:393-402, (Aug 1989). (3.4.1).

37. Apers, P.M.: Recursive Samplesort; BIT, 18(2):125-132, (1978). (4.1.3).

38. Apostolico, A. and Giancarlo, R.: The Boyer-Moore-Galil String Searching Strategies Revisited; SIAM J on Computing, 15:98-105, (1986). (7.1.3).

39. Apostolico, A. and Guerra, C.: The Longest Common Subsequence Problem Revisited; Algorithmica, 2:315-336, (1987). (7.3.1).

40. Apostolico, A. and Preparata, F.P.: Structural Properties of the String Statistics Problem; JCSS, 31:394-411, (1985). (7.2.2).

41. Apostolico, A.: Improving the Worst-Case Performance of the Hunt-Szymanski Strategy for the Longest Common Subsequence of two Strings; Inf. Proc. Letters, 23:63-69, (1986). (7.3.1).

42. Apostolico, A.: Remark on the Hsu-Du New Algorithm for the Longest Common Subsequence Problem; Inf. Proc. Letters, 25:235-236, (1987). (7.3.1).

43. Apostolico, A.: The Myriad Virtues of Subword Trees; Combinatorial Algorithms on Words, NATO ASI Series, Springer-Verlag, F12:85-96, (1985). (7.2.2).

44. Aragon, C. and Seidel, R.: Randomized Search Trees; Proceedings FOCS, Research Triangle Park, NC, 30:540-545, (1989). (3.4.1).

45. Arazi, B.: A Binary Search with a Parallel Recovery of the Bits; SIAM J on Computing, 15(3):851-855, (Aug 1986). (3.2.1).

46. Arnow, D. and Tenenbaum, A.M.: An Empirical Comparison of B-Trees, Compact B-Trees and Multiway Trees; Proceedings ACM SIGMOD, Boston, Mass, 14:33-46, (June 1984). (3.4.2, 3.4.1.10).

47. Arora, S.R. and Dent, W.T.: Randomized Binary Search Technique; C.ACM, 12(2):77-80, (1969). (3.2.1, 3.4.1).

48. Artzy, E., Hinds, J.A. and Saal, H.J.: A Fast Technique for Constant Divisors; C.ACM, 19(2):98-101, (Feb 1976). (6.1).

49. Atkinson, M.D., Sack, J.R., Santoro, N. and Strothotte, T.: Min-Max Heaps and Generalized Priority Queues; C.ACM, 29(10):996-1000, (Oct 1986). (5.1.3, 5.1.6).

50. Atkinson, M.D. and Santoro, N.: A Practical Algorithm for Boolean Matrix Multiplication; Inf. Proc. Letters, 29(1):37-38, (Sep 1988). (6.3).

51. Aviad, Z. and Shamir, E.: A Direct Dynamic Solution to Range Search and Related Problems for Product Regions; Proceedings FOCS, Nashville TN, 22:123-126, (Oct 1981). (3.6.2).

52. Badley, J.: Use of Mean distance between overflow records to compute average search lengths in hash files with open addressing; Computer Journal, 29(2):167-170, (Apr 1986). (3.3).

53. Baer, J.L. and Schwab, B.: A Comparison of Tree-Balancing Algorithms; C.ACM, 20(5):322-330, (May 1977). (3.4.1.3, 3.4.1.4, 3.4.1.6).

54. Baer, J.L.: Weight-Balanced Trees; Proceedings AFIPS, Anaheim CA, 44:467-472, (1975). (3.4.1.5).

55. Baeza-Yates, R.A., Gonnet, G.H. and Regnier, M.: Analysis of Boyer-Moore-type String Searching Algorithms; Proceedings SODA, San Francisco CA, 1:328-343, (Jan 1990). (7.1.3).

56. Baeza-Yates, R.A., Gonnet, G.H. and Ziviani, N.: Expected Behaviour Analysis of AVL Trees; Proceedings Scandinavian Workshop in Algorithmic Theory, SWAT'90, Lecture Notes in Computer Science 447, Springer-Verlag, Bergen, Norway, 2:143-159, (July 1990). (3.4.1.3).

57. Baeza-Yates, R.A. and Gonnet, G.H.: A New Approach to Text Searching; Proceedings ACM SIGIR, Cambridge, Mass., 12:168-175, (June 1989). (7.1.7, 7.1.8).

58. Baeza-Yates, R.A. and Gonnet, G.H.: Efficient Text Searching of Regular Expressions; Proceedings ICALP, Lecture Notes in Computer Science 372, Springer-Verlag, Stresa, Italy, 16:46-62, (July 1989). (7.2.3).

59. Baeza-Yates, R.A. and Gonnet, G.H.: Average Case Analysis of Algorithms using Matrix Recurrences; Proceedings ICCI, Niagara Falls, Canada, 2:47-51, (May 1990). (3.4.2, 7.2.3).

60. Baeza-Yates, R.A. and Larson, P.: Performance of B^+-trees with Partial Expansions; IEEE Trans. on Knowledge and Data Engineering, 1(2):248-257, (June 1989). (3.4.2).

61. Baeza-Yates, R.A. and Poblete, P.V.: Reduction of the Transition Matrix of a Fringe Analysis and Its Application to the Analysis of 2-3 Trees; Proceedings SCCC Int. Conf. in Computer Science, Santiago, Chile, 5:56-82, (1985). (3.4.2.1).

62. Baeza-Yates, R.A. and Regnier, M.: Fast Algorithms for Two Dimensional and Multiple Pattern Matching; Proceedings Scandinavian Workshop in Algorithmic Theory, SWAT'90, Lecture Notes in Computer Science 447, Springer-Verlag, Bergen, Norway, 2:332-347, (July 1990). (7.1.4, 7.3.2).

63. Baeza-Yates, R.A.: Efficient Text Searching; PhD Dissertation, Department of Computer Science, University of Waterloo, (May 1989). (7.1, 7.1.1, 7.1.2, 7.1.3, 7.1.5, 7.1.7, 7.1.8, 7.2.2, 7.2.3).

64. Baeza-Yates, R.A.: A Trivial Algorithm Whose Analysis Isn't: A Continuation; BIT, 29:88-113, (1989). (3.4.1.9).

65. Baeza-Yates, R.A.: An Adaptive Overflow Technique for the B-tree; Proceedings Extending Data Base Technology Conference, Lecture Notes in Computer Science 416, Springer-Verlag, Venice, :16-28, (Mar 1990). (3.4.2).

66. Baeza-Yates, R.A.: Expected Behaviour of B^+-trees under Random Insertions; Acta Informatica, 26(5):439-472, (1989). (3.4.2).

67. Baeza-Yates, R.A.: Improved String Searching; Software - Practice and Experience, 19(3):257-271, (1989). (7.1.3).

68. Baeza-Yates, R.A.: Modeling Splits in File Structures; Acta Informatica, 26(4):349-362, (1989). (3.3.14, 3.4.2, 3.4.2.5).

69. Baeza-Yates, R.A.: Some Average Measures in m-ary Search Trees; Inf. Proc. Letters, 25:375-381, (July 1987). (3.4.1.10).

70. Baeza-Yates, R.A.: String Searching Algorithms Revisited; Proceedings Workshop in Algorithms and Data Structures, Lecture Notes in Computer Science 382, Springer-Verlag, Ottawa, Canada, 1:75-96, (Aug 1989). (7.1, 7.1.1, 7.1.2, 7.1.3).

71. Baeza-Yates, R.A.: A Storage Allocation Algorithm suitable for File Structures; Inform. Systems, 15(5):515-521, (1990). (3.4.2.5).

72. Baeza-Yates, R.A.: Searching Subsequences; Theoretical Computer Science, to app.. (7.2.5, 7.3.1).

73. Bagchi, A. and Pal, A.K.: Asymptotic normality in the generalized Polya-Eggenberger urn model, with an application to computer data structures; SIAM J Alg Disc Methods, 6:394-405, (1985). (3.4.2, 3.4.2.1).

74. Bagchi, A. and Reingold, E.M.: Aspects of Insertion in Random Trees; Computing, 29:11-29, (1982). (3.4.1.4, 3.4.1.1).

75. Bagchi, A. and Roy, J.K.: On V-Optimal Trees; SIAM J on Computing, 8(4):524-541, (Nov 1979). (3.4.1.7).

76. Bailey, T.A. and Dromey, R.G.: Fast String Searching by Finding Subkeys in Subtext; Inf. Proc. Letters, 11:130-133, (1980). (7.1.3, 7.1.4).

77. Baker, T.: A Technique for Extending Rapid Exact String Matching to Arrays of More than One Dimension; SIAM J on Computing, 7:533-541, (1978). (7.3.2).

78. Bandyopadhyay, S.K.: Comment on Weighted Increment Linear Search for Scatter Tables; C.ACM, 20(4):262-263, (Apr 1977). (3.3.4).

79. Banerjee, J. and Ramaraman, V.: A Dual Link Data Structure for Random File Organization; Inf. Proc. Letters, 4(3):64-69, (Dec 1975). (3.3.12).

80. Barnett, J.K.R.: A Technique for Reducing Comparison Times in Certain Applications of the Merging Method of Sorting; Inf. Proc. Letters, 2(5):127-128, (Dec 1973). (4.4).

81. Barstow, D.R.: Remarks on A Synthesis of Several Sorting Algorithms; Acta Informatica, 13(3):225-227, (1980). (2.2.2).

82. Barth, G.: An Alternative for the Implementation of Knuth-Morris-Pratt Algorithm; Inf. Proc. Letters, 13:134-137, (1981). (7.1.2).

83. Barth, G.: An Analytical Comparison of two String Searching Algorithms; Inf. Proc. Letters, 18:249-256, (1984). (7.1.1, 7.1.2).

84. Batagelj, V.: The Quadratic Hash Method When the Table Size is Not a Prime Number; C.ACM, 18(4):216-217, (Apr 1975). (3.3.6).

85. Batory, D.S.: B+Trees and Indexed Sequential Files: A Performance Comparison; Proceedings ACM SIGMOD, Ann Arbor MI, 11:30-39, (Apr 1981). (3.4.3, 3.4.2).

86. Batson, A.: The Organization of Symbol Tables; C.ACM, 8(2):111-112, (1965). (3.3, 3.4.1).

87. Bayer, R. and McCreight, E.M.: Organization and Maintenance of Large Ordered Indexes; Acta Informatica, 1(3):173-189, (1972). (3.4.2).

88. Bayer, R. and Metzger, J.K.: On the Encipherment of Search Trees and Random Access Files; ACM TODS, 1(1):37-52, (1976). (3.4.2).

89. Bayer, R. and Unterauer, K.: Prefix B-trees; ACM TODS, 2(1):11-26, (Mar 1977). (3.4.2).

90. Bayer, R.: Binary B-trees for virtual memory; Proceedings ACM SIGFIDET Workshop on Data Description, Access and Control, San Diego CA, :219-235, (Nov 1971). (3.4.2).

91. Bayer, R.: Symmetric Binary B-trees: Data Structure and Maintenance Algorithms; Acta Informatica, 1(4):290-306, (1972). (3.4.2.2).

92. Bayer, R.: Storage Characteristics and Methods for Searching and Addressing; Proceedings Information Processing 74, North-Holland, Stockholm, Sweden, :440-444, (1974). (3.3, 3.4.2).

93. Bays, C.: A Note on When to Chain Overflow Items Within a Direct-Access Table; C.ACM, 16(1):46-47, (Jan 1973). (3.3.11).

94. Bays, C.: Some Techniques for Structuring Chained Hash Tables; Computer Journal, 16(2):126-131, (May 1973). (3.3.12).

95. Bays, C.: The Reallocation of Hash-Coded Tables; C.ACM, 16(1):11-14, (Jan 1973). (3.3).

96. Bechtald, U. and Kuspert, K.: On the use of extendible Hashing without hashing; Inf. Proc. Letters, 19(1):21-26, (July 1984). (3.3.13).

97. Beck, I. and Krogdahl, S.: A select and insert sorting algorithm; BIT, 28(4):726-735, (1988). (4.1).

98. Beckley, D.A., Evans, M.W. and Raman, V.K.: Multikey Retrieval from K-d Trees and Quad Trees; Proceedings ACM SIGMOD, Austin TX, 14:291-303, (1985). (3.5.1, 3.5.2).

99. Behymer, J.A., Ogilive, R.A. and Merten, A.G.: Analysis of Indexed Sequential and Direct Access File Organization; Proceedings ACM SIGMOD Workshop on Data Description, Access and Control, Ann Arbor MI, :389-417, (May 1974). (3.3.11, 3.4.3).

100. Belaga, E.G.: Some Problems Involved in the Computation of Polynomials; Dokladi Akademia Nauk SSSR, 123:775-777, (1958). (6.4).

101. Bell, C.: An Investigation into the Principles of the Classification and Analysis of Data on an Automatic Digital Computer; PhD Dissertation, Leeds University, (1965). (3.4.1).

102. Bell, D.A. and Deen, S.M.: Hash trees vs. B-trees; Computer Journal, 27(3):218-224, (Aug 1984). (3.4.2).

103. Bell, J.R. and Kaman, C.H.: The Linear Quotient Hash Code; C.ACM, 13(11):675-677, (Nov 1970). (3.3.5).

104. Bell, J.R.: The Quadratic Quotient Method: A Hash Code Eliminating Secondary Clustering; C.ACM, 13(2):107-109, (Feb 1970). (3.3.6).

105. Bell, R.C. and Floyd, B.: A Monte Carlo Study of Cichelli Hash-Function Solvability; C.ACM, 26(11):924-925, (Nov 1983). (3.3.16).

106. Bender, E.A., Praeger, C.E. and Wornald, C.N.: Optimal worst case trees; Acta Informatica, 24(4):475-489, (1987). (3.4.1.7).

107. Bender, E.A.: Asymptotic methods in enumeration; SIAM Review, 16:485-515, (1974). (II).

108. Bent, S.W. and John, J.W.: Finding the median requires $2n$ comparisons; Proceedings STOC SIGACT, Providence, RI, 17:213-216, (May 1985). (5.2).

109. Bent, S.W., Sleator, D.D. and Tarjan, R.E.: Biased 2-3 Trees; Proceedings FOCS, Syracuse NY, 21:248-254, (Oct 1980). (3.4.2.1).

110. Bent, S.W., Sleator, D.D. and Tarjan, R.E.: Biased Search Trees; SIAM J on Computing, 14(3):545-568, (Aug 1985). (3.4.1.6).

111. Bent, S.W.: Ranking Trees Generated by Rotations; Proceedings Scandinavian Workshop in Algorithmic Theory, SWAT'90, Lecture Notes in Computer Science 447, Springer-Verlag, Bergen, Norway, 2:132-142, (July 1990). (3.4.1.8).

112. Bentley, J.L. and Brown, D.J.: A General Class of Resource Tradeoffs; Proceedings FOCS, Syracuse NY, 21:217-228, (Oct 1980). (2.2).

113. Bentley, J.L. and Friedman, J.H.: Data Structures for Range Searching; ACM C. Surveys, 11(4):397-409, (Dec 1979). (3.6).

114. Bentley, J.L. and Maurer, H.A.: A Note on Euclidean Near Neighbor Searching in the Plane; Inf. Proc. Letters, 8(3):133-136, (Mar 1979). (3.5).

115. Bentley, J.L. and Maurer, H.A.: Efficient Worst-Case Data Structures for Range Searching; Acta Informatica, 13(2):155-168, (1980). (3.6).

116. Bentley, J.L. and McGeoch, C.C.: Amortized Analyses of Self-Organizing Sequential Search Heuristics; C.ACM, 28(4):404-411, (Apr 1985). (3.1.2, 3.1.3).

117. Bentley, J.L. and Saxe, J.B.: Decomposable Searching Problems. I. Static-to-Dynamic Transformation; J of Algorithms, 1(4):301-358, (Dec 1980). (2.2).

118. Bentley, J.L. and Saxe, J.B.: Generating Sorted Lists of Random Numbers; ACM TOMS, 6(3):359-364, (Sep 1980). (4.2).

119. Bentley, J.L. and Shamos, M.I.: Divide and Conquer for Linear Expected Time; Inf. Proc. Letters, 7(2):87-91, (Feb 1978). (2.2.2.1).

120. Bentley, J.L. and Shamos, M.I.: Divide and Conquer in Multidimensional Space; Proceedings STOC-SIGACT, Hershey PA, 8:220-230, (May 1976). (2.2.2.1).

121. Bentley, J.L. and Stanat, D.F.: Analysis of Range Searches in Quad Trees; Inf. Proc. Letters, 3(6):170-173, (July 1975). (3.5.1).

122. Bentley, J.L. and Yao, A.C-C.: An Almost Optimal Algorithm for Unbounded Searching; Inf. Proc. Letters, 5(3):82-87, (Aug 1976). (3.2.1).

123. Bentley, J.L.: An Introduction to Algorithm Design; IEEE Computer, 12(2):66-78, (Feb 1979). (2.2).

124. Bentley, J.L.: Decomposable Searching Problems; Inf. Proc. Letters, 8(5):244-251, (June 1979). (2.2).

125. Bentley, J.L.: Multidimensional Binary Search Trees in Database Applications; IEEE Trans. Software Engineering, 5(4):333-340, (July 1979). (3.5.2).

126. Bentley, J.L.: Multidimensional Binary Search Trees Used for Associative Searching; C.ACM, 18(9):509-517, (Sep 1975). (3.5.2).

127. Bentley, J.L.: Multidimensional Divide-and-Conquer; C.ACM, 23(4):214-229, (Apr 1980). (3.5).

128. Bentley, J.L.: Programming Pearls: Selection; C.ACM, 28(11):1121-1127, (Nov 1985). (5.2.2).

129. Berman, F., Bock, M.E., Dittert, E., O'Donell, M.J. and Plank, P.: Collections of functions for perfect hashing; SIAM J on Computing, 15(2):604-618, (May 1986). (3.3.16).

130. Berman, G. and Colijn, A.W.: A Modified List Technique Allowing Binary Search; J.ACM, 21(2):227-232, (Apr 1974). (3.1.1, 3.2.1).

131. Bing-Chao, H. and Knuth, D.E.: A one-way, stackless quicksort algorithm; BIT, 26(1):127-130, (1986). (4.1.3).

132. Bini, D., Capovani, M., Romani, F. and Lotti, G.: O(n**2.7799) Complexity for n x n Approximate Matrix Multiplication; Inf. Proc. Letters, 8(5):234-235, (June 1979). (6.3).

133. Bird, R.: Two Dimensional Pattern Matching; Inf. Proc. Letters, 6:168-170, (1977). (7.3.2).

134. Bitner, J.R. and Huang, S-H.S.: Key Comparison Optimal 2-3 Trees with Maximum Utilization; SIAM J on Computing, 10(3):558-570, (Aug 1981). (3.4.2.1).

135. Bitner, J.R.: Heuristics that Dynamically Organize Data Structures; SIAM J on Computing, 8(1):82-110, (Feb 1979). (3.1.2, 3.1.3).

136. Bjork, H.: A Bi-Unique Transformation into Integers of Identifiers and Other Variable-Length Items; BIT, 11(1):16-20, (1971). (3.3.1).

137. Blake, I.F. and Konheim, A.G.: Big Buckets Are (Are Not) Better!; J.ACM, 24(4):591-606, (Oct 1977). (3.3.4).

138. Bloom, B.H.: Space/Time Trade-offs in Hash Coding with Allowable Errors; C.ACM, 13(7):422-426, (1970). (3.3).

139. Blum, N., Floyd, R.W., Pratt, V., Rivest, R.L. and Tarjan, R.E.: Time Bounds for Selection; JCSS, 7(4):448-461, (Aug 1973). (5.2).

140. Blum, N. and Mehlhorn, K.: On the Average Number of Rebalancing Operations in Weight-Balanced Trees; Theoretical Computer Science, 11(3):303-320, (July 1980). (3.4.1.4).

141. Blumer, A., Blumer, J., Haussler, D., Ehrenfeucht, A., Chen, M.T. and Seiferas, J.: The Smallest Automaton Recognizing the Subwords of a Text; Theoretical Computer Science, 40:31-55, (1985). (7.2.5).

142. Blumer, A., Blumer, J., Haussler, D., McConnell, R. and Ehrenfeucht, A.: Complete Inverted Files for Efficient Text Retrieval and Analysis; J.ACM, 34(3):578-595, (July 1987). (7.2.1, 7.2.5).

143. Bobrow, D.G. and Clark, D.W.: Compact Encodings of List Structure; ACM TOPLAS, 1(2):266-286, (Oct 1979). (2.1).

144. Bobrow, D.G.: A Note on Hash Linking; C.ACM, 18(7):413-415, (July 1975). (3.3).

145. Bollobas, B. and Simon, I.: Repeated Random Insertion in a Priority Queue; J of Algorithms, 6(4):466-477, (Dec 1985). (5.1.3).

146. Bolour, A.: Optimal Retrieval Algorithms for Small Region Queries; SIAM J on Computing, 10(4):721-741, (Nov 1981). (3.3).

147. Bolour, A.: Optimality Properties of Multiple-Key Hashing Functions; J.ACM, 26(2):196-210, (Apr 1979). (3.3.1, 3.5.4).

148. Bookstein, A.: Double Hashing; J American Society of Information Science, 23(6):402-405, (1972). (3.3.5, 3.3.11).

149. Bookstein, A.: On Harrison's Substring Testing Technique; C.ACM, 16:180-181, (1973). (7.2.6).

150. Boothroyd, J.: Algorithm 201, Shellsort; C.ACM, 6(8):445, (Aug 1963). (4.1.4).

151. Boothroyd, J.: Algorithm 207, Stringsort; C.ACM, 6(10):615, (Oct 1963). (4.1).

152. Borodin, A. and Cook, S.: A Time-Space Tradeoff for Sorting on a General Sequential Model of Computation; SIAM J on Computing, 11(2):287-297, (May 1982). (4.1, 4.2).

153. Borodin, A. and Cook, S.: On the Number of Additions to Compute Specific Polynomials; SIAM J on Computing, 5(1):146-157, (Mar 1976). (6.4).

154. Borodin, A., Fischer, M.J., Kirkpatrick, D.G., Lynch, N.A. and Tompa, M.P.: A Time-Space Tradeoff for Sorting on Non-Oblivious Machines; Proceedings FOCS, San Juan PR, 20:319-327, (Oct 1979). (4.1, 4.2).

155. Borwein, J.M. and Borwein, P.M.: The Arithmetic-Geometric Mean and Fast Computation of Elementary Functions; SIAM Review, 26(3):351-366, (1984). (6.2).

156. Boyer, R. and Moore, S.: A Fast String Searching Algorithm; C.ACM, 20:762-772, (1977). (7.1.3).

157. Bradford, J.: Sequence Matching with Binary Codes; Inf. Proc. Letters, 34(4):193-196, (July 1983). (7.1.8).

158. Brain, M.D. and Tharp, A.L.: Perfect Hashing Using Sparse Matrix Packing; Inform. Systems, 15(3):281-290, (1990). (3.3.16).

159. Brent, R.P.: Fast Multiple-Precision Evaluation of Elementary Functions; J.ACM, 23(2):242-251, (1976). (6.1, 6.2).

160. Brent, R.P.: Multiple-Precision Zero-Finding Methods and the Complexity of Elementary Function Evaluation; Analytic Computational Complexity, Academic Press, :151-176, (1976). (6.1, 6.2).

161. Brent, R.P.: Reducing the Retrieval Time of Scatter Storage Techniques; C.ACM, 16(2):105-109, (Feb 1973). (3.3.8.1).

162. Brinck, K. and Foo, N.Y.: Analysis of Algorithms on Threaded Trees; Computer Journal, 24(2):148-155, (May 1981). (3.4.1.1).

163. Brinck, K.: Computing parent nodes in threaded binary trees; BIT, 26(4):402-409, (1986). (3.4.1).

164. Brinck, K.: On deletion in threaded binary trees; J of Algorithms, 7(3):395-411, (Sep 1986). (3.4.1.9).

165. Brinck, K.: The expected performance of traversal algorithms in binary trees; Computer Journal, 28(4):426-432, (Aug 1985). (3.4.1).

166. Brockett, R.W. and Dobkin, D.: On the Number of Multiplications Required for Matrix Multiplication; SIAM J on Computing, 5(4):624-628, (Dec 1976). (6.3).

167. Broder, A.Z. and Karlin, A.R.: Multilevel Adaptive Hashing; Proceedings SODA, San Francisco CA, 1:43-53, (Jan 1990). (3.3).

168. Bron, C.: Algorithm 426: Merge Sort Algorithm (M1); C.ACM, 15(5):357-358, (May 1972). (4.2.1).

169. Brown, G.G. and Shubert, B.O.: On random binary trees; Math. Operations Research, 9:43-65, (1984). (3.4.1).

170. Brown, M.R. and Dobkin, D.: An Improved Lower Bound on Polynomial Multiplication; IEEE Trans. on Computers, 29(5):337-340, (May 1980). (6.4).

171. Brown, M.R. and Tarjan, R.E.: A Fast Merging Algorithm; J.ACM, 26(2):211-226, (Apr 1979). (4.3, 5.1).

172. Brown, M.R. and Tarjan, R.E.: A Representation for Linear Lists with Movable Fingers; Proceedings STOC-SIGACT, San Diego CA, 10:19-29, (May 1978). (3.4.2.1).

173. Brown, M.R. and Tarjan, R.E.: Design and Analysis of a Data Structure for Representing Sorted Lists; SIAM J on Computing, 9(3):594-614, (Aug 1980). (3.4.2.1).

174. Brown, M.R.: A Partial Analysis of Random Height-Balanced Trees; SIAM J on Computing, 8(1):33-41, (Feb 1979). (3.4.1.3).

175. Brown, M.R.: A Storage Scheme for Height-Balanced Trees; Inf. Proc. Letters, 7(5):231-232, (Aug 1978). (3.4.1.3, 3.4.2.1).

176. Brown, M.R.: Implementation and Analysis of Binomial Queue Algorithms; SIAM J on Computing, 7(3):298-319, (Aug 1978). (5.1.7).

177. Brown, M.R.: Some Observations on Random 2-3 Trees; Inf. Proc. Letters, 9(2):57-59, (Aug 1979). (3.4.2.1).

178. Brown, M.R.: The Complexity of Priority Queue Maintenance; Proceedings STOC-SIGACT, Boulder CO, 9:42-48, (May 1977). (5.1.7).

179. Bruno, J. and Coffman, E.G.: Nearly Optimal Binary Search Trees; Proceedings Information Processing 71, Ljubjana, Yugoslavia, :99-103, (Aug 1971). (3.4.1.7).

180. Bruss, A.R. and Meyer, A.R.: On Time-Space Classes and their Relation to the Theory of Real Addition; Theoretical Computer Science, 11(1):59-69, (1980). (6.1).

181. Buchholz, W.: File Organization and Addressing; IBM Systems J, 2(2):86-111, (June 1963). (3.3.4).

182. Bui, T.D. and Thanh, M.: Significant improvements to the Ford-Johnson algorithm; BIT, 25(1):70-759, (1985). (4.1).

183. Burgdorff, H.A., Jajodia, S., Sprigstell, N.F. and Zalcstein, Y.: Alternative methods for the reconstruction of trees from their traversals; BIT, 27(2):134-140, (1987). (3.4.1).

184. Burge, W.H.: An Analysis of Binary Search Trees Formed from Sequences of Nondistinct Keys; J.ACM, 23(3):451-454, (July 1976). (3.4.1).

185. Burkhard, W.A.: Full Table Quadratic Quotient Searching; Computer Journal, 18(1):161-163, (Feb 1975). (3.3.6).

186. Burkhard, W.A.: Hashing and Trie Algorithms for Partial Match Retrieval; ACM TODS, 1(2):175-187, (June 1976). (3.4.4, 3.5.4).

187. Burkhard, W.A.: Interpolation-based index maintenance; BIT, 23(3):274-294, (1983). (3.2.2, 3.3.13, 3.4.3).

188. Burkhard, W.A.: Non-uniform partial-match file designs; Theoretical Computer Science, 5(1):1-23, (1977). (3.6.2).

189. Burkhard, W.A.: Nonrecursive Traversals of Trees; Computer Journal, 18(3):227-230, (1975). (3.4.1).

190. Burkhard, W.A.: Partial-Match Hash Coding: Benefits of Redundancy; ACM TODS, 4(2):228-239, (June 1979). (3.5.4, 7.2.6).

191. Burkhard, W.A.: Associative Retrieval Trie Hash-Coding; JCSS, 15(3):280-299, (Dec 1977). (3.4.4, 3.5.4).

192. Burton, F.W., Kollias, J.G., Matsakis, D.G. and Kollias, V.G.: Implementation of Overlapping B-trees for Time and Space Efficient Representation of Collection of Similar Files; Computer Journal, 33(3):279-280, (June 1989). (3.4.2).

193. Burton, F.W. and Lewis, G.N.: A Robust Variation of Interpolation Search; Inf. Proc. Letters, 10(4):198-201, (July 1980). (3.2.2).

194. Burton, F.W.: Generalized Recursive Data Structures; Acta Informatica, 12(2):95-108, (1979). (2.1).

195. Cardenas, A.F. and Sagamang, J.P.: Doubly-Chained Tree Data Base Organization - Analysis and Design Strategies; Computer Journal, 20(1):15-26, (1977). (3.4.3).

196. Cardenas, A.F.: Evaluation and Selection of File Organization - A Model and a System; C.ACM, 16(9):540-548, (Sep 1973). (3.4.3).

197. Carlsson, S., Chen, J. and Strothotte, T.: A note on the construction of the data structure deap; Inf. Proc. Letters, 31(6):315-317, (June 1989). (5.1.3).

198. Carlsson, S. and Mattsson, C.: An Extrapolation on the Interpolation Search; Proceedings SWAT 88, Halmstad, Sweden, 1:24-33, (1988). (3.2.2).

199. Carlsson, S., Munro, J.I. and Poblete, P.V.: An Implicit Binomial Queue with Constant Insertion Time; Proceedings SWAT 88, Halmstad, Sweden, 1:1-13, (1988). (5.1.7).

200. Carlsson, S.: Average-case results on heapsort; BIT, 27(1):2-16, (1987). (4.1.5).

201. Carlsson, S.: Improving worst-case behavior of heaps; BIT, 24(1):14-18, (1984). (5.1.3).

202. Carlsson, S.: Split Merge-A Fast Stable Merging Algorithm; Inf. Proc. Letters, 22(4):189-192, (Apr 1986). (4.3.2).

203. Carlsson, S.: The Deap - A double-ended heap to implement double-ended priority queues; Inf. Proc. Letters, 26(1):33-36, (Sep 1987). (5.1.3).

204. Carter, J.L. and Wegman, M.N.: Universal Classes of Hash Functions; JCSS, 18(2):143-154, (Apr 1979). (3.3.1).

205. Casey, R.G.: Design of Tree Structures for Efficient Querying; C.ACM, 16(9):549-556, (Sep 1973). (3.4.3).

206. Celis, P., Larson, P. and Munro, J.I.: Robin Hood Hashing; Proceedings FOCS, Portland OR, 26:281-288, (Oct 1985). (3.3.3, 3.3.8.4).

207. Celis, P.: External Robin Hood Hashing; Proceedings SCCC Int. Conf. in Computer Science, Santiago, Chile, 6:185-200, (July 1986). (3.3.3, 3.3.8.4).

208. Celis, P.: Robin Hood Hashing; PhD Dissertation, University of Waterloo, (1985). (3.3.3, 3.3.8.4).

209. Cercone, N., Boates, J. and Krause, M.: An Interactive System for Finding Perfect Hashing Functions; IEEE Software, 2(6):38-53, (1985). (3.3.16).

210. Cesarini, F. and Sada, G.: An algorithm to construct a compact B-tree in case of ordered keys; Inf. Proc. Letters, 17(1):13-16, (July 1983). (3.4.2).

211. Cesarini, F. and Soda, G.: Binary Trees Paging; Inform. Systems, 7:337-334, (1982). (3.4.1).

212. Chang, C.C. and Lee, R.C.T.: A Letter-oriented minimal perfect hashing; Computer Journal, 29(3):277-281, (June 1986). (3.3.16).

213. Chang, C.C.: The Study of an Ordered Minimal Perfect Hashing Scheme; C.ACM, 27(4):384-387, (Apr 1984). (3.3.16).

214. Chang, H. and Iyengar, S.S.: Efficient Algorithms to Globally Balance a Binary Search Tree; C.ACM, 27(7):695-702, (July 1984). (3.4.1.6).

215. Chapin, N.: A Comparison of File Organization Techniques; Proceedings ACM-NCC, New York NY, 24:273-283, (Sep 1969). (3.3, 3.4.3).

216. Chapin, N.: Common File Organization Techniques Compared; Proceedings AFIPS Fall JCC, Las Vegas NE, :413-422, (Nov 1969). (3.3, 3.4.3).

217. Chazelle, B. and Guibas, L.J.: Fractional Cascading: I. A Data Structuring technique; Algorithmica, 1(2):133-162, (1986). (2.2).

218. Chazelle, B.: Filtering Search: A New Approach to Query-Answering; Proceedings FOCS, Tucson AZ, 24:122-132, (Nov 1983). (2.2.2.1).

219. Chazelle, B.: Lower Bounds in the Complexity of Multidimensional Searching; Proceedings FOCS, Toronto, Canada, 27:87-96, (Oct 1986). (3.6.2).

220. Chazelle, B.: Polytope Range Searching and Integral Geometry; Proceedings FOCS, Los Angeles CA, 28:1-10, (Oct 1987). (3.6).

221. Chen, L.: Space complexity deletion for AVL-trees; Inf. Proc. Letters, 22(3):147-149, (Mar 1986). (3.4.1.3).

222. Chen, M.T. and Seiferas, J.: Efficient and Elegant Subword Tree Construction; Combinatorial Algorithms on Words, NATO ASI Series, Springer-Verlag, F12:97-107, (1985). (7.2.2).

223. Chen, W-C. and Vitter, J.S.: Analysis of Early-Insertion Standard Coalescing Hashing; SIAM J on Computing, 12(4):667-676, (Nov 1983). (3.3.12).

224. Chen, W-C. and Vitter, J.S.: Deletion algorithms for coalesced hashing; Computer Journal, 29(5):436-450, (Oct 1986). (3.3.12).

225. Chen, W-C. and Vitter, J.S.: Analysis of New Variants of Coalesced Hashing; ACM TODS, 9(4):616-645, (1984). (3.3.12).

226. Chin, F.Y. and Fok, K.S.: Fast Sorting Algorithms on Uniform Ladders (Multiple Shift-Register Loops); IEEE Trans. on Computers, C29(7):618-631, (July 1980). (4.2).

227. Chin, F.Y.: A Generalized Asymptotic Upper Bound on Fast Polynomial Evaluation and Interpolation; SIAM J on Computing, 5(4):682-690, (Dec 1976). (6.4).

228. Choy, D.M. and Wong, C.K.: Bounds for Optimal $\alpha - \beta$ Binary Trees; BIT, 17(1):1-15, (1977). (3.4.1.7).

229. Choy, D.M. and Wong, C.K.: Optimal $\alpha - \beta$ trees with Capacity Constraint; Acta Informatica, 10(3):273-296, (1978). (3.4.1.7).

230. Christen, C.: Improving the Bounds on Optimal Merging; Proceedings FOCS, Ann Arbor MI, 19:259-266, (Oct 1978). (4.3.3).

231. Christodoulakis, S. and Ford, D.A.: File Organizations and Access Methods for CLV optical disks; Proceedings ACM SIGIR, Cambridge, Mass., 12:152-159, (June 1989). (3.3, 3.4.2.5).

232. Chung, F.R.K., Hajela, D.J. and Seymour, P.D.: Self-Organizing Sequential search and Hilbert's Inequalities; JCSS, 36(2):148-157, (Apr 1988). (3.1.2).

233. Cichelli, R.J.: Minimal Perfect Hash Functions Made Simple; C.ACM, 23(1):17-19, (Jan 1980). (3.3.16).

234. Clapson, P.: Improving the Access Time for Random Access Files; C.ACM, 20(3):127-135, (Mar 1977). (3.3).

235. Clark, D.W.: An Efficient List-Moving Algorithm Using Constant Workspace; C.ACM, 19(6):352-354, (June 1976). (3.1.1).

236. Clark, K.L. and Darlington, J.: Algorithm Classification Through Synthesis; Computer Journal, 23(1):61-65, (Feb 1980). (2.2.2).

237. Claybrook, B.G. and Yang, C-S.: Efficient Algorithms for Answering Queries with Unsorted Multilists; Inform. Systems, 3:93-57, (1978). (3.1).

238. Claybrook, B.G.: A Facility for Defining and Manipulating Generalized Data Structures; ACM TODS, 2(4):370-406, (Dec 1977). (2.1).

239. Coffman, E.G. and Bruno, J.: On File Structuring for Non-Uniform Access Frequencies; BIT, 10(4):443-456, (1970). (3.4.1).

240. Coffman, E.G. and Eve, J.: File Structures Using Hashing Functions; C.ACM, 13(7):427-436, (1970). (3.3).

241. Cohen, J. and Roth, M.: On the Implementation of Strassen's Fast Multiplication Algorithm; Acta Informatica, 6:341-355, (1976). (6.3).

242. Cohen, J.: A Note on a Fast Algorithm for Sparse Matrix Multiplication; Inf. Proc. Letters, 16(5):247-248, (June 1983). (6.3).

243. Cole, R.: On the Dynamic Finger Conjecture for Splay Trees; Proceedings STOC-SIGACT, Baltimore MD, 22:8-17, (May 1990). (3.4.1.6).

244. Cole, R.: Searching and Storing similar lists; J of Algorithms, 7(2):202-220, (June 1986). (3.5).

245. Colin, A.J.T., McGettrick, A.D. and Smith, P.D.: Sorting Trains; Computer Journal, 23(3):270-273, (Aug 1980). (4.2, 4.4.4).

246. Collins, G.E. and Musser, D.R.: Analysis of the Pope-Stein Division Algorithm; Inf. Proc. Letters, 6(5):151-155, (Oct 1977). (6.1).

247. Collmeyer, A.J. and Shemer, J.E.: Analysis of Retrieval Performance for Selected File Organization Techniques; Proceedings AFIPS, Houston TX, 37:201-210, (1970). (3.3, 3.4.3).

248. Comer, D. and Sethi, R.: The Complexity of Trie Index Construction; J.ACM, 24(3):428-440, (July 1977). (3.4.4).

249. Comer, D. and Shen, V.: Hash-Bucket Search: A Fast Technique for Searching an English Spelling Dictionary; Software - Practice and Experience, 12:669-682, (1982). (7.2.2, 7.2.6).

250. Comer, D.: A Note on Median Split Trees; ACM TOPLAS, 2(1):129-133, (Jan 1980). (3.4.1.6).

251. Comer, D.: Analysis of a Heuristic for Full Trie Minimization; ACM TODS, 6(3):513-537, (Sep 1981). (3.4.4).

252. Comer, D.: Effects of Updates on Optimality in Tries; JCSS, 26(1):1-13, (Feb 1983). (3.4.4).

253. Comer, D.: Heuristics for Trie Index Minimization; ACM TODS, 4(3):383-395, (Sep 1979). (3.4.4).

254. Comer, D.: The Ubiquitous B-tree; ACM C. Surveys, 11(2):121-137, (June 1979). (3.4.2).

255. Commentz-Walter, B.: A String Matching Algorithm Fast on the Average; Proceedings ICALP, Lecture Notes in Computer Science 71, Springer-Verlag, Graz, Austria, 6:118-132, (July 1979). (7.1.4).

256. Cook, C.R. and Kim, D.J.: Best Sorting Algorithm for Nearly Sorted Lists; C.ACM, 23(11):620-624, (Nov 1980). (4.1).

257. Cooper, D., Dicker, M.E. and Lynch, F.: Sorting of Textual Data Bases: A Variety Generation Approach to Distribution Sorting; Inf. Processing and Manag., 16:49-56, (1980). (4.2.3).

258. Cooper, R.B. and Solomon, M.K.: The Average Time until Bucket Overflow; ACM TODS, 9(3):392-408, (1984). (3.4.3).

259. Coppersmith, D. and Winograd, S.: Matrix Multiplication via Arithmetic Progressions; Proceedings STOC-SIGACT, New York, 19:1-6, (1987). (6.3).

260. Coppersmith, D. and Winograd, S.: On the Asymptotic Complexity of Matrix Multiplication; SIAM J on Computing, 11(3):472-492, (Aug 1982). (6.3).

261. Coppersmith, D.: Rapid Multiplication of Rectangular Matrices; SIAM J on Computing, 11(3):467-471, (Aug 1982). (6.3).

262. Cormack, G.V., Horspool, R.N.S. and Kaiserswerth, M.: Practical perfect hashing; Computer Journal, 28(1):54-55, (Feb 1985). (3.3.16).

263. Coulbourn, C.J. and van Oorshot, P.C.: Applications of Combinatorial Designs in Computer Science; ACM C. Surveys, 21(2):223-250, (June 1989). (3.6, 7.2.1).

264. Cowan, R. and Griss, M.: Hashing: the key to rapid pattern matching; Proceedings EUROSAM, Lecture Notes in Computer Science 72, Springer-Verlag, Marseille, France, :266-278, (June 1979). (7.2.6).

265. Cremers, A.B. and Hibbard, T.N.: Orthogonality of Information Structures; Acta Informatica, 9(3):243-261, (1978). (2.1).

266. Crochemore, M.: An Optimal Algorithm for Computing the Repetitions in a Word; Inf. Proc. Letters, 12:244-250, (1981). (7.1).

267. Crochemore, M.: Computing LCF in linear time; Bulletin EATCS, 30:57-61, (1986). (7.3.1).

268. Crochemore, M.: Optimal Factor transducers; Combinatorial Algorithms on Words, NATO ASI Series, Springer-Verlag, F12:31-44, (1985). (7.2.5).

269. Culberson, J.C. and Munro, J.I.: Analysis of the standard deletion algorithm in exact fit domain binary search trees; Algorithmica, 5(3):295-312, (1990). (3.4.1.9).

270. Culberson, J.C. and Munro, J.I.: Explaining the behavior of Binary Search Trees under Prolonged Updates: A Model and Simulations; Computer Journal, 32(1):68-75, (Feb 1989). (3.4.1.9).

271. Culberson, J.C.: The Effect of Asymmetric Deletions on Binary Search Trees; PhD Dissertation, Department of Computer Science, University of Waterloo, (May 1986). (3.4.1).

272. Culik II, K., Ottmann, T. and Wood, D.: Dense Multiway Trees; ACM TODS, 6(3):486-512, (Sep 1981). (3.4.2, 3.4.1.10).

273. Cunto, W. and Gascon, J.L.: Improving Time and Space Efficiency in Generalized Binary Search Trees; Acta Informatica, 24(5):583-594, (1987). (3.4.1.1).

274. Cunto, W., Gonnet, G.H. and Munro, J.I.: EXTQUICK: An In Situ Distributive External Sorting Algorithm; Information and Computation, to app.. (4.4.6).

275. Cunto, W., Lau, G. and Flajolet, P.: Analysis of KDT-Trees: KD-Trees improved by Local Reorganizations; Proceedings Workshop in Algorithms and Data Structures, Lecture Notes in Computer Science 382, Springer-Verlag, Ottawa, Canada, 1:24-38, (Aug 1989). (3.5.1).

276. Cunto, W. and Munro, J.I.: Average Case Selection; J.ACM, 36(2):270-279, (Apr 1989). (5.2).

277. Cunto, W. and Poblete, P.V.: Transforming Unbalanced Multiway trees into a Practical External Data structure; Acta Informatica, 26(3):193-212, (1988). (3.4.1.10).

278. Cunto, W. and Poblete, P.V.: Two Hybrid Methods for Collision Resolution in Open Addressing Hashing; Proceedings SWAT 88, Halmstad, Sweden, 1:113-119, (1988). (3.3.8.3).

279. Cunto, W.: Lower Bounds in Selection and Multiple Selection Problems; PhD Dissertation, University of Waterloo, (Dec 1983). (5.2).

280. Darlington, J.: A Synthesis of Several Sorting Algorithms; Acta Informatica, 11(1):1-30, (1978). (2.2.2).

281. Dasarathy, B. and Yang, C.: A Transformation on Ordered Trees; Computer Journal, 23(2):161-164, (Feb 1980). (3.4.1).

282. Davis, L.S. and Roussopoulos, N.: Approximate Pattern Matching in a Pattern Database System; Inform. Systems, 5:107-120, (1980). (7.3.2).

283. Day, A.C.: Full Table Quadratic Searching for Scatter Storage; C.ACM, 13(8):481-482, (1970). (3.3.6).

284. de la Brandais, R.: File Searching Using Variable Length Keys; Proceedings AFIPS Western JCC, San Francisco CA, :295-298, (Mar 1959). (3.4.4).

285. de la Torre, P.: Analysis of Tries; PhD Dissertation, University of Maryland, (July 1987). (3.4.4).

286. Deutscher, R.F., Sorenson, P.G. and Tremblay, J.P.: Distribution dependent hashing functions and their characteristics; Proceedings ACM SIGMOD, Ann Arbor MI, 11:224-236, (1975). (3.3).

287. Devillers, R. and Louchard, G.: Hashing Techniques, a Global Approach; BIT, 19(4):302-311, (1979). (3.3.4, 3.3.11, 3.3.1).

288. Devroye, L. and Klincsek, T.: Average Time Behavior of Distributive Sorting Algorithms; Computing, 26(1):1-7, (1981). (4.2.3).

289. Devroye, L.: A Note on the Average Depth of Tries; Computing, 28:367-371, (1982). (3.4.4).

290. Devroye, L.: A Note on the Height of Binary Search Trees; J.ACM, 33(3):489-498, (July 1986). (3.4.1.1).

291. Devroye, L.: A Probabilistic Analysis of the Height of Tries and of the Complexity of Triesort; Acta Informatica, 21(3):229-237, (1984). (3.4.1.1, 3.4.4, 4.2.4).

292. Devroye, L.: Applications of the theory of records in the study of random trees; Acta Informatica, 26(1-2):123-130, (1988). (3.4.1.1).

293. Devroye, L.: Branching Processes in the Analysis of the Heights of Trees; Acta Informatica, 24(3):277-298, (1987). (3.4.1.1).

294. Devroye, L.: Exponential Bounds for the Running Time of a Selection Algorithm; JCSS, 29(1):1-7, (Aug 1984). (5.2).

295. Devroye, L.: The expected length of the longest probe sequence for bucket searching when the distribution is not uniform; J of Algorithms, 6(1):1-9, (Mar 1985). (3.3).

296. Dewar, R.B.K.: A Stable Minimum Storage Sorting Algorithm; Inf. Proc. Letters, 2(6):162-164, (Apr 1974). (4.2.1).

297. Dhawan, A.K. and Srivastava, V.K.: On a New Division Algorithm; BIT, 17(4):481-485, (1977). (6.1).

298. Diehr, G. and Faaland, B.: Optimal Pagination of B-Trees with Variable-Length Items; C.ACM, 27(3):241-247, (Mar 1984). (3.4.2).

299. Dietzfelbinger, M., Karlin, A.R., Mehlhorn, K., Meyer auf der Heide, F., Rohnert, H. and Tarjan, R.E.: Dynamic Perfect Hashing; Proceedings FOCS, White Plains NY, 29:524-531, (Oct 1988). (3.3.16).

300. Dijkstra, E.W. and Gasteren, A.J.M.: An Introduction to Three Algorithms for Sorting in Situ; Inf. Proc. Letters, 15(3):129-134, (Oct 1982). (4.1.2, 4.1.5).

301. Dijkstra, E.W.: Smoothsort, an Alternative for Sorting In Situ; Science of Computer Programming, 1(3):223-233, (May 1982). (4.1.5).

302. Dinsmore, R.J.: Longer Strings from Sorting; C.ACM, 8(1):48, (Jan 1965). (4.4.1).

303. Doberkat, E.E.: An average case analysis of Floyd's Algorithm to construct heaps; Information and Control, 61(2):114-131, (May 1984). (4.1.5, 5.1.3).

304. Doberkat, E.E.: Asymptotic Estimates for the Higher Moments of the Expected Behavior of Straight Insertion Sort; Inf. Proc. Letters, 14(4):179-182, (June 1982). (4.1.2).

305. Doberkat, E.E.: Deleting the Root of a Heap; Acta Informatica, 17(3):245-265, (1982). (5.1.3).

306. Doberkat, E.E.: Inserting a New Element in a Heap; BIT, 21(3):255-269, (1981). (5.1.3).

307. Doberkat, E.E.: Some Observations on the Average Behavior of Heapsort; Proceedings FOCS, Syracuse NY, 21:229-237, (Oct 1980). (4.1.5).

308. Dobkin, D. and Lipton, R.J.: Addition Chain Methods for the Evaluation of Specific Polynomials; SIAM J on Computing, 9(1):121-125, (Feb 1980). (6.4).

309. Dobkin, D. and Lipton, R.J.: Multidimensional Searching Problems; SIAM J on Computing, 5(2):181-186, (June 1976). (3.5).

310. Dobkin, D. and Lipton, R.J.: Some Generalizations of Binary Search; Proceedings STOC-SIGACT, Seattle WA, 6:310-316, (Apr 1974). (3.5).

311. Dobkin, D. and Munro, J.I.: Determining the Mode; Theoretical Computer Science, 12(3):255-263, (Nov 1980). (5.2.3).

312. Dobkin, D. and Munro, J.I.: Optimal Time Minimal Space Selection Algorithms; J.ACM, 28(3):454-461, (July 1981). (5.2).

313. Dobkin, D. and van Leeuwen, J.: The Complexity of Vector-Products; Inf. Proc. Letters, 4(6):149-154, (Mar 1976). (6.3).

314. Dobkin, D.: On the Optimal Evaluation of a Set of N-Linear Forms; Proceedings SWAT (FOCS), Iowa City IO, 14:92-102, (Oct 1973). (6.3).

315. Dobosiewicz, W.: A Note on natural selection; Inf. Proc. Letters, 21(5):239-243, (Nov 1985). (4.4.1).

316. Dobosiewicz, W.: An Efficient Variation of Bubble Sort; Inf. Proc. Letters, 11(1):5-6, (Aug 1980). (4.1.1).

317. Dobosiewicz, W.: Sorting by Distributive Partitioning; Inf. Proc. Letters, 7(1):1-6, (Jan 1978). (4.2.5).

318. Dobosiewicz, W.: The Practical Significance of D.P. Sort Revisited; Inf. Proc. Letters, 8(4):170-172, (Apr 1979). (4.2.5).

319. Douglas, C.C. and Miranker, W.L.: The multilevel principle applied to sorting; BIT, 30(2):178-195, (1990). (4.1).

320. Downey, P., Leong, B.L. and Sethi, R.: Computing Sequences with Addition Chains; SIAM J on Computing, 10(3):638-646, (Aug 1981). (6.2).

321. Draws, L., Eriksson, P., Forslund, E., Hoglund, L., Vallner, S. and Strothotte, T.: Two New Algorithms for Constructing Min-Max Heaps; Proceedings SWAT 88, Halmstad, Sweden, 1:43-50, (1988). (5.1.3).

322. Driscoll, J.R., Gabow, H.N., Shrairman, R. and Tarjan, R.E.: Relaxed Heaps: an alternative to Fibonacci heaps with applications to parallel computations; C.ACM, 31(11):1343-1354, (Nov 1988). (5.1.3).

323. Driscoll, J.R., Lang, S.D. and Bratman, S.M.: Achieving Minimum Height for Block Split Tree Structured Files; Inform. Systems, 12:115-124, (1987). (3.4.2).

324. Driscoll, J.R. and Lien, Y.E.: A Selective Traversal Algorithm for Binary Search Trees; C.ACM, 21(6):445-447, (June 1978). (3.4.1).

325. Dromey, R.G.: A Fast Algorithm for Text Comparison; Australian Computer J, 11:63-67, (1979). (7.3.1).

326. Du, M.W., Hsieh, T.M., Jea, K.F. and Shieh, D.W.: The Study of a New Perfect Hash Scheme; IEEE Trans. Software Engineering, SE-9(3):305-313, (Mar 1983). (3.3.16).

327. Ducoin, F.: Tri par Adressage Direct; RAIRO Informatique, 13(3):225-237, (1979). (4.1.6).

328. Dudzinski, K. and Dydek, A.: On a Stable Minimum Storage Merging Algorithm; Inf. Proc. Letters, 12(1):5-8, (Feb 1981). (4.3.2).

329. Dvorak, S. and Durian, B.: Merging by decomposition revisited; Computer Journal, 31(6):553-556, (Dec 1988). (4.3.2).

330. Dvorak, S. and Durian, B.: Stable linear time sublinear space merging; Computer Journal, 30(4):372-374, (Aug 1987). (4.3.2).

331. Dvorak, S. and Durian, B.: Unstable linear time 0(1) space merging; Computer Journal, 31(3):279-282, (June 1988). (4.3.2).

332. Dwyer, B.: One More Time-How to Update a Master File; C.ACM, 24(1):3-8, (Jan 1981). (2.2.2.1).

333. Eades, P. and Staples, J.: On Optimal Trees; J of Algorithms, 2(4):369-384, (Dec 1981). (3.4.1.6).

334. Eastman, C.M. and Weiss, S.F.: Tree Structures for High Dimensionality Nearest Neighbor Searching; Inform. Systems, 7:115-122, (1982). (3.5).

335. Eastman, C.M. and Zemankova, M.: Partially Specified Nearest Neighbor Searches Using k-d Trees; Inf. Proc. Letters, 15(2):53-56, (Sep 1982). (3.5.2).

336. Eastman, C.M.: Optimal Bucket Size for Nearest Neighbor Searching in k-d Trees; Inf. Proc. Letters, 12(4):165-167, (Aug 1981). (3.5.2).

337. Eberlein, P.J.: A Note on Median Selection and Spider Production; Inf. Proc. Letters, 9(1):19-22, (July 1979). (5.2).

338. Ecker, A.: The Period of Search for the Quadratic and Related Hash Methods; Computer Journal, 17(4):340-343, (Nov 1974). (3.3.6).

339. Ehrenfeucht, A. and Haussler, D.: A new distance metric on strings computable in linear time; Discr App Math, 20:191-203, (1988). (7.1.8).

340. Ehrlich, G.: Searching and Sorting Real Numbers; J of Algorithms, 2(1):1-12, (Mar 1981). (3.2.2, 4.1.6).

341. Eisenbarth, B., Ziviani, N., Gonnet, G.H., Mehlhorn, K. and Wood, D.: The Theory of Fringe Analysis and Its Application to 2-3 Trees and B-Trees; Information and Control, 55(1):125-174, (Oct 1982). (3.4.2, 3.4.2.1).

342. Enbody, R.J. and Du, H.C.: Dynamic Hashing Schemes; ACM C. Surveys, 20(2):85-114, (June 1988). (3.3.13, 3.3.14).

343. Eppinger, J.L.: An Empirical Study of Insertion and Deletion in Binary Search Trees; C.ACM, 26(9):663-669, (Sep 1983). (3.4.1.1).

344. Eppstein, D., Galil, Z., Giancarlo, R. and Italiano, G.: Sparse Dynamic Programming; Proceedings SODA, San Francisco CA, 1:513-522, (Jan 1990). (7.1.8, 7.3.1).

345. Er, M.C. and Lowden, B.G.T.: The Theory and Practice of Constructing an Optimal Polyphase Sort; Computer Journal, 25(1):93-101, (Feb 1982). (4.4.4).

346. Erkio, H.: A Heuristic Approximation of the Worst Case of Shellsort; BIT, 20(2):130-136, (1980). (4.1.4).

347. Erkio, H.: Internal Merge Sorting with Delayed Selection; Inf. Proc. Letters, 11(3):137-140, (Nov 1980). (4.2.1).

348. Erkio, H.: Speeding Sort Algorithms by Special Instructions; BIT, 21(1):2-19, (1981). (4.1).

349. Erkio, H.: The worst case permutation for median-of-three quicksort; Computer Journal, 27(3):276-277, (Aug 1984). (4.1.3).

350. Erkioe, H. and Terkki, R.: Binary Search with Variable-Length Keys Within an Index Page; Inform. Systems, 8:137-140, (1983). (3.2.1).

351. Espelid, T.O.: Analysis of a Shellsort Algorithm; BIT, 13(4):394-400, (1973). (4.1.4).

352. Espelid, T.O.: On Replacement Selection and Dinsmore's Improvement; BIT, 16(2):133-142, (1976). (4.4.1).

353. Estivill-Castro, V. and Wood, D.: A new measure of presortedness; Information and Computation, 83(1):111-119, (Oct 1989). (4.1.8).

354. Eve, J.: The Evaluation of Polynomials; Numer Math, 6:17-21, (1974). (6.4).

355. Fabbrini, F. and Montani, C.: Autumnal Quadtrees; Computer Journal, 29(5):472-474, (Oct 1986). (3.5.1).

356. Fabri, J.: Some Remarks on p-Way Merging; SIAM J on Computing, 6(2):268-271, (June 1977). (4.3).

357. Fagin, R., Nievergelt, J., Pippenger, N. and Strong, H.R.: Extendible Hashing-A Fast Access Method for Dynamic Files; ACM TODS, 4(3):315-344, (Sep 1979). (3.3.13).

358. Faloutsos, C. and Christodoulakis, S.: Description and Performance Analysis of Signature File Methods; ACM TOOIS, 5(3):237-257, (1987). (7.2.6).

359. Faloutsos, C. and Christodoulakis, S.: Signature Files: An Access Method for Documents and Its Analytical Performance Evaluation; ACM TOOIS, 2(4):267-288, (Oct 1984). (7.2.6).

360. Faloutsos, C., Sellis, T. and Roussopoulos, N.: Analysis of Object Oriented Spatial Access Methods; Proceedings ACM SIGMOD, San Francisco CA, 16:426-439, (May 1987). (3.5).

361. Faloutsos, C.: Access Methods for Text; ACM C. Surveys, 17:49-74, (1985). (7.2).

362. Faloutsos, C. and Roseman, S.: Fractals for Secondary Key Retrieval; Proceedings ACM PODS, Philadelfia PA, 8, (Mar 1989). (3.5.4).

363. Faloutsos, C.: Multiattribute Hashing using Gray Codes; Proceedings ACM SIGMOD, Washington DC, 15:227-238, (May 1986). (3.5.4).

364. Faloutsos, C.: Signature Files : an integrated access method for text and attributes suitable for optical disk storage; BIT, 28(4):736-754, (1988). (7.2.6).

365. Feig, E.: Minimal Algorithms for Bilinear Forms May Have Divisions; J of Algorithms, 4(1):81-84, (Mar 1983). (6.3).

366. Feig, E.: On Systems of Bilinear Forms Whose Minimal Division-Free Algorithms are all Bilinear; J of Algorithms, 2(3):261-281, (Sep 1981). (6.3).

367. Feldman, J.A. and Low, J.R.: Comment on Brent's Scatter Storage Algorithm; C.ACM, 16(11):703, (Nov 1973). (3.3.8.1).

368. Felician, L.: Linked-hashing: an Improvement of Open Addressing Techniques for Large Secondary Storage Files; Inform. Systems, 12(4):385-390, (1987). (3.3).

369. Fiat, A., Naor, M., Schaffer, A., Schmidt, J.P. and Siegel, A.: Storing and Searching a Multikey Table; Proceedings STOC-SIGACT, Chicago IL, 20:344-353, (May 1988). (3.5).

370. Fiat, A., Naor, M., Schmidt, J.P. and Siegel, A.: Non-Oblivious Hashing; Proceedings STOC-SIGACT, Chicago IL, 20:367-376, (May 1988). (3.3.1).

371. Fiat, A. and Naor, M.: Implicit $O(1)$ Probe Search; Proceedings STOC-SIGACT, Seattle, Washington, 21:336-344, (May 1989). (3.3.1).

372. Finkel, R.A. and Bentley, J.L.: Quad Trees: A Data Structure for Retrieval on Composite Keys; Acta Informatica, 4(1):1-9, (1974). (3.5.1).

373. Fischer, M.J. and Paterson, M.S.: Fishpear: A priority queue algorithm; Proceedings FOCS, Singer Island FL, 25:375-386, (Oct 1984). (5.1).

374. Fischer, M.J. and Paterson, M.S.: String Matching and Other Products; Complexity of Computation (SIAM-AMS Proceedings 7), American Mathematical Society, Providence, RI, 7:113-125, (1974). (7.1).

375. Fisher, M.T.R.: On universal binary search trees; Fundamenta Informaticae, 4(1):173-184, (1981). (3.4.1).

376. Flajolet, P., Francon, J. and Vuillemin, J.: Computing Integrated Costs of Sequences of Operations with Applications to Dictionaries; Proceedings STOC-SIGACT, Atlanta GA, 11:49-61, (Apr 1979). (3.1.1, 3.2.1, 3.4.1).

377. Flajolet, P., Francon, J. and Vuillemin, J.: Sequence of Operations Analysis for Dynamic Data Structures; J of Algorithms, 1(2):111-141, (June 1980). (3.1.1, 3.2, 3.4.1, 5.1).

378. Flajolet, P., Francon, J. and Vuillemin, J.: Towards Analysing Sequences of Operations for Dynamic Data Structures; Proceedings FOCS, San Juan PR, 20:183-195, (Oct 1979). (3.1.1, 3.2, 3.4.1, 5.1).

379. Flajolet, P. and Martin, N.G.: Probabilistic Counting Algorithms for Data Base Applications; JCSS, 31(2):182-209, (Oct 1985). (6.1).

380. Flajolet, P. and Odlyzko, A.M.: Exploring Binary Trees and Other Simple Trees; Proceedings FOCS, Syracuse NY, 21:207-216, (Oct 1980). (3.4.1.2).

381. Flajolet, P. and Odlyzko, A.M.: Limit Distributions for Coefficients of Iterates of Polynomials with Applications to Combinatorial Enumerations; Math Proc Camb Phil Soc, 96:237-253, (1984). (3.4.1.2).

382. Flajolet, P. and Odlyzko, A.M.: The Average Height of Binary Trees and Other Simple Trees; JCSS, 25(2):171-213, (Oct 1982). (3.4.1.2).

383. Flajolet, P., Ottmann, T. and Wood, D.: Search Trees and Bubble Memories; RAIRO Informatique Theorique, 19(2):137-164, (1985). (3.4.1.1).

384. Flajolet, P. and Puech, C.: Partial Match Retrieval of Multidimensional Data; J.ACM, 33(2):371-407, (Apr 1986). (3.5.2, 3.6.2).

385. Flajolet, P. and Puech, C.: Tree Structures for Partial Match Retrieval; Proceedings FOCS, Tucson AZ, 24:282-288, (Nov 1983). (3.5.1, 3.5.2, 3.6.2).

386. Flajolet, P., Gonnet, G.H., Puech, C. and Robson, M.: The Analysis of Multidimensional Searching in Quad-Trees; Proceedings SODA'91, San Francisco CA, 2, (Jan 1991). (3.5.1).

387. Flajolet, P., Regnier, M. and Sotteau, D.: Algebraic Methods for Trie Statistics; Annals of Discrete Mathematics, 25:145-188, (1985). (3.4.4, 3.5.1).

388. Flajolet, P. and Saheb, N.: Digital Search Trees and the Generation of an Exponentially Distributed Variate; Proceedings CAAP, L'Aquila, Italy, 10:221-235, (1983). (3.4.4).

389. Flajolet, P. and Sedgewick, R.: Digital Search Trees Revisited; SIAM J on Computing, 15:748-767, (1986). (3.4.4).

390. Flajolet, P. and Steyaert, J.M.: A Branching Process Arising in Dynamic Hashing, Trie Searching and Polynomial Factorization; Proceedings ICALP, Aarhus, 9:239-251, (July 1982). (3.3.13, 3.4.4).

391. Flajolet, P.: Approximate Counting: A Detailed Analysis; BIT, 25:113-134, (1985). (6.1).

392. Flajolet, P.: On the Performance Evaluation of Extendible Hashing and Trie Search; Acta Informatica, 20(4):345-369, (1983). (3.3.13, 3.4.4).

393. Flores, I. and Madpis, G.: Average Binary Search Length for Dense Ordered Lists; C.ACM, 14(9):602-603, (Sep 1971). (3.2.1).

394. Flores, I.: Analysis of Internal Computer Sorting; J.ACM, 8(1):41-80, (Jan 1961). (4.1).

395. Flores, I.: Computer Time for Address Calculation Sorting; J.ACM, 7(4):389-409, (Oct 1960). (4.1.6, 4.2.3).

396. Floyd, R.W. and Rivest, R.L.: Expected Time Bounds for Selection; C.ACM, 18(3):165-172, (Mar 1975). (5.2).

397. Floyd, R.W. and Smith, A.J.: A Linear Time Two Tape Merge; Inf. Proc. Letters, 2(5):123-125, (Dec 1973). (4.3).

398. Floyd, R.W.: Algorithm 245, Treesort3; C.ACM, 7(12):701, (Dec 1964). (4.1.5, 5.1.3).

399. Floyd, R.W.: The Exact Time Required to Perform Generalized Addition; Proceedings FOCS, Berkeley CA, 16:3-5, (Oct 1975). (6.1).

400. Forbes, K.: Random Files and Subroutine for Creating a Random Address; Australian Computer J, 4(1):35-40, (1972). (3.3.1).

401. Foster, C.C.: A Generalization of AVL Trees; C.ACM, 16(8):513-517, (Aug 1973). (3.4.1.3).

402. Foster, C.C.: Information Storage and Retrieval Using AVL Trees; Proceedings ACM-NCC, Cleveland OH, 20:192-205, (1965). (3.4.1.3).

403. Francon, J., Randrianarimanana, B. and Schott, R.: Analysis of dynamic algorithms in Knuth's model; Theoretical Computer Science, 72(2/3):147-168, (May 1990). (3.4.1).

404. Francon, J., Viennot, G. and Vuillemin, J.: Description and Analysis of an Efficient Priority Queue Representation; Proceedings FOCS, Ann Arbor MI, 19:1-7, (Oct 1978). (5.1.5).

405. Francon, J.: On the analysis of algorithms for trees; Theoretical Computer Science, 4(2):155-169, (1977). (3.4.1.1).

406. Franklin, W.R.: Padded Lists: Set Operations in Expected O(log log N) Time; Inf. Proc. Letters, 9(4):161-166, (Nov 1979). (3.2.2).

407. Frazer, W.D. and Bennett, B.T.: Bounds of Optimal Merge Performance, and a Strategy for Optimality; J.ACM, 19(4):641-648, (Oct 1972). (4.4).

408. Frazer, W.D. and McKellar, A.C.: Samplesort: A Sampling Approach to Minimal Storage Tree Sorting; J.ACM, 17(3):496-507, (July 1970). (4.1.3, 4.2.6).

409. Frazer, W.D. and Wong, C.K.: Sorting by Natural Selection; C.ACM, 15(10):910-913, (Oct 1972). (4.4.1).

410. Frederickson, G.N. and Johnson, D.B.: Generalized Selection and Ranking; Proceedings STOC-SIGACT, Los Angeles CA, 12:420-428, (Apr 1980). (5.2).

411. Frederickson, G.N.: Improving Storage Utilization in Balanced Trees; Proceedings Allerton Conference, Monticello, IL, 17:255-264, (1979). (3.4.2).

412. Frederickson, G.N.: The Information Theory Bound is Tight for Selection in a Heap; Proceedings STOC-SIGACT, Baltimore MD, 22:26-33, (May 1990). (5.1.3, 5.2).

413. Fredkin, E.: Trie Memory; C.ACM, 3(9):490-499, (Sep 1960). (3.4.4, 7.2.2).

414. Fredman, M.L., Komlos, J. and Szemeredi, E.: Storing a Sparse Table with O(1) Worst Case Access Time; J.ACM, 31(3):538-544, (July 1984). (3.3.16).

415. Fredman, M.L. and Komlos, J.: On the Size of Separating Systems and Families of Perfect Hash Functions; SIAM J Alg Disc Methods, 5(1):61-68, (Mar 1984). (3.3.16).

416. Fredman, M.L., Sedgewick, R., Sleator, D.D. and Tarjan, R.E.: The Pairing Heap: A New Form of Self-Adjusting Heap; Algorithmica, 1(1):111-129, (Mar 1986). (5.1.3).

417. Fredman, M.L. and Spencer, T.H.: Refined complexity analysis for heap operations; JCSS, 35(3):269-284, (Dec 1987). (5.1.3).

418. Fredman, M.L. and Tarjan, R.E.: Fibonacci Heaps and Their Uses in Improved Network Optimization Algorithms; J.ACM, 34(3):596-615, (July 1987). (5.1.3).

419. Fredman, M.L. and Willard, D.E.: Blasting Through the Information Theoretic Barrier with Fusion Trees; Proceedings STOC-SIGACT, Baltimore MD, 22:1-7, (May 1990). (3.4.1, 3.5.3, 4.1).

420. Fredman, M.L.: A Lower Bound on the Complexity of Orthogonal Range Queries; J.ACM, 28(4):696-705, (Oct 1981). (3.6.2).

421. Fredman, M.L.: A Near Optimal Data Structure for a Type of Range Query Problem; Proceedings STOC-SIGACT, Atlanta GA, 11:62-66, (Apr 1979). (3.6.2).

422. Fredman, M.L.: How good is the information theory bound in sorting?; Theoretical Computer Science, 1(4):355-361, (1976). (4.1).

423. Fredman, M.L.: The Inherent Complexity of Dynamic Data Structures Which Accommodate Range Queries; Proceedings FOCS, Syracuse NY, 21:191-199, (Oct 1980). (3.6.2).

424. Fredman, M.L.: Two Applications of a Probabilistic Search Technique: Sorting X+Y and Building Balanced Search Trees; Proceedings STOC-SIGACT, Albuquerque NM, 7:240-244, (May 1975). (3.4.1.6).

425. Freeston, M.: Advances in the design of the BANG file; Proceedings Foundations of Data Organisation and Algorithms, Lecture Notes in Computer Science 367, Springer-Verlag, Paris, France, 3:322-338, (June 1989). (3.5.4).

426. Freeston, M.: The Bang file: a new kind of grid file; Proceedings ACM SIGMOD, San Francisco CA, 16:260-269, (May 1987). (3.5.4).

427. Friedman, J.H., Bentley, J.L. and Finkel, R.A.: An Algorithm for Finding Best Matches in Logarithmic Expected Time; ACM TOMS, 3(3):209-226, (Sep 1977). (3.5.2, 3.6).

428. Friend, E.H.: Sorting on Electronic Computer Systems; J.ACM, 3(3):134-168, (July 1956). (4.1, 4.2, 4.4).

429. Frieze, A.M.: On the random construction of heaps; Inf. Proc. Letters, 27(2):103-109, (Feb 1988). (5.1.3).

430. Furukawa, K.: Hash Addressing with Conflict Flag; Information Proc. in Japan, 13(1):13-18, (1973). (3.3.2, 3.3.3).

431. Fussenegger, F. and Gabow, H.N.: A Counting Approach to Lower Bounds for Selection Problems; J.ACM, 26(2):227-238, (Apr 1979). (5.2).

432. Fussenegger, F. and Gabow, H.N.: Using Comparison Trees to Derive Lower Bounds for Selection Problems; Proceedings FOCS, Houston TX, 17:178-182, (Oct 1976). (5.2).

433. Gairola, B.K. and Rajaraman, V.: A Distributed Index Sequential Access Method; Inf. Proc. Letters, 5(1):1-5, (May 1976). (3.4.3).

434. Gajewska, H. and Tarjan, R.E.: Dequeues with Heap Order; Inf. Proc. Letters, 22(4):197-200, (Apr 1986). (5.1.3).

435. Galil, Z. and Giancarlo, R.: Improved String Matching with k Mismatches; SIGACT News, 17:52-54, (1986). (7.1.8).

436. Galil, Z. and Megiddo, N.: A Fast Selection Algorithm and the Problem of Optimum Distribution of Effort; J.ACM, 26(1):58-64, (Jan 1979). (5.2).

437. Galil, Z. and Park, K.: An Improved Algorithm for Approximate String Matching; Proceedings ICALP, Stressa, Italy, 16:394-404, (July 1989). (7.1.8).

438. Galil, Z. and Seiferas, J.: A linear-time on-line recognition algorithm for Palstar; J.ACM, 25:102-111, (1978). (7.1).

439. Galil, Z. and Seiferas, J.: Linear-Time String Matching Using Only a Fixed Number of Local Storage Locations; Theoretical Computer Science, 13:331-336, (1981). (7.1).

440. Galil, Z. and Seiferas, J.: Saving Space in Fast String-Matching; SIAM J on Computing, 9:417-438, (1980). (7.1).

441. Galil, Z. and Seiferas, J.: Time-Space-Optimal String Matching; JCSS, 26:280-294, (1983). (7.1).

442. Galil, Z.: On Improving the Worst Case Running Time of the Boyer-Moore String Matching Algorithm; C.ACM, 22:505-508, (1979). (7.1.3).

443. Galil, Z.: Open Problems in Stringology; Combinatorial Algorithms on Words, NATO ASI Series, Springer-Verlag, F12:1-8, (1985). (7.1, 7.2).

444. Galil, Z.: Real-Time Algorithms for String-Matching and Palindrome Recognition; Proceedings STOC-SIGACT, Hershey, PA, 8:161-173, (1976). (7.1).

445. Galil, Z.: String Matching in Real Time; J.ACM, 28:134-149, (1981). (7.1).

446. Gamzon, E. and Picard, C.F.: Algorithme de Tri par Adressage Direct; C.R. Academie Sc. Paris, 269A, :38-41, (July 1969). (4.1.6).

447. Gardy, D., Flajolet, P. and Puech, C.: On the performance of orthogonal range queries in multiattribute and double chained trees; Proceedings Workshop in Algorithms and Data Structures, Lecture Notes in Computer Science 382, Springer-Verlag, Ottawa, Canada, 1:218-229, (Aug 1989). (3.6.2).

448. Garey, M.R.: Optimal Binary Search Trees with Restricted Maximal Depth; SIAM J on Computing, 3(2):101-110, (June 1974). (3.4.1.7).

449. Gargantini, I.: An Effective Way to Represent Quadtrees; C.ACM, 25(12):905-910, (Dec 1982). (3.5.1.1).

450. Garsia, A.M. and Wachs, M.L.: A New Algorithm for Minimum Cost Binary Trees; SIAM J on Computing, 6(4):622-642, (Dec 1977). (3.4.1.7).

451. Gassner, B.J.: Sorting by Replacement Selecting; C.ACM, 10(2):89-93, (Feb 1967). (4.4.1).

452. Gerash, T.E.: An Insertion Algorithm for a Minimal Internal Path Length Binary Search Tree; C.ACM, 31(5):579-585, (May 1988). (3.4.1.5).

453. Ghosh, S.P. and Lum, V.Y.: Analysis of Collisions when Hashing by Division; Inform. Systems, 1(1):15-22, (1975). (3.3).

454. Ghosh, S.P. and Senko, M.E.: File Organization: On the Selection of Random Access Index Points for Sequential Files; J.ACM, 16(4):569-579, (Oct 1969). (3.4.3).

455. Ghoshdastidar, D. and Roy, M.K.: A Study on the Evaluation of Shell's Sorting Technique; Computer Journal, 18(3):234-235, (Aug 1975). (4.1.4).

456. Gil, J., Meyer auf der Heide, F. and Wigderson, A.: Not all Keys can be Hashed in Constant Time; Proceedings STOC-SIGACT, Baltimore MD, 22:244-253, (May 1990). (3.3).

457. Gill, A.: Hierarchical Binary Search; C.ACM, 23(5):294-300, (May 1980). (3.4.1).

458. Gilstad, R.L.: Polyphase Merge Sorting - an Advanced Technique; Proceedings AFIPS Eastern JCC, New York NY, 18:143-148, (Dec 1960). (4.4.4).

459. Gilstad, R.L.: Read-Backward Polyphase Sorting; C.ACM, 6(5):220-223, (May 1963). (4.4.4).

460. Goetz, M.A. and Toth, G.S.: A Comparison Between the Polyphase and Oscillating Sort Techniques; C.ACM, 6(5):223-225, (May 1963). (4.4.5).

461. Goetz, M.A.: Internal and Tape Sorting Using the Replacement-Selection Technique; C.ACM, 6(5):201-206, (May 1963). (4.4.1).

462. Gonnet, G.H. and Baeza-Yates, R.A.: An Analysis of the Karp-Rabin String Matching Algorithm; Inf. Proc. Letters, 34:271-274, (1990). (7.1.5).

463. Gonnet, G.H. and Larson, P.: External Hashing with Limited Internal Storage; J.ACM, 35(1):161-184, (Jan 1988). (3.3.15).

464. Gonnet, G.H., Munro, J.I. and Suwanda, H.: Exegesis of Self-Organizing Linear Search; SIAM J on Computing, 10(3):613-637, (Aug 1981). (3.1.2, 3.1.3).

465. Gonnet, G.H., Munro, J.I. and Suwanda, H.: Toward Self-Organizing Linear Search; Proceedings FOCS, San Juan PR, 20:169-174, (Oct 1979). (3.1.2, 3.1.3).

466. Gonnet, G.H. and Munro, J.I.: A Linear Probing Sort and its Analysis; Proceedings STOC-SIGACT, Milwaukee WI, 13:90-95, (May 1981). (4.1.7).

467. Gonnet, G.H. and Munro, J.I.: Efficient Ordering of Hash Tables; SIAM J on Computing, 8(3):463-478, (Aug 1979). (3.3.9, 3.3.8.2).

468. Gonnet, G.H. and Munro, J.I.: Heaps on Heaps; SIAM J on Computing, 15(4):964-971, (Nov 1986). (5.1.3).

469. Gonnet, G.H. and Munro, J.I.: The Analysis of an Improved Hashing Technique; Proceedings STOC-SIGACT, Boulder CO, 9:113-121, (May 1977). (3.3.8.2, 3.3.9).

470. Gonnet, G.H. and Munro, J.I.: The Analysis of Linear Probing by the Use of a New Mathematical Transform; J of Algorithms, 5:451-470, (1984). (4.1.7).

471. Gonnet, G.H., Olivie, H.J. and Wood, D.: Height-Ratio-Balanced Trees; Computer Journal, 26(2):106-108, (May 1983). (3.4.1.3).

472. Gonnet, G.H., Rogers, L.D. and George, J.A.: An Algorithmic and Complexity Analysis of Interpolation Search; Acta Informatica, 13(1):39-52, (Jan 1980). (3.2.2).

473. Gonnet, G.H. and Rogers, L.D.: The Interpolation-Sequential Search Algorithm; Inf. Proc. Letters, 6(4):136-139, (Aug 1977). (3.2.3).

474. Gonnet, G.H. and Tompa, F.W.: A Constructive Approach to the Design of Algorithms and Their Data Structures; C.ACM, 26(11):912-920, (Nov 1983). (2.1, 2.2).

475. Gonnet, G.H. and Tompa, F.W.: Mind your Grammar: A New Approach to Modelling Text; Proceedings VLDB, Brighton, England, 13:339-346, (Aug 1987). (7.2.7).

476. Gonnet, G.H.: Average Lower Bounds for Open Addressing Hash Coding; Proceedings Theoretical Computer Science, Waterloo, Ont, :159-162, (Aug 1977). (3.3.9).

477. Gonnet, G.H.: Balancing Binary Trees by Internal Path Reduction; C.ACM, 26(12):1074-1081, (Dec 1983). (3.4.1.5).

478. Gonnet, G.H.: Efficient Searching of Text and Pictures; (Technical Report OED-88-02)(1988). (7.2.2, 7.2.3, 7.3.2).

479. Gonnet, G.H.: Expected Length of the Longest Probe Sequence in Hash Code Searching; J.ACM, 28(2):289-304, (Apr 1981). (3.3.2, 3.3.9, 3.3.10).

480. Gonnet, G.H.: Heaps Applied to Event Driven Mechanisms; C.ACM, 19(7):417-418, (July 1976). (5.1.3).

481. Gonnet, G.H.: Interpolation and Interpolation-Hash Searching; PhD Dissertation, University of Waterloo, (Feb 1977). (3.2.2).

482. Gonnet, G.H.: Notes on the Derivation of Asymptotic Expressions from Summations; Inf. Proc. Letters, 7(4):165-169, (June 1978). (II).

483. Gonnet, G.H.: On Direct Addressing Sort; RAIRO TSI, 3(2):123-127, (Mar 1984). (4.1.6).

484. Gonnet, G.H.: Open Addressing Hashing with Unequal Probability Keys; JCSS, 21(3):354-367, (Dec 1980). (3.3.2).

485. Gonnet, G.H.: PAT Implementation; (1986). (7.2.4).

486. Gonnet, G.H.: Unstructured Data Bases or Very Efficient Text Searching; Proceedings ACM PODS, Atlanta, GA, 2:117-124, (Mar 1983). (7.2, 7.2.2).

487. Gonzalez, T.F. and Johnson, D.B.: Sorting Numbers in Linear Expected Time and Optimal Extra Space; Inf. Proc. Letters, 15(3):119-124, (Oct 1982). (4.1.8).

488. Goodman, J.E. and Pollack, R.: Multidimensional Sorting; SIAM J on Computing, 12(3):484-507, (Aug 1983). (4.2).

489. Gordon, D.: Eliminating the flag in threaded binary search trees; Inf. Proc. Letters, 23(4):209-214, (Apr 1986). (3.4.1).

490. Gori, M. and Soda, G.: An algebraic approach to Cichelli's perfect hashing; BIT, 29(1):2-13, (1989). (3.3.16).

491. Gotlieb, C.C. and Walker, W.A.: A Top-Down Algorithm for Constructing Nearly Optimal Lexicographical Trees; Graph Theory and Computing, Academic Press, :303-323, (1972). (3.4.1.6).

492. Gotlieb, C.C.: Sorting on Computers; C.ACM, 6(5):194-201, (May 1963). (4.4).

493. Gotlieb, L.R.: Optimal Multi-Way Search Trees; SIAM J on Computing, 10(3):422-433, (Aug 1981). (3.4.2).

494. Goto, E. and Kanada, Y.: Hashing Lemmas on Time Complexity; Proceedings ACM Symp. on Algebr. and Symbolic Comp., Yorktown Heights NY, :154-158, (Aug 1976). (3.3).

495. Greene, D.H.: Labelled Formal Languages and Their Uses; PhD Dissertation, Stanford University, (June 1983). (3.4.1.6).

496. Grimson, J.B. and Stacey, G.M.: A Performance Study of Some Directory Structures for Large Files; Inf. Storage and Retrieval, 10(11/12):357-364, (1974). (3.4.3, 7.2.1).

497. Grossi, R. and Luccio, F.: Simple and Efficient string matching with k mismatches; Inf. Proc. Letters, 33(3):113-120, (July 1989). (7.1.8).

498. Guibas, L.J., McCreight, E.M., Plass, M.F. and Roberts, J.R.: A New Representation for Linear Lists; Proceedings STOC-SIGACT, Boulder CO, 9:49-60, (May 1977). (3.2, 3.4.2).

499. Guibas, L.J. and Odlyzko, A.M.: A New Proof of the Linearity of the Boyer-Moore String Searching Algorithm; SIAM J on Computing, 9:672-682, (1980). (7.1.3).

500. Guibas, L.J. and Sedgewick, R.: A Dichromatic Framework for Balanced Trees; Proceedings FOCS, Ann Arbor MI, 19:8-21, (Oct 1978). (3.4.1.3, 3.4.2.4).

501. Guibas, L.J. and Szemeredi, E.: The Analysis of Double Hashing; JCSS, 16(2):226-274, (Apr 1978). (3.3.5).

502. Guibas, L.J.: A Principle of Independence for Binary Tree Searching; Acta Informatica, 4:293-298, (1975). (3.4.1.1).

503. Guibas, L.J.: The Analysis of Hashing Algorithms; PhD Dissertation, Stanford University, (Aug 1976). (3.3.5, 3.3).

504. Guibas, L.J.: The Analysis of Hashing Techniques that Exhibit k-ary Clustering; J.ACM, 25(4):544-555, (Oct 1978). (3.3).

505. Guntzer, U. and Paul, M.C.: Jump interpolation search trees and symmetric binary numbers; Inf. Proc. Letters, 26(4):193-204, (Dec 1987). (3.1.5).

506. Gupta, G.K. and Srinivasan, B.: Approximate Storage utilization of B-trees; Inf. Proc. Letters, 22(5):243-246, (Apr 1986). (3.4.2).

507. Gupta, U.I., Lee, D.T. and Wong, C.K.: Ranking and Unranking of 2-3 Trees; SIAM J on Computing, 11(3):582-590, (Aug 1982). (3.4.2.1).

508. Gupta, U.I., Lee, D.T. and Wong, C.K.: Ranking and Unranking of B-trees; J of Algorithms, 4(1):51-60, (Mar 1983). (3.4.2).

509. Gurski, A.: A Note on Analysis of Keys for Use in Hashing; BIT, 13(1):120-122, (1973). (3.3.1).

510. Guting, R.H. and Kriegel, H.P.: Dynamic k-dimensional Multiway Search Under Time-varying Access Frequencies; Lecture Notes in Computer Science 104, Springer-Verlag, :135-145, (1981). (3.5).

511. Guting, R.H. and Kriegel, H.P.: Multidimensional B-tree: An Efficient Dynamic File Structure for Exact Match Queries; Informatik Fachberichte, 33:375-388, (1980). (3.5).

512. Guttman, A.: R-Trees: A Dynamic Index Structure for Spatial Searching; Proceedings ACM SIGMOD, Boston, Mass, 14:47-57, (June 1984). (3.5).

513. Gwatking, J.C.: Random Index File Design; Australian Computer J, 5(1):29-34, (1973). (3.3.11).

514. Halatsis, C. and Philokypru, G.: Pseudo Chaining in Hash Tables; C.ACM, 21(7):554-557, (July 1978). (3.3).

515. Hall, P.A.V. and Dowling, G.R.: Approximate String Matching; ACM C. Surveys, 12:381-402, (1980). (7.1.8).

516. Handley, C.: An in-situ distributive sort; Inf. Proc. Letters, 23(5):265-270, (Apr 1986). (4.2.5).

517. Hansen, E.R., Patrick, M.L. and Wong, R.L.C.: Polynomial evaluation with scaling; ACM TOMS, 16(1):86-93, (Mar 1990). (6.4).

518. Hansen, W.J.: A Cost Model for the Internal Organization of B+ Tree Nodes; ACM TOPLAS, 3(4):508-532, (Oct 1981). (3.4.2).

519. Hansen, W.J.: A Predecessor Algorithm for Ordered Lists; Inf. Proc. Letters, 7(3):137-138, (Apr 1978). (3.1.1).

520. Harper, L.H., Payne, T.H., Savage, J.E. and Straus, E.: Sorting X+Y; C.ACM, 18(6):347-349, (June 1975). (4.2, 4.3).

521. Harrison, M.C.: Implementation of the Substring Test by Hashing; C.ACM, 14:777-779, (1971). (7.1.5, 7.2.6).

522. Hasham, A. and Sack, J.R.: Bounds for min-max heaps; BIT, 27(3):315-323, (1987). (5.1.3).

523. Head, A.K.: Multiplication Modulo n; BIT, 20(1):115-116, (1980). (6.1).

524. Heintz, J. and Schnorr, C.P.: Testing Polynomials Which are Easy to Compute; Proceedings STOC-SIGACT, Los Angeles CA, 12:262-272, (Apr 1980). (6.4).

525. Heintz, J. and Sieveking, M.: Lower Bounds for Polynomials with Algebraic Coefficients; Theoretical Computer Science, 11:321-330, (1980). (6.4).

526. Heising, W.P.: Note on Random Addressing Techniques; IBM Systems J, 2(2):112-116, (June 1963). (I.4).

527. Held, G. and Stonebraker, M.: B-trees re-examined; C.ACM, 21(2):139-143, (Feb 1978). (3.4.2).

528. Hendricks, W.J.: An account of self-organizing systems; SIAM J on Computing, 5(4):715-723, (Dec 1976). (3.1.2, 3.1.3).

529. Henrich, A., Six, H. and Widmayer, P.: The LSD tree: spatial access to multidimensional point- and non-point objects; Proceedings VLDB, Amsterdam, Netherlands, 15:45-54, (Aug 1989). (3.3.13, 3.5).

530. Hermosilla, L. and Olivos, J.: A Bijective Approach to Single rotation trees; Proceedings SCCC Int. Conf. in Computer Science, Santiago, Chile, 5:22-30, (1985). (3.4.1.6).

531. Hertel, S.: Smoothsort's Behavior on Presorted Sequences; Inf. Proc. Letters, 16(4):165-170, (May 1983). (4.1.5).

532. Hester, J.H., Hirschberg, D.S., Huang, S-H.S. and Wong, C.K.: Faster construction of optimal binary split trees; J of Algorithms, 7(3):412-424, (Sep 1986). (3.4.1.6).

533. Hester, J.H., Hirschberg, D.S. and Larmore, L.L.: Construction of optimal Binary Split trees in the presence of bounded access probabilities; J of Algorithms, 9(22):245-253, (June 1988). (3.4.1.6).

534. Hester, J.H. and Hirschberg, D.S.: Self-Organizing Linear Search; ACM C. Surveys, 17(3):295-311, (Sep 1985). (3.1.2, 3.1.3).

535. Hester, J.H. and Hirschberg, D.S.: Self-Organizing Search Lists Using Probabilistic Back-Pointers; C.ACM, 30(12):1074-1079, (Dec 1987). (3.1.2, 3.1.3).

536. Hibbard, T.N.: An Empirical Study of Minimal Storage Sorting; C.ACM, 6(5):206-213, (May 1963). (4.1, 4.2.4).

537. Hibbard, T.N.: Some Combinatorial Properties of Certain Trees with Applications to Searching and Sorting; J.ACM, 9(1):13-28, (Jan 1962). (3.4.1).

538. Hinrichs, K.: Implementation of the grid file: design concepts and experience; BIT, 25(4):569-592, (1985). (3.5.4).

539. Hirschberg, D.S.: A linear space algorithm for computing maximal common subsequences; C.ACM, 18:341-343, (1975). (7.3.1).

540. Hirschberg, D.S.: Algorithms for the longest common subsequence problem; J.ACM, 24:664-675, (1977). (7.3.1).

541. Hirschberg, D.S.: An information-theoretic lower bound for the longest common subsequence problem; Inf. Proc. Letters, 7:40-41, (1978). (7.3.1).

542. Hirschberg, D.S.: An Insertion Technique for One-Sided Height-Balanced Trees; C.ACM, 19(8):471-473, (Aug 1976). (3.4.1.3).

543. Hirschberg, D.S.: On the Complexity of Searching a Set of Vectors; SIAM J on Computing, 9(1):126-129, (Feb 1980). (3.5).

544. Hoare, C.A.R.: Algorithm 63 and 64; C.ACM, 4(7):321, (July 1961). (4.1.3).

545. Hoare, C.A.R.: Algorithm 65 (FIND); C.ACM, 4(7):321-322, (July 1961). (5.2).

546. Hoare, C.A.R.: Quicksort; Computer Journal, 5(4):10-15, (Apr 1962). (4.1.3).

547. Hollander, C.R.: Remark on Uniform Insertions in Structured Data Structures; C.ACM, 20(4):261-262, (1977). (2.1).

548. Honig, W.L. and Carlson, C.R.: Toward an Understanding of (actual) Data Structures; Computer Journal, 21(2):98-104, (1977). (2.1).

549. Hopgood, F.R.A. and Davenport, J.: The Quadratic Hash Method when the Table Size is a Power of 2; Computer Journal, 15(4):314-315, (1972). (3.3.6).

550. Horibe, Y. and Nemetz, T.O.H.: On the Max-Entropy Rule for a Binary Search Tree; Acta Informatica, 12(1):63-72, (1979). (3.4.1.6).

551. Horibe, Y.: An Improved Bound for Weight-Balanced Tree; Information and Control, 34(2):148-151, (June 1977). (3.4.1.7).

552. Horibe, Y.: Weight Sequences and Individual Path Length in a Balanced Binary Tree; J. of Combinatorics, Information and System Sciences, 4(1):19-22, (1979). (3.4.1.7).

553. Horowitz, E.: A Unified View of the Complexity of Evaluation and Interpolation; Acta Informatica, 3(2):123-133, (1974). (6.4).

554. Horowitz, E.: The Efficient Calculation of Powers of Polynomials; JCSS, 7(5):469-480, (Oct 1973). (6.2, 6.4).

555. Horspool, R.N.S.: Practical Fast Searching in Strings; Software - Practice and Experience, 10:501-506, (1980). (7.1.3).

556. Horvath, E.C.: Some Efficient Stable Sorting Algorithms; Proceedings STOC-SIGACT, Seattle WA, 6:194-215, (Apr 1974). (4.3.2).

557. Horvath, E.C.: Stable Sorting in Asymptotically Optimal Time and Extra Space; J.ACM, 25(2):177-199, (Apr 1978). (4.1, 4.3.2).

558. Hoshi, M. and Yuba, T.: A Counter Example to a Monotonicity Property of k-d Trees; Inf. Proc. Letters, 15(4):169-173, (Oct 1982). (3.5.2).

559. Hosken, W.H.: Optimum Partitions of Tree Addressing Structures; SIAM J on Computing, 4(3):341-347, (Sep 1975). (3.4.1.7).

560. Hsiao, Y-S. and Tharp, A.L.: Adaptive Hashing; Inform. Systems, 13(1):111-128, (1988). (3.4.2.5).

561. Hsu, W.J. and Du, M.W.: Computing a Longest Common Subsequence for A Set of Strings; BIT, 24:45-59, (1984). (7.3.1).

562. Hsu, W.J. and Du, M.W.: New algorithms for the longest common subsequence problem; JCSS, 29:133-152, (1984). (7.3.1).

563. Hu, T.C., Kleitman, D.J. and Tamaki, J.K.: Binary Trees Optimum Under Various Criteria; SIAM J Appl Math, 37(2):246-256, (Oct 1979). (3.4.1.7).

564. Hu, T.C. and Shing, M.T.: Computation of Matrix Chain Products. Part I; SIAM J on Computing, 11(2):362-373, (May 1982). (6.3).

565. Hu, T.C. and Tan, K.C.: Least Upper Bound on the Cost of Optimum Binary Search Trees; Acta Informatica, 1(4):307-310, (1972). (3.4.1.7).

566. Hu, T.C. and Tucker, A.C.: Optimal Computer Search Trees and Variable-Length Alphabetical Codes; SIAM J Appl Math, 21(4):514-532, (Dec 1971). (3.4.1.7).

567. Hu, T.C.: A New Proof of the T-C Algorithm; SIAM J Appl Math, 25(1):83-94, (July 1973). (3.4.1.7).

568. Huang, B. and Langston, M.A.: Practical In-Place Merging; C.ACM, 31(3):348-352, (Mar 1988). (4.3, 4.3.1, 4.3.2).

569. Huang, B. and Langston, M.A.: Fast Stable Merging and Sorting in Constant Extra Space; Proceedings ICCI'89, 71-80, (1989). (4.3, 4.3.1, 4.3.2).

570. Huang, B. and Langston, M.A.: Stable Duplicate-key Extraction with Optimal Time and Space bounds; Acta Informatica, 26(5):473-484, (1989). (4.1).

571. Huang, S-H.S. and Viswanathan, V.: On the construction of weighted time-optimal B-trees; BIT, 30(2):207-215, (1990). (3.4.2).

572. Huang, S-H.S. and Wong, C.K.: Binary search trees with limited rotation; BIT, 23(4):436-455, (1983). (3.4.1.6).

573. Huang, S-H.S. and Wong, C.K.: Generalized Binary Split Trees; Acta Informatica, 21(1):113-123, (1984). (3.4.1.6).

574. Huang, S-H.S. and Wong, C.K.: Optimal Binary Split Trees; J of Algorithms, 5(1):65-79, (Mar 1984). (3.4.1.6).

575. Huang, S-H.S. and Wong, C.K.: Average Number of rotation and access cost in iR-trees; BIT, 24(3):387-390, (1984). (3.4.1.6).

576. Huang, S-H.S.: Height-balanced trees of order (β, γ, δ); ACM TODS, 10(2):261-284, (1985). (3.4.2).

577. Huang, S-H.S.: Optimal Multiway split trees; J of Algorithms, 8(1):146-156, (Mar 1987). (3.4.1.6, 3.4.1.10).

578. Huang, S-H.S.: Ordered priority queues; BIT, 26(4):442-450, (1986). (5.1).

579. Huddleston, S. and Mehlhorn, K.: A New Data Structure for Representing Sorted Lists; Acta Informatica, 17(2):157-184, (1982). (3.4.2.1).

580. Huddleston, S. and Mehlhorn, K.: Robust Balancing in B-Trees; Lecture Notes in Computer Science 104, Springer-Verlag, :234-244, (1981). (3.4.2).

581. Huits, M. and Kumar, V.: The Practical Significance of Distributive Partitioning Sort; Inf. Proc. Letters, 8(4):168-169, (Apr 1979). (4.2.5).

582. Hunt, J. and Szymanski, T.G.: A fast algorithm for computing longest common subsequences; C.ACM, 20:350-353, (1977). (7.3.1).

583. Hutflesz, A., Six, H. and Widmayer, P.: Globally Order Preserving Multidimensional Linear Hashing; Proceedings IEEE Conf. on Data Eng., Los Angeles CA, 4:572-579, (1988). (3.5.4).

584. Hutflesz, A., Six, H. and Widmayer, P.: Twin Grid Files: Space Optimizing Access Schemes; Proceedings ACM SIGMOD, Chicago IL, 17:183-190, (June 1988). (3.5.4).

585. Hwang, F.K. and Lin, S.: A Simple Algorithm for Merging Two Disjoint Linearly Ordered Sets; SIAM J on Computing, 1(1):31-39, (Mar 1972). (4.3.3).

586. Hwang, F.K. and Lin, S.: Optimal Merging of 2 Elements with n Elements; Acta Informatica, 1(2):145-158, (1971). (4.3.3).

587. Hwang, F.K.: Optimal Merging of 3 Elements with n Elements; SIAM J on Computing, 9(2):298-320, (May 1980). (4.3.3).

588. Hyafil, L., Prusker, F. and Vuillemin, J.: An Efficient Algorithm for Computing Optimal Disk Merge Patterns; Proceedings STOC-SIGACT, Seattle WA, 6:216-229, (Apr 1974). (4.3, 4.4).

589. Hyafil, L. and van de Wiele, J.P.: On the Additive Complexity of Specific Polynomials; Inf. Proc. Letters, 4(2):45-47, (Nov 1975). (6.4).

590. Hyafil, L.: Bounds for Selection; SIAM J on Computing, 5(1):109-114, (Mar 1976). (5.2).

591. Incerpi, J. and Sedgewick, R.: Improved Upper Bounds on Shellsort; JCSS, 31(2):210-224, (Oct 1985). (4.1.4).

592. Incerpi, J. and Sedgewick, R.: Practical Variations of Shellsort; Inf. Proc. Letters, 26(1):37-43, (Sep 1987). (4.1.4).

593. Isaac, E.J. and Singleton, R.C.: Sorting by Address Calculation; J.ACM, 3(3):169-174, (July 1956). (4.1.6, 4.2.3).

594. Itai, A., Konheim, A.G. and Rodeh, M.: A Sparse Table Implementation of Priority Queues; Proceedings ICALP, Lecture Notes in Computer Science 115, Springer-Verlag, Acre, 8:417-430, (July 1981). (5.1).

595. Itai, A.: Optimal Alphabetic Trees; SIAM J on Computing, 5(1):9-18, (Mar 1976). (3.4.1.7).

596. Ja'Ja', J. and Takche, J.: Improved Lower Bounds for some matrix multiplication problems; Inf. Proc. Letters, 21(3):123-127, (Sep 1985). (6.3).

597. Ja'Ja', J.: On the Complexity of Bilinear Forms with Commutativity; SIAM J on Computing, 9(4):713-728, (Nov 1980). (6.3).

598. Ja'Ja', J.: On the Computational Complexity of the Permanent; Proceedings FOCS, Tucson AZ, 24:312-319, (Nov 1983). (6.3).

599. Ja'Ja', J.: Optimal Evaluation of Pairs of Bilinear Forms; SIAM J on Computing, 8(3):443-462, (Aug 1979). (6.1, 6.3).

600. Jackowski, B.L., Kubiak, R. and Sokolowski, S.: Complexity of Sorting by Distributive Partitioning; Inf. Proc. Letters, 9(2):100, (Aug 1979). (4.2.5).

601. Jacobs, D. and Feather, M.: Corrections to A synthesis of Several Sorting algorithms; Acta Informatica, 26(1-2):19-24, (1988). (2.2).

602. Jacobs, M.C.T. and van Emde-Boas, P.: Two results on Tables; Inf. Proc. Letters, 22(1):43-48, (Jan 1986). (3.3).

603. Jacquet, P. and Regnier, M.: Trie Partitioning Process: Limiting Distributions; Proceedings CAAP, Nice, 13:196-210, (1986). (3.4.4).

604. Jaeschke, G. and Osterburg, G.: On Cichelli's Minimal Perfect Hash Functions Method; C.ACM, 23(12):728-729, (Dec 1980). (3.3.16).

605. Jaeschke, G.: Reciprocal Hashing: A Method for Generating Minimal Perfect Hashing Functions; C.ACM, 24(12):829-833, (Dec 1981). (3.3.16).

606. Jakobsson, M.: Reducing Block Accesses in Inverted Files by Partial Clustering; Inform. Systems, 5(1):1-5, (1980). (7.2.1).

607. Janko, W.: A List Insertion Sort for Keys with Arbitrary Key Distribution; ACM TOMS, 2(2):143-153, (1976). (4.1.2).

608. Janko, W.: Variable Jump Search: The Algorithm and its Efficiency; Angewandte Informatik, 23(1):6-11, (Jan 1981). (3.1.5).

609. Johnson, D.B. and Mizoguchi, T.: Selecting the Kth Element in X+Y and X1+X2+...+Xm; SIAM J on Computing, 7(2):147-153, (May 1978). (5.2).

610. Johnson, D.B.: Priority Queues with Update and Finding Minimum Spanning Trees; Inf. Proc. Letters, 4(3):53-57, (Dec 1975). (5.1).

611. Johnson, J.H.: Formal Models for String Similarity; PhD Dissertation, University of Waterloo, Waterloo, Ontario, Canada, (1983). (7.1.8).

612. Johnson, L.R.: An Indirect Chaining Method for Addressing on Secondary Keys; C.ACM, 4(5):218-222, (May 1961). (3.3.11).

613. Johnson, T. and Shasha, D.: Utilization of B-trees with Inserts, Deletes and Modifies; Proceedings ACM PODS, Philadelphia PN, 8:235-246, (Mar 1989). (3.4.2).

614. Jonassen, A.T. and Dahl, O-J.: Analysis of an Algorithm for Priority Queue Administration; BIT, 15(4):409-422, (1975). (5.1.2).

615. Jonassen, A.T. and Knuth, D.E.: A Trivial Algorithm Whose Analysis Isn't; JCSS, 16(3):301-322, (June 1978). (3.4.1.9).

616. Jones, B.: A Variation on Sorting by Address Calculation; C.ACM, 13(2):105-107, (Feb 1970). (4.1.6, 4.2.1).

617. Jones, D.W.: An Empirical Comparison of Priority-Queue and Event-Set Implementations; C.ACM, 29(4):300-311, (Apr 1986). (5.1).

618. Jones, P.R.: Comment on Average Binary Search Length; C.ACM, 15(8):774, (Aug 1972). (3.2.1).

619. Kahaner, D.K.: Algorithm 561-Fortran Implementation of Heap Programs for Efficient Table Maintenance; ACM TOMS, 6(3):444-449, (Sep 1980). (5.1.3).

620. Kaminski, M.: A Linear Time Algorithm for Residue Computation and a Fast Algorithm for Division with a Sparse Divisor; J.ACM, 34(4):968-984, (Oct 1987). (6.1).

621. Karlsson, R.G. and Overmars, M.H.: Normalized Divide-and-Conquer: A scaling technique for solving multi-dimensional problems; Inf. Proc. Letters, 26(6):307-312, (Jan 1987). (2.2.2.1, 3.5).

622. Karlton, P.L., Fuller, S.H., Scroggs, R.E. and Kaehler, E.B.: Performance of Height-Balanced Trees; C.ACM, 19(1):23-28, (Jan 1976). (3.4.1.3).

623. Karp, R.M., Miller, R. and Rosenberg, A.L.: Rapid Identification of Repeated Patterns in Strings, Trees, and Arrays; Proceedings STOC-SIGACT, Boulder CO, 4:125-136, (May 1972). (7.1).

624. Karp, R.M. and Rabin, M.O.: Efficient Randomized Pattern-Matching Algorithms; IBM J Res. Development, 31(2):249-260, (Mar 1987). (7.1.5, 7.2.6, 7.3.2).

625. Katz, M.D. and Volper, D.J.: Data structures for retrieval on square grids; SIAM J on Computing, 15(4):919-931, (Nov 1986). (3.6.2).

626. Kawagoe, K.: Modified Dynamic Hashing; Proceedings ACM SIGMOD, Austin TX, 14:201-213, (1985). (3.3.13, 3.3.14).
627. Kedem, Z.M.: Combining Dimensionality and Rate of Growth Arguments for Establishing Lower Bounds on the Number of Multiplications; Proceedings STOC-SIGACT, Seattle WA, 6:334-341, (Apr 1974). (6.2).
628. Keehn, D.G. and Jacy, J.O.: VSAM Data Set Design Parameters; IBM Systems J, 13(3):186-212, (1974). (3.4.3).
629. Kemp, M., Bayer, R. and Guntzer, U.: Time optimal Left to Right construction of position Trees; Acta Informatica, 24(4):461-474, (1987). (7.2.2).
630. Kemp, R.: A Note on the Stack Size of Regularly Distributed Binary Trees; BIT, 20(2):157-163, (1980). (3.4.1.2).
631. Kemp, R.: The Average Number of Registers Needed to Evaluate a Binary Tree Optimally; Acta Informatica, 11(4):363-372, (1979). (3.4.1.2).
632. Kemp, R.: The Expected additive weight of trees; Acta Informatica, 26(8):711-740, (1989). (3.4.1.2).
633. Kennedy, S.: A Note on Optimal Doubly-Chained Trees; C.ACM, 15(11):997-998, (Nov 1972). (3.4.1.7).
634. Kent, P.: An efficient new way to represent multidimensional data; Computer Journal, 28(2):184-190, (May 1985). (3.5).
635. Kingston, J.H.: A new proof of the Garsia-Wachs algorithm; J of Algorithms, 9(1):129-136, (Mar 1988). (3.4.1.7).
636. Kirkpatrick, D.G. and Reisch, S.: Upper bounds for sorting integers on random access machines; Theoretical Computer Science, 28(3):263-276, (Feb 1984). (4.2.3).
637. Kirkpatrick, D.G.: A Unified Lower Bound for Selection and Set Partitioning Problems; J.ACM, 28(1):150-165, (Jan 1981). (5.2).
638. Kirschenhofer, P., Prodinger, H. and Szpankowski, W.: Do we Really Need to Balance Patricia Tries; Proceedings ICALP, Lecture Notes in Computer Science 317, Springer-Verlag, Tampere, Finland, 15:302-316, (1988). (3.4.4.5).
639. Kirschenhofer, P., Prodinger, H. and Szpankowski, W.: On the balance property of Patricia trees: External path length view point; Theoretical Computer Science, 68(1):1-18, (Oct 1989). (3.4.4.5).
640. Kirschenhofer, P. and Prodinger, H.: Further results on digital search trees; Theoretical Computer Science, 58(1-3):143-154, (1988). (3.4.4).
641. Kirschenhofer, P. and Prodinger, H.: On the recursion depth of special Tree traversal algorithms; Information and Computation, 74(1):15-32, (July 1987). (3.4.1.2).
642. Kirschenhofer, P. and Prodinger, H.: Some Further Results on Digital Trees; Proceedings ICALP, Lecture Notes in Computer Science 226, Springer-Verlag, Rennes, France, 13:177-185, (1986). (3.4.4).
643. Kirschenhofer, P.: On the Height of Leaves in Binary Trees; J. of Combinatorics, Information and System Sciences, 8(1):44-60, (1983). (3.4.1).
644. Kjellberg, P. and Zahle, T.U.: Cascade Hashing; Proceedings VLDB, Singapore, 10:481-492, (Aug 1984). (3.3.14).
645. Klein, R. and Wood, D.: A tight upper bound for the path length of AVL trees; Theoretical Computer Science, 72(2/3):251-264, (May 1990). (3.4.1.3).
646. Klein, R. and Wood, D.: The Node Visit Cost of Brother Trees; Information and Computation, 75(2):107-129, (Nov 1987). (3.4.1.3, 3.4.2.1, 3.4.2.3).

647. Klein, R. and Wood, D.: On the Path Length of Binary Trees; J.ACM, 36(2):280-289, (Apr 1989). (3.4.1).

648. Kleitman, D.J., Meyer, A.R., Rivest, R.L., Spencer, J. and Winklmann, K.: Coping with Errors in Binary Search Procedures; JCSS, 20(3):396-404, (June 1980). (3.4.1).

649. Kleitman, D.J. and Saks, M.E.: Set Orderings Requiring Costliest Alphabetic Binary Trees; SIAM J Alg Disc Methods, 2(2):142-146, (June 1981). (3.4.1.7).

650. Knott, G.D. and de la Torre, P.: Hash table collision resolution with direct chaining; J of Algorithms, 10(1):20-34, (Mar 1989). (3.3.10).

651. Knott, G.D.: A Balanced Tree Storage and Retrieval Algorithm; Proceedings ACM Symposium of Information Storage and Retrieval, College Park MD, 175-196, (1971). (3.4.1.3).

652. Knott, G.D.: A Numbering System for Binary Trees; C.ACM, 20(2):113-115, (Feb 1977). (3.4.1).

653. Knott, G.D.: Deletions in Binary Storage Trees; PhD Dissertation, Computer Science Department, Stanford University, (May 1975). (3.4.1.9).

654. Knott, G.D.: Direct-chaining with coalescing lists; J of Algorithms, 5(1):7-21, (Mar 1984). (3.3.10, 3.3.12).

655. Knott, G.D.: Fixed-Bucket Binary Storage Trees; J of Algorithms, 3(3):276-287, (Sep 1982). (3.4.1.1, 3.4.4).

656. Knott, G.D.: Hashing Functions; Computer Journal, 18(3):265-278, (Aug 1975). (3.3.1).

657. Knott, G.D.: Linear open addressing and Peterson's theorem rehashed; BIT, 28(2):364-371, (1988). (3.3.4).

658. Knott, G.D.: Expandable Open Addressing Hash Table Storage and Retrieval; Proceedings ACM SIGFIDET Workshop on Data Description, Access and Control, San Diego CA, :186-206, (Nov 1971). (3.3).

659. Knuth, D.E., Morris, J. and Pratt, V.: Fast Pattern Matching in Strings; SIAM J on Computing, 6:323-350, (1977). (7.1.2).

660. Knuth, D.E.: Deletions that Preserve Randomness; IEEE Trans. Software Engineering, 3:351-359, (1977). (3.4.1.9).

661. Knuth, D.E.: Evaluating Polynomials by Computers; C.ACM, 5:595-599, (1962). (6.4).

662. Knuth, D.E.: Length of Strings for a Merge Sort; C.ACM, 6(11):685-688, (Nov 1963). (4.4.1).

663. Knuth, D.E.: Optimum Binary Search Trees; Acta Informatica, 1(1):14-25, (1971). (3.4.1.7).

664. Knuth, D.E.: Structured Programming with Go To Statements; ACM C. Surveys, 6(4):261-301, (Dec 1974). (3.1.1, 3.4.1.1, 4.1, 4.1.3).

665. Knuth, D.E.: The Average Time for Carry Propagation; P. Kon Ned A, 81(2):238-242, (1978). (6.1).

666. Kollias, J.G.: An Estimate of Seek Time for Batched Searching of Random or Index Sequential Structured Files; Computer Journal, 21(2):132-133, (1978). (3.3, 3.4.3).

667. Konheim, A.G. and Weiss, B.: An Occupancy Discipline and Applications; SIAM J Appl Math, 14:1266-1274, (1966). (3.3.4).

668. Korsh, J.F.: Greedy Binary Search Trees are Nearly Optimal; Inf. Proc. Letters, 13(1):16-19, (Oct 1981). (3.4.1.6).

669. Korsh, J.F.: Growing Nearly Optimal Binary Search Trees; Inf. Proc. Letters, 14(3):139-143, (May 1982). (3.4.1.6).

670. Kosaraju, S.R.: Insertions and Deletions in One-Sided Height-Balanced Trees; C.ACM, 21(3):226-227, (Mar 1978). (3.4.1.3).

671. Kosaraju, S.R.: Localized Search in Sorted Lists; Proceedings STOC-SIGACT, Milwaukee WI, 13:62-69, (May 1981). (3.4.2.1).

672. Kosaraju, S.R.: On a Multidimensional Search Problem; Proceedings STOC-SIGACT, Atlanta GA, 11:67-73, (Apr 1979). (3.5).

673. Kosaraju, S.R.: Efficient Tree Pattern Matching; Proceedings FOCS, Research Triangle Park, NC, 30:178-183, (1989). (7.1.7).

674. Kral, J.: Some Properties of the Scatter Storage Technique with Linear Probing; Computer Journal, 14(2):145-149, (1971). (3.3.4).

675. Krichersky, R.E.: Optimal Hashing; Information and Control, 62(1):64-92, (July 1984). (3.3.9).

676. Kriegel, H.P. and Kwong, Y.S.: Insertion-Safeness in Balanced Trees; Inf. Proc. Letters, 16(5):259-264, (June 1983). (3.4.2.1).

677. Kriegel, H.P. and Seeger, B.: Multidimensional Order Preserving Linear Hashing with Partial Expansions; Proceedings Int. Conf. on Database Theory, Lecture Notes in Computer Science, Springer-Verlag, Rome, 243:203-220, (1986). (3.5.4).

678. Kriegel, H.P. and Seeger, B.: PLOP-Hashing: A Grid File without Directory; Proceedings IEEE Conf. on Data Eng., Los Angeles, CA, 4:369-376, (1988). (3.5.4).

679. Kriegel, H.P., Vaishnavi, V.K. and Wood, D.: 2-3 Brother Trees; BIT, 18(4):425-435, (1978). (3.4.2.1).

680. Krithivasan, K. and Sitalakshmi, R.: Efficient Two-Dimensional Pattern Matching in the Presence of Errors; Information Sciences, 43:169-184, (1987). (7.1.8, 7.3.2).

681. Kritzinger, P.S. and Graham, J.W.: A Theorem in the Theory of Compromise Merge Methods; J.ACM, 21(1):157-160, (Jan 1974). (4.4.4, 4.4.3).

682. Kronmal, R.A. and Tarter, M.E.: Cumulative Polygon Address Calculation Sorting; Proceedings ACM-NCC, Cleveland OH, 20:376-384, (1965). (4.1.6).

683. Kronrod, M.A.: An Optimal Ordering Algorithm Without a Field of Operation; Dokladi Akademia Nauk SSSR, 186:1256-1258, (1969). (4.3.2).

684. Kruijer, H.S.M.: The Interpolated File Search Method; Informatie, 16(11):612-615, (Nov 1974). (3.2.2).

685. Kumar, S.K. and Ranzon, C.P.: A linear space algorithm for the LCS problem; Acta Informatica, 24(3):353-362, (1987). (7.3.1).

686. Kung, H.T.: A New Upper Bound on the Complexity of Derivative Evaluation; Inf. Proc. Letters, 2(5):146-147, (Dec 1973). (6.4).

687. Kuspert, K.: Storage Utilization in B^*-trees with a Generalized Overflow Technique; Acta Informatica, 29(1):35-56, (1983). (3.4.2).

688. Ladi, E., Luccio, F., Mugnai, C. and Pagli, L.: On two dimensional data organization I; Fundamenta Informaticae, 3(2):211-226, (1979). (3.5).

689. Lai, T.W. and Wood, D.: Implicit Selection; Proceedings SWAT 88, Halmstad, Sweden, 1:14-23, (1988). (5.2).

690. Lan, K.K.: A note on synthesis and Classification of Sorting Algorithms; Acta Informatica, 27(1):73-80, (1989). (2.2).

691. Landau, G.M. and Vishkin, U.: Efficient String Matching in the Presence of Errors; Proceedings FOCS, Portland OR, 26:126-136, (Oct 1985). (7.1.8).

692. Landau, G.M. and Vishkin, U.: Efficient String Matching with k Mismatches; Theoretical Computer Science, 43:239-249, (1986). (7.1.8).

693. Landau, G.M. and Vishkin, U.: Introducing efficient parallelism into approximate string matching and a new serial algorithm; Proceedings STOC-SIGACT, Berkeley CA, 18:220-230, (May 1986). (7.1.8).

694. Landau, G.M.: String Matching in Erroneous Input; PhD Dissertation, Tel Aviv University, Tel Aviv, Israel, (1986). (7.1.8).

695. Lang, S.D.: Analysis of recursive batched interpolation sort; BIT, 30(1):42-50, (1990). (4.1.6).

696. Langenhop, C.E. and Wright, W.E.: A model of the Dynamic Behavior of B-trees; Acta Informatica, 27(1):41-60, (1989). (3.4.2).

697. Langenhop, C.E. and Wright, W.E.: An Efficient Model for Representing and Analyzing B-Trees; Proceedings ACM-NCC, Denver CO, 40:35-40, (1985). (3.4.2).

698. Langenhop, C.E. and Wright, W.E.: Probabilities related to Father-Son Distances in Binary search; SIAM J on Computing, 15(2):520-530, (May 1986). (3.4.1).

699. Larmore, L.L.: A Subquadratic algorithm for constructing approximately optimal binary search trees; J of Algorithms, 8(4):579-591, (Dec 1987). (3.4.1.7).

700. Larson, J.A. and Walden, W.E.: Comparing Insertion Schemes Used to Update 3-2 Trees; Inform. Systems, 4:127-136, (1979). (3.4.2.1).

701. Larson, P. and Kajla, A.: File Organization: Implementation of a Method Guaranteeing Retrieval in one Access; C.ACM, 27(7):670-677, (July 1984). (3.3.15).

702. Larson, P. and Ramakrishna, M.V.: External Perfect Hashing; Proceedings ACM SIGMOD, Austin TX, 14:190-200, (June 1985). (3.3.16).

703. Larson, P.: A Method for Speeding up Text Retrieval; Proceedings ACM SIGMOD, San Jose CA, 12:117-123, (May 1983). (7.2, 7.2.6).

704. Larson, P.: A Single-File Version of Linear Hashing with Partial Expansions; Proceedings VLDB, Mexico City, 8:300-309, (Sep 1982). (3.3.14).

705. Larson, P.: Analysis of Hashing with Chaining in the Prime Area; J of Algorithms, 5(1):36-47, (1984). (3.3).

706. Larson, P.: Analysis of Index-Sequential Files with Overflow Chaining; ACM TODS, 6(4):671-680, (Dec 1981). (3.4.3).

707. Larson, P.: Analysis of Repeated Hashing; BIT, 20(1):25-32, (1980). (3.3).

708. Larson, P.: Analysis of Uniform Hashing; J.ACM, 30(4):805-819, (Oct 1983). (3.3.2).

709. Larson, P.: Dynamic Hash Tables; C.ACM, 31(4):446-457, (Apr 1988). (3.3.14).

710. Larson, P.: Dynamic Hashing; BIT, 18(2):184-201, (1978). (3.3.14).

711. Larson, P.: Expected Worst-Case Performance of Hash Files; Computer Journal, 25(3):347-352, (Aug 1982). (3.3.3, 3.3.4, 3.3.11).

712. Larson, P.: Frequency Loading and Linear Probing; BIT, 19(2):223-228, (1979). (3.3.4).

713. Larson, P.: Linear Hashing with Overflow-Handling by Linear Probing; ACM TODS, 10(1):75-89, (Mar 1985). (3.3.14).

714. Larson, P.: Linear Hashing with Partial Expansions; Proceedings VLDB, Montreal, 6:224-232, (1980). (3.3.14).

715. Larson, P.: Linear Hashing with Separators - A Dynamic Hashing Scheme Achieving One-Access Retrieval; ACM TODS, 13(3):366-388, (1988). (3.3.14, 3.3.15).

716. Larson, P.: Performance Analysis of a Single-File Version of Linear Hashing; Computer Journal, 28(3):319-329, (1985). (3.3.14).

717. Larson, P.: Performance Analysis of Linear Hashing with Partial Expansions; ACM TODS, 7(4):566-587, (Dec 1982). (3.3.14).

718. Lea, D.: Digital and Hilbert K-D trees; Inf. Proc. Letters, 27(1):35-41, (Feb 1988). (3.5.2).

719. Lee, C.C., Lee, D.T. and Wong, C.K.: Generating Binary Trees of Bounded Height; Acta Informatica, 23(5):529-544, (1986). (3.4.1).

720. Lee, D.T. and Wong, C.K.: Quintary Trees: A File Structure for Multidimensional Database System; ACM TODS, 5(3):339-353, (Sep 1980). (3.5).

721. Lee, D.T. and Wong, C.K.: Worst-Case Analysis for Region and Partial Region Searches in Multidimensional Binary Search Trees and Balanced Quad Trees; Acta Informatica, 9(1):23-29, (1977). (3.5.1, 3.5.2, 3.6.2).

722. Lee, K.P.: A Linear Algorithm for Copying Binary Trees Using Bounded Workspace; C.ACM, 23(3):159-162, (Mar 1980). (3.4.1).

723. Leipala, T.: On a Generalization of Binary Search; Inf. Proc. Letters, 8(5):230-233, (June 1979). (3.2.1).

724. Leipala, T.: On Optimal Multilevel Indexed Sequential Files; Inf. Proc. Letters, 15(5):191-195, (Dec 1982). (3.4.3).

725. Leipala, T.: On the Design of One-Level Indexed Sequential Files; Int. J of Comp and Inf Sciences, 10(3):177-186, (June 1981). (3.1.5, 3.4.3).

726. Lentfert, P. and Overmars, M.H.: Data structures in a real time environment; Inf. Proc. Letters, 31(3):151-155, (May 1989). (3.4.1, 5.1).

727. Lescarne, P. and Steyaert, J.M.: On the Study of Data Structures: Binary Tournaments with Repeated Keys; Proceedings ICALP, Lecture Notes in Computer Science 154, Springer-Verlag, Barcelona, Spain, 10:466-477, (July 1983). (3.4.1).

728. Lesuisse, R.: Some Lessons Drawn from the History of the Binary Search Algorithm; Computer Journal, 26(2):154-163, (May 1983). (3.2.1, 2.2.2.1).

729. Leung, H.C.: Approximate storage utilization of B-trees: A simple derivation and generalizations; Inf. Proc. Letters, 19(4):199-201, (Nov 1984). (3.4.2).

730. Levcopoulos, C., Lingas, A. and Sack, J.R.: Heuristics for Optimum Binary Search Trees and Minimum Weight Trangulation problems; Theoretical Computer Science, 66(2):181-204, (1989). (3.4.1.7).

731. Levcopoulos, C., Lingas, A. and Sack, J.R.: Nearly Optimal heuristics for Binary Search Trees with Geometric Applications; Proceedings ICALP, Lecture Notes in Computer Science 267, Springer-Verlag, Karslruhe, West Germany, 14:376-385, (1987). (3.4.1.6, 3.4.1.7).

732. Levcopoulos, C. and Overmars, M.H.: A balanced search tree with O(1) worst case update time; Acta Informatica, 26(3):269-278, (1988). (3.4.1).

733. Levcopoulos, C. and Petersson, O.: Heapsort - adapted for presorted files; Proceedings Workshop in Algorithms and Data Structures, Lecture Notes in Computer Science 382, Springer-Verlag, Ottawa, Canada, 1:499-509, (Aug 1989). (4.1.8).

734. Levcopoulos, C. and Petersson, O.: Sorting shuffled monotone sequences; Proceedings Scandinavian Workshop in Algorithmic Theory, SWAT'90, Lecture Notes in Computer Science 447, Springer-Verlag, Bergen, Norway, 2:181-191, (July 1990). (4.1.8).

735. Levenshtein, V.: Binary Codes capable of correcting deletions, insertions and reversals; Soviet Phys. Dokl, 6:126-136, (1966). (7.1.8).

736. Levenshtein, V.: Binary codes capable of correcting spurious insertions and deletions of ones; Problems of Information Transmission, 1:8-17, (1965). (7.1.8).

737. Lewis, G.N., Boynton, N.J. and Burton, F.W.: Expected Complexity of Fast Search with Uniformly Distributed Data; Inf. Proc. Letters, 13(1):4-7, (Oct 1981). (3.2.2).

738. Li, L.: Ranking and Unranking AVL-Trees; SIAM J on Computing, 15(4):1025-1035, (Nov 1986). (3.4.1.3).

739. Li, M. and Yesha, Y.: String matching cannot be done by a two-head one way deterministic finite automaton; Inf. Proc. Letters, 22:231-235, (1986). (7.1).

740. Li, S. and Loew, M.H.: Adjacency Detection Using Quadcodes; C.ACM, 30(7):627-631, (July 1987). (3.5.1.1).

741. Li, S. and Loew, M.H.: The Quadcode and its Arithmetic; C.ACM, 30(7):621-626, (July 1987). (3.5.1.1).

742. Linial, N. and Saks, M.E.: Searching ordered structures; J of Algorithms, 6(1):86-103, (Mar 1985). (3.2).

743. Linnainmaa, S.: Software for Doubled-Precision Floating-Point Computations; ACM TOMS, 7(3):272-283, (Sep 1981). (6.1).

744. Lipski, Jr., W., Ladi, E., Luccio, F., Mugnai, C. and Pagli, L.: On two dimensional data organization II; Fundamenta Informaticae, 3(3):245-260, (1979). (3.5).

745. Lipton, R.J. and Dobkin, D.: Complexity Measures and Hierarchies for the Evaluation of Integers, Polynomials and N-Linear Forms; Proceedings STOC-SIGACT, Albuquerque NM, 7:1-5, (May 1975). (6.4).

746. Lipton, R.J., Rosenberg, A.L. and Yao, A.C-C.: External Hashing Schemes for Collection of Data Structures; J.ACM, 27(1):81-95, (Jan 1980). (3.3).

747. Lipton, R.J. and Stockmeyer, L.J.: Evaluation of Polynomials with Super-Preconditioning; Proceedings STOC-SIGACT, Hershey PA, 8:174-180, (May 1976). (6.4).

748. Lipton, R.J.: Polynomials With 0-1 Coefficients That are Hard to Evaluate; SIAM J on Computing, 7(1):61-69, (Feb 1978). (6.4).

749. Litwin, W. and Lomet, D.B.: A New Method for Fast Data Searches with Keys; IEEE Software, 4(2):16-24, (Mar 1987). (3.3.14, 3.4.2).

750. Litwin, W. and Lomet, D.B.: The Bounded Disorder Access Method; Proceedings IEEE Conf. on Data Eng., Los Angeles CA, 2:38-48, (1986). (3.3.14, 3.4.2.5, 3.4.4).

751. Litwin, W.: Linear Hashing: A New Tool for File and Table Addressing; Proceedings VLDB, Montreal, 6:212-223, (1980). (3.3.14).

752. Litwin, W.: Linear Virtual Hashing: A New Tool for Files and Tables Implementation; Proceedings IFIP TC-2 Conference, Venice, Italy, (1979). (3.3.14).

753. Litwin, W.: Trie Hashing; Proceedings ACM SIGMOD, Ann Arbor MI, 11:19-29, (Apr 1981). (3.4.4, 3.3).

754. Litwin, W.: Virtual Hashing: A Dynamically Changing Hashing; Proceedings VLDB, Berlin, 4:517-523, (Sep 1978). (3.3.14).

755. Lloyd, J.W. and Ramamohanarao, K.: Partial-Match Retrieval for Dynamic Files; BIT, 22(2):150-168, (1982). (3.3.13, 3.3.14, 3.6.2).

756. Lloyd, J.W.: Optimal Partial-Match Retrieval; BIT, 20(4):406-413, (1980). (3.6.2).

757. Lodi, E., Luccio, F., Pagli, L. and Santoro, N.: Random Access in a List Environment; Inform. Systems, 2:11-17, (1976). (3.1).

758. Lodi, E. and Luccio, F.: Split sequence hash search; Inf. Proc. Letters, 20(3):131-136, (Apr 1985). (3.3.7).

759. Loeser, R.: Some Performance Tests of Quicksort and Descendants; C.ACM, 17(3):143-152, (Mar 1974). (4.1.3).

760. Lomet, D.B. and Salzberg, B.: Access Methods for Multiversion Data; Proceedings ACM SIGMOD, Portland OR, 18:315-324, (May 1989). (3.4.2.5).

761. Lomet, D.B. and Salzberg, B.: The hB-tree: A robust multiattribute search structure; Proceedings IEEE Conf. on Data Eng., Los Angeles CA, 5, (Feb 1989). (3.5).

762. Lomet, D.B. and Salzberg, B.: The Performance of a Multiversion Access Method; Proceedings ACM SIGMOD, Atlantic City NJ, 19:353-363, (May 1990). (3.4.2.5).

763. Lomet, D.B.: A High Performance, Universal, Key Associative Access Method; Proceedings ACM SIGMOD, San Jose CA, 13:120-133, (May 1983). (3.3.13, 3.4.2.5).

764. Lomet, D.B.: A Simple Bounded Disorder File Organization with Good Performance; ACM TODS, 13(4):525-551, (1988). (3.3.14, 3.4.4).

765. Lomet, D.B.: Bounded Index Exponential Hashing; ACM TODS, 8(1):136-165, (Mar 1983). (3.3.13).

766. Lomet, D.B.: Digital B-Trees; Proceedings VLDB, Cannes, 7:333-344, (Sep 1981). (3.4.2.5, 3.4.4).

767. Lomet, D.B.: Partial Expansions for file organizations with an index; ACM TODS, 12:65-84, (1987). (3.4.2).

768. Lotka, A.J.: The Frequency Distribution of Scientific Production; J of the Washington Academy of Sciences, 16(12):317-323, (1926). (I.3).

769. Lotti, G. and Romani, F.: Application of Approximating Algorithms to Boolean Matrix Multiplication; IEEE Trans. on Computers, C29(10):927-928, (Oct 1980). (6.3).

770. Lowden, B.G.T.: A Note on the Oscillating Sort; Computer Journal, 20(1):92, (Feb 1977). (4.4.5).

771. Luccio, F. and Pagli, L.: Comment on Generalized AVL Trees; C.ACM, 23(7):394-395, (July 1980). (3.4.1.3).

772. Luccio, F. and Pagli, L.: On the Height of Height-Balanced Trees; IEEE Trans. on Computers, C25(1):87-90, (Jan 1976). (3.4.1.3).

773. Luccio, F. and Pagli, L.: Power Trees; C.ACM, 21(11):941-947, (Nov 1978). (3.4.1.3).

774. Luccio, F. and Pagli, L.: Rebalancing Height Balanced Trees; IEEE Trans. on Computers, C27(5):386-396, (May 1978). (3.4.1.3).

775. Luccio, F., Regnier, M. and Schott, R.: Disc and other related data structures; Proceedings Workshop in Algorithms and Data Structures, Lecture Notes in Computer Science 382, Springer-Verlag, Ottawa, Canada, 1:192-205, (Aug 1989). (3.4.4).

776. Luccio, F.: Weighted Increment Linear Search for Scatter Tables; C.ACM, 15(12):1045-1047, (Dec 1972). (3.3.5).

777. Lueker, G.S. and Molodowitch, M.: More Analysis of Double Hashing; Proceedings STOC-SIGACT, Chicago IL, 20:354-359, (May 1988). (3.3.5).

778. Lueker, G.S. and Willard, D.E.: A Data Structure for Dynamic Range Queries; Inf. Proc. Letters, 15(5):209-213, (Dec 1982). (3.6.2).

779. Lueker, G.S.: A Data Structure for Orthogonal Range Queries; Proceedings FOCS, Ann Arbor MI, 19:28-34, (Oct 1978). (3.6.2).

780. Lum, V.Y., Yuen, P.S.T. and Dodd, M.: Key-to-Address Transform Techniques: a Fundamental Performance Study on Large Existing Formatted Files; C.ACM, 14(4):228-239, (1971). (3.3.1).

781. Lum, V.Y. and Yuen, P.S.T.: Additional Results on Key-to-Address Transform Techniques: A Fundamental Performance Study on Large Existing Formatted Files; C.ACM, 15(11):996-997, (Nov 1972). (3.3.1).

782. Lum, V.Y.: General Performance Analysis of Key-to-Address Transformation Methods Using an Abstract File Concept; C.ACM, 16(10):603-612, (Oct 1973). (3.3.1).

783. Lum, V.Y.: Multi-Attribute Retrieval with Combined Indexes; C.ACM, 13(11):660-665, (Nov 1970). (3.4.3, 3.5).

784. Lynch, W.C.: More combinatorial problems on certain trees; Computer Journal, 7:299-302, (1965). (3.4.1).

785. Lyon, G.E.: Batch Scheduling From Short Lists; Inf. Proc. Letters, 8(2):57-59, (Feb 1979). (3.3.8.2).

786. Lyon, G.E.: Hashing with Linear Probing and Frequency Ordering; J Res. Nat. Bureau of Standards, 83(5):445-447, (Sep 1978). (3.3.4).

787. Lyon, G.E.: Packed Scatter Tables; C.ACM, 21(10):857-865, (Oct 1978). (3.3.9).

788. MacCallum, I.R.: A Simple Analysis of the nth Order Polyphase Sort; Computer Journal, 16(1):16-18, (Feb 1973). (4.4.4).

789. MacLaren, M.D.: Internal Sorting by Radix Plus Shifting; J.ACM, 13(3):404-411, (July 1966). (4.2.4).

790. MacVeigh, D.T.: Effect of Data Representation on Cost of Sparse Matrix Operations; Acta Informatica, 7:361-394, (1977). (2.1).

791. Madhavan, C.E.V.: Secondary attribute retrieval using tree data structures; Theoretical Computer Science, 33(1):107-116, (1984). (3.5).

792. Madison, J.A.T.: Fast Lookup in Hash Tables with Direct Rehashing; Computer Journal, 23(2):188-189, (Feb 1980). (3.3.8.2).

793. Mahmoud, H.M. and Pittel, B.: Analysis of the space of search trees under the random insertion algorithm; J of Algorithms, 10(1):52-75, (Mar 1989). (3.4.1.10).

794. Mahmoud, H.M. and Pittel, B.: On the Most Probable Shape of a Search Tree Grown from a Random Permutation; SIAM J Alg Disc Methods, 5(1):69-81, (Mar 1984). (3.4.1.1).

795. Mahmoud, H.M.: On the Average Internal Path length of m-ary search trees; Acta Informatica, 23(1):111-117, (1986). (3.4.1.10).

796. Mahmoud, H.M.: The expected distribution of degrees in random binary search trees; Computer Journal, 29(1):36-37, (Feb 1986). (3.4.1.1).

797. Maier, D. and Salveter, S.C.: Hysterical B-Trees; Inf. Proc. Letters, 12(4):199-202, (Aug 1981). (3.4.2.1).

798. Maier, D.: The Complexity of some Problems on Subsequences and Supersequences; J.ACM, 25:322-336, (1978). (7.3.1, 7.3).

799. Main, M. and Lorentz, R.: An $O(n \log n)$ Algorithm for Finding all Repetitions in a String; J of Algorithms, 1:359-373, (1980). (7.1).

800. Mairson, H.G.: Average Case Lower Bounds on the Construction and Searching of Partial Orders; Proceedings FOCS, Portland OR, 26:303-311, (Oct 1985). (5.1).

801. Mairson, H.G.: The Program Complexity of Searching a Table; Proceedings FOCS, Tucson AZ, 24:40-47, (Nov 1983). (3.3.16).

802. Majster, M. and Reiser, A.: Efficient On-Line Construction and Correction of Position Trees; SIAM J on Computing, 9:785-807, (1980). (7.2.2).

803. Makarov, O.M.: Using Duality for the Synthesis of an Optimal Algorithm Involving Matrix Multiplication; Inf. Proc. Letters, 13(2):48-49, (Nov 1981). (6.3).

804. Makinen, E.: Constructing a binary tree from its traversals; BIT, 29(3):572-575, (1989). (3.4.1).

805. Makinen, E.: On Linear Search Heuristics; Inf. Proc. Letters, 29(1):35-36, (Sep 1988). (3.1.2, 3.1.3).

806. Makinen, E.: On top-down splaying; BIT, 27(3):330-339, (1987). (3.4.1.6).

807. Malcolm, W.D.: String Distribution for the Polyphase Sort; C.ACM, 6(5):217-220, (May 1963). (4.4.4).

808. Mallach, E.G.: Scatter Storage Techniques: A Unifying Viewpoint and a Method for Reducing Retrieval Times; Computer Journal, 20(2):137-140, (May 1977). (3.3.8.2).

809. Maly, K.: A Note on Virtual Memory Indexes; C.ACM, 21(9):786-787, (Sep 1978). (3.4.2).

810. Maly, K.: Compressed Tries; C.ACM, 19(7):409-415, (July 1976). (3.4.4).

811. Manacher, G.K., Bui, T.D. and Mai, T.: Optimum Combinations of Sorting and Merging; J.ACM, 36(2):290-334, (Apr 1989). (4.3.3).

812. Manacher, G.K.: Significant Improvements to the Hwang-Lin Merging Algorithm; J.ACM, 26(3):434-440, (July 1979). (4.3.3).

813. Manacher, G.K.: The Ford-Johnson Sorting Algorithm is Not Optimal; J.ACM, 26(3):441-456, (July 1979). (4.1).

814. Manber, U. and Baeza-Yates, R.A.: An Algorithm for String Matching with a Sequence of Don't Cares; Inf. Proc. Letters, to app.. (7.2.4, 7.3).

815. Manber, U. and Myers, G.: Suffix Arrays: A new method for on-line string searches; Proceedings SODA, San Francisco CA, 1:319-327, (Jan 1990). (7.2.4).

816. Manber, U.: Using Induction to Design Algorithms; C.ACM, 31(11):1300-1313, (1988). (2.2).

817. Manker, H.H.: Multiphase Sorting; C.ACM, 6(5):214-217, (May 1963). (4.4.4).

818. Mannila, H. and Ukkonen, E.: A Simple Linear-time algorithm for in-situ merging; Inf. Proc. Letters, 18(4):203-208, (May 1984). (4.3.2).

819. Mannila, H.: Measures of Presortedness and Optimal Sorting Algorithms; Proceedings ICALP, Lecture Notes in Computer Science 267, Springer-Verlag, Antwerp, Belgium, 11:324-336, (1984). (4.1.8).

820. Manolopoulos, Y.P., Kollias, J.G. and Burton, F.W.: Batched interpolation search; Computer Journal, 30(6):565-568, (Dec 1987). (3.2.2).

821. Manolopoulos, Y.P., Kollias, J.G. and Hatzupoulos, M.: Sequential vs. Binary Batched searching; Computer Journal, 29(4):368-372, (Aug 1986). (3.1, 3.2).

822. Manolopoulos, Y.P.: Batched search of index sequential files; Inf. Proc. Letters, 22(5):267-272, (Apr 1986). (3.4.3).

823. Mansour, Y., Nisan, N. and Tiwari, P.: The Computational Complexity of Universal Hashing; Proceedings STOC-SIGACT, Baltimore MD, 22:235-243, (May 1990). (3.3.1).

824. Martin, W.A. and Ness, D.N.: Optimizing Binary Trees Grown with a Sorting Algorithm; C.ACM, 15(2):88-93, (Feb 1972). (3.4.1.6).

825. Martin, W.A.: Sorting; Computing Surveys, 3(4):147-174, (Dec 1971). (4.1, 4.4).

826. Maruyama, K. and Smith, S.E.: Analysis of Design Alternatives for Virtual Memory Indexes; C.ACM, 20(4):245-254, (Apr 1977). (3.4.3).

827. Maurer, H.A., Ottmann, T. and Six, H.: Implementing Dictionaries Using Binary Trees of Very Small Height; Inf. Proc. Letters, 5(1):11-14, (May 1976). (3.4.2.3).

828. Maurer, W.D. and Lewis, T.E.: Hash table methods; ACM C. Surveys, 7(1):5-19, (Mar 1975). (3.3).

829. Maurer, W.D.: An Improved Hash Code for Scatter Storage; C.ACM, 11(1):35-38, (Jan 1968). (3.3.1, 3.3.6).

830. McAllester, R.L.: Polyphase Sorting with Overlapped Rewind; C.ACM, 7(3):158-159, (Mar 1964). (4.4.4).

831. McCabe, J.: On serial files with relocatable records; Operations Research, 13(4):609-618, (1965). (3.1.2).

832. McCreight, E.M.: Pagination of B*-trees with variable-length records; C.ACM, 20(9):670-674, (Sep 1977). (3.4.2).

833. McCreight, E.M.: Priority search trees; SIAM J on Computing, 14(2):257-276, (May 1985). (5.1.6).

834. McCulloch, C.M.: Quickshunt - A Distributive Sorting Algorithm; Computer Journal, 25(1):102-104, (Feb 1982). (4.2.4, 4.4).

835. McDiarmid, C.J.H. and Reed, B.A.: Building Heaps Fast; J of Algorithms, 10(3):352-365, (Sep 1989). (5.1.3).

836. McDonell, K.J.: An Inverted Index Implementation; Computer Journal, 20(2):116-123, (1977). (7.2.1, 7.2.2).

837. McKellar, A.C. and Wong, C.K.: Bounds on Algorithms for String Generation; Acta Informatica, 1(4):311-319, (1972). (4.4.1).

838. McKellar, A.C. and Wong, C.K.: Dynamic Placement of Records in Linear Storage; J.ACM, 25(3):421-434, (July 1978). (3.1).

839. Mehlhorn, K. and Naher, S.: Dynamic Fractional cascading; Algorithmica, 5(2):215-141, (1990). (2.2).

840. Mehlhorn, K. and Overmars, M.H.: Optimal Dynamization of Decomposable Searching Problems; Inf. Proc. Letters, 12(2):93-98, (Apr 1981). (2.2).

841. Mehlhorn, K. and Tsakalidis, A.K.: An Amortized Analysis of Insertions into AVL-Trees; SIAM J on Computing, 15(1):22-33, (Feb 1986). (3.4.1.3).

842. Mehlhorn, K. and Tsakalidis, A.K.: Dynamic Interpolation Search; Proceedings ICALP, Lecture Notes in Computer Science 194, Springer-Verlag, Nafplion, Greece, 12:424-434, (1985). (3.2.2).

843. Mehlhorn, K.: A Best Possible Bound for the Weighted Path Length of Binary Search Trees; SIAM J on Computing, 6(2):235-239, (June 1977). (3.4.1.6).

844. Mehlhorn, K.: A Partial Analysis of Height-Balanced Trees Under Random Insertions and Deletions; SIAM J on Computing, 11(4):748-760, (Nov 1982). (3.4.1.3, 3.4.2.1, 3.4.2.3).

845. Mehlhorn, K.: Dynamic Binary Search; SIAM J on Computing, 8(2):175-198, (May 1979). (3.4.1.6, 3.4.4).

846. Mehlhorn, K.: Nearly Optimal Binary Search Trees; Acta Informatica, 5:287-295, (1975). (3.4.1.6).

847. Mehlhorn, K.: On the Program Size of Perfect and Universal Hash Functions; Proceedings FOCS, Chicago IL, 23:170-175, (Oct 1982). (3.3.16, 3.3.1).

848. Mehlhorn, K.: Sorting Presorted Files; Proceedings GI Conference on Theoretical Computer Science, Lecture Notes in Computer Science 67, Springer-Verlag, Aachen, Germany, 4:199-212, (1979). (4.1).

849. Meijer, H. and Akl, S.G.: The Design and Analysis of a New Hybrid Sorting Algorithm; Inf. Proc. Letters, 10(4):213-218, (July 1980). (4.1.1, 4.1.8, 4.2.5).

850. Meir, A. and Moon, J.W.: On the Altitude of Nodes in Random Trees; Canad J Math, 30(5):997-1015, (1978). (3.4.1.1).

851. Melville, R. and Gries, D.: Controlled Density Sorting; Inf. Proc. Letters, 10(4):169-172, (July 1980). (4.1.2, 4.1.7).

852. Mendelson, H. and Yechiali, U.: A New Approach to the Analysis of Linear Probing Schemes; J.ACM, 27(2):474-483, (July 1980). (3.3.4).

853. Mendelson, H. and Yechiali, U.: Performance Measures for Ordered Lists in Random-Access Files; J.ACM, 26(4):654-667, (Oct 1979). (3.3).

854. Mendelson, H.: Analysis of Linear Probing with Buckets; Inform. Systems, 8:207-216, (1983). (3.3.4).

855. Merrett, T.H. and Fayerman, B.: Dynamic Patricia; Proceedings Int. Conf. on Foundations of Data Organization, Kyoto, Japan, :13-20, (1985). (3.4.4.5, 7.2.2).

856. Merritt, S.M.: An Inverted Taxonomy of Sorting Algorithms; C.ACM, 28(1):96-99, (Jan 1985). (2.2.2, 4.1).

857. Mescheder, B.: On the Number of Active *-Operations Needed to Compute the Discrete Fourier Transform; Acta Informatica, 13(4):383-408, (1980). (6.4).

858. Mesztenyi, C. and Witzgall, C.: Stable Evaluation of Polynomials; J Res. Nat. Bureau of Standards, 71B(1):11-17, (Jan 1967). (6.4).

859. Meyer, B.: Incremental String Matching; Inf. Proc. Letters, 21:219-227, (1985). (7.1.2, 7.1.4).

860. Miller, R., Pippenger, N., Rosenberg, A.L. and Snyder, L.: Optimal 2-3 trees; SIAM J on Computing, 8(1):42-59, (Feb 1979). (3.4.2.1).

861. Miyakawa, M., Yuba, T., Sugito, Y. and Hoshi, M.: Optimum Sequence Trees; SIAM J on Computing, 6(2):201-234, (June 1977). (3.4.4).

862. Mizoguchi, T.: On Required Space for Random Split Trees; Proceedings Allerton Conference, Monticello, IL, 17:265-273, (1979). (3.4.3).

863. Moenk, R. and Borodin, A.: Fast Modular Transforms Via Division; Proceedings FOCS, College Park Md, 13:90-96, (Oct 1972). (6.4).

864. Moffat, A. and Port, G.: A fast algorithm for melding splay trees; Proceedings Workshop in Algorithms and Data Structures, Lecture Notes in Computer Science 382, Springer-Verlag, Ottawa, Canada, 1:450-459, (Aug 1989). (3.4.1.6).

865. Moller-Nielsen, P. and Staunstrup, J.: Experiments with a Fast String Searching Algorithm; Inf. Proc. Letters, 18:129-135, (1984). (7.1.3).

866. Monard, M.C.: Design and Analysis of External Quicksort Algorithms; PhD Dissertation, PUC University of Rio de Janeiro, (Feb 1980). (4.4.6).

867. Montgomery, A.Y.: Algorithms and Performance Evaluation of a New Type of Random Access File Organisation; Australian Computer J, 6(1):3-11, (1974). (3.3).

868. Moran, S.: On the complexity of designing optimal partial-match retrieval systems; ACM TODS, 8(4):543-551, (1983). (3.6).

869. Morris, R.: Counting Large Numbers of Events in Small Registers; C.ACM, 21(10):840-842, (Oct 1978). (6.1).

870. Morris, R.: Scatter Storage Techniques; C.ACM, 11(1):38-44, (Jan 1968). (3.3.3, 3.3.4, 3.3.10, 3.3.11).

871. Morrison, D.R.: PATRICIA - Practical Algorithm to Retrieve Information Coded in Alphanumeric; J.ACM, 15(4):514-534, (Oct 1968). (3.4.4.5, 7.2.2).

872. Motoki, T.: A Note on Upper Bounds for the Selection Problem; Inf. Proc. Letters, 15(5):214-219, (Dec 1982). (5.2).

873. Motzkin, D.: A Stable Quicksort; Software - Practice and Experience, 11:607-611, (1981). (4.2.2).

874. Motzkin, D.: Meansort; C.ACM, 26(4):250-251, (Apr 1983). (4.1.3).

875. Motzkin, T.S.: Evaluation of Polynomials and Evaluation of Rational Functions; Bull of Amer Math Soc, 61:163, (1965). (6.4).

876. Mukhopadhay, A.: A Fast Algorithm for the Longest-Common-Subsequence Problem; Information Sciences, 20:69-82, (1980). (7.3.1).

877. Mullen, J.: Unified Dynamic Hashing; Proceedings VLDB, Singapore, 10:473-480, (1984). (3.3.13, 3.3.14).

878. Mullin, J.K.: An Improved Index Sequential Access Method Using Hashed Overflow; C.ACM, 15(5):301-307, (May 1972). (3.4.3).

879. Mullin, J.K.: Retrieval-Update Speed Tradeoffs Using Combined Indices; C.ACM, 14(12):775-776, (1971). (3.4.3).

880. Mullin, J.K.: Spiral Storage: Efficient Dynamic Hashing with Constant Performance; Computer Journal, 28(3):330-334, (1985). (3.3.13).

881. Mullin, J.K.: Tightly Controlled Linear Hashing Without Separate Overflow Storage; BIT, 21(4):390-400, (1981). (3.3.14).

882. Munro, J.I. and Paterson, M.S.: Selection and Sorting with Limited Storage; Theoretical Computer Science, 12(3):315-323, (1980). (4.4, 5.2).

883. Munro, J.I. and Poblete, P.V.: A Discipline for Robustness or Storage Reduction in Binary Search Trees; Proceedings ACM PODS, Atlanta GA, 2:70-75, (Mar 1983). (3.4.1).

884. Munro, J.I. and Poblete, P.V.: Fault Tolerance and Storage reduction in Binary search trees; Information and Control, 62(2-3):210-218, (Aug 1984). (3.4.1).

885. Munro, J.I. and Poblete, P.V.: Searchability in merging and implicit data structures; BIT, 27(3):324-329, (1987). (4.3).

886. Munro, J.I., Raman, V.K. and Salowe, J.S.: Stable in-situ sorting and minimum data movement; BIT, 30(2):220-234, (1990). (4.1).

887. Munro, J.I. and Raman, V.K.: Sorting with minimum data movement; Proceedings Workshop in Algorithms and Data Structures, Lecture Notes in Computer Science 382, Springer-Verlag, Ottawa, Canada, 1:552-562, (Aug 1989). (4.1).

888. Munro, J.I. and Spira, P.M.: Sorting and Searching in Multisets; SIAM J on Computing, 5(1):1-8, (Mar 1976). (4.2).

889. Munro, J.I.: Searching a Two Key Table Under a Single Key; Proceedings STOC-SIGACT, New York, 19:383-387, (May 1987). (3.5, 3.6.2).

890. Murphy, L.J.: Lotka's Law in the Humanities; J American Society of Information Science, 24(6):461-462, (1973). (I.3).

891. Murphy, O.J. and Selkow, S.M.: The efficiency of using k-d trees for finding rearest neighbours in discrete space; Inf. Proc. Letters, 23(4):215-218, (Apr 1986). (3.5.2).

892. Murphy, O.J.: A Unifying Frame work for Trie Design Heuristics; Inf. Proc. Letters, 34:243-249, (1990). (3.4.4).

893. Murphy, P.E. and Paul, M.C.: Minimum Comparison Merging of sets of approximately equal size; Information and Control, 42(1):87-96, (July 1979). (4.3.2).

894. Murthy, D. and Srimani, P.K.: Split Sequence Coalesced Hashing; Inform. Systems, 13(2):211-218, (1988). (3.3.12).

895. Murthy, Y.D., Bhattacharjee, G.P. and Seetaramanath, M.N.: Time- and Space-Optimal Height Balanced 2-3 Trees; J. of Combinatorics, Information and System Sciences, 8(2):127-141, (1983). (3.4.2.1).

896. Myers, E. and Miller, W.: Approximate Matching of Regular Expressions; Bulletin of Mathematical Biology, 51(1):5-37, (1989). (7.1.6, 7.3).

897. Myers, E.: An $O(ND)$ Difference Algorithm and Its Variations; Algorithmica, 1:251-266, (1986). (7.3.1).

898. Myers, E.: Incremental Alignment Algorithms and Their Applications; SIAM J on Computing, to app.. (7.3.1).

899. Nakamura, T. and Mizoguchi, T.: An Analysis of Storage Utilization Factor in Block Split Data Structuring Scheme; Proceedings VLDB, Berlin, 4:489-495, (Sep 1978). (3.4.3).

900. Nakatsu, N., Kambayashi, Y. and Yajima, S.: A Longest Common Subsequence Algorithm Suitable for Similar Text Strings; Acta Informatica, 18:171-179, (1982). (7.3.1).

901. Naor, M. and Yung, M.: Universal One-Way Hash Functions and their Cryptographic Applications; Proceedings STOC-SIGACT, Seattle WA, 21:33-43, (May 1989). (3.3.1).

902. Nelson, R.C. and Samet, H.: A Population Analysis for Hierarchical Data Structures; Proceedings ACM SIGMOD, San Francisco CA, 16:270-277, (May 1987). (3.5.1).

903. Nevalainen, O. and Teuhola, J.: Priority Queue Administration by Sublist Index; Computer Journal, 22(3):220-225, (Mar 1979). (5.1.1).

904. Nevalainen, O. and Teuhola, J.: The Efficiency of Two Indexed Priority Queue Algorithms; BIT, 18(3):320-333, (1978). (5.1.2).

905. Nevalainen, O. and Vesterinen, M.: Determining Blocking Factors for Sequential Files by Heuristic Methods; Computer Journal, 20(3):245-247, (1977). (3.1).

906. Nicklas, B.M. and Schlageter, G.: Index Structuring in Inverted Data Bases by Tries; Computer Journal, 20(4):321-324, (Nov 1977). (3.4.4, 7.2.1, 7.2.2).

907. Nievergelt, J., Hinterberger, H. and Sevcik, K.: The Grid File: An Adaptable, Symmetric Multikey File Structure; ACM TODS, 9(1):38-71, (Mar 1984). (3.5.4).

908. Nievergelt, J. and Reingold, E.M.: Binary Search Trees of Bounded Balance; SIAM J on Computing, 2(1):33-43, (1973). (3.4.1.4).

909. Nievergelt, J. and Wong, C.K.: On Binary Search Trees; Proceedings Information Processing 71, Ljubjana, Yugoslavia, :91-98, (Aug 1971). (3.4.1).

910. Nievergelt, J. and Wong, C.K.: Upper bounds for the total path length of binary trees; J.ACM, 20(1):1-6, (Jan 1973). (3.4.1).

911. Nievergelt, J.: Binary Search Trees and File Organization; ACM C. Surveys, 6(3):195-207, (Sep 1974). (3.4.1).

912. Nijssen, G.M.: Efficient Batch Updating of a Random File; Proceedings ACM SIGFIDET Workshop an Data Description, Access and Control, San Diego CA, :174-186, (Nov 1971). (3.3).

913. Nijssen, G.M.: Indexed Sequential versus Random; IAG Journal, 4:29-37, (1971). (3.3, 3.4.3).

914. Nishihara, S. and Hagiwara, H.: A Full Table Quadratic Search Method Eliminating Secondary Clustering; Int. J of Comp and Inf Sciences, 3(2):123-128, (1974). (3.3.6).

915. Nishihara, S. and Ikeda, K.: Reducing the Retrieval Time of Hashing Method by Using Predictors; C.ACM, 26(12):1082-1088, (Dec 1983). (3.3).

916. Noga, M.T. and Allison, D.C.S.: Sorting in linear expected time; BIT, 25(3):451-465, (1985). (4.2.5).

917. Norton, R.M. and Yeager, D.P.: A Probability Model for Overflow Sufficiency in Small Hash Tables; C.ACM, 28(10):1068-1075, (Oct 1985). (3.3.11).

918. Noshita, K.: Median Selection of 9 Elements in 14 Comparisons; Inf. Proc. Letters, 3(1):8-12, (July 1974). (5.2).

919. Nozaki, A.: A Note on the Complexity of Approximative Evaluation of Polynomials; Inf. Proc. Letters, 9(2):73-75, (Aug 1979). (6.4).

920. Nozaki, A.: Sorting Using Networks of Deques; JCSS, 19(3):309-315, (Dec 1979). (4.2).

921. Nozaki, A.: Two Entropies of a Generalized Sorting Problem; JCSS, 7(5):615-621, (Oct 1973). (4.1, 5.2).

922. O'Dunlaing, C. and Yap, C.K.: Generic Transformation of Data Structures; Proceedings FOCS, Chicago IL, 23:186-195, (Oct 1982). (2.1).

923. Odlyzko, A.M.: Periodic Oscillations of Coefficients of Power Series that Satisfy Functional Equations; Advances in Mathematics, to app.. (3.4.2).

924. Olivie, H.J.: On a Relationship Between 2-3 Brother Trees and Dense Ternary Trees; Int. J Computer Math, 8:233-245, (1980). (3.4.2.1).

925. Olivie, H.J.: On Random Son-trees; Int. J Computer Math, 9:287-303, (1981). (3.4.2.3).

926. Olivie, H.J.: On the Relationship Between Son-trees and Symmetric Binary B-trees; Inf. Proc. Letters, 10(1):4-8, (Feb 1980). (3.4.2.2, 3.4.2.3).

927. Olson, C.A.: Random Access File Organization for Indirectly Addressed Records; Proceedings ACM-NCC, New York NY, 24:539-549, (Sep 1969). (3.3.11).

928. Orenstein, J.A.: Multidimensional Tries Used for Associative Searching; Inf. Proc. Letters, 14(4):150-157, (June 1982). (3.4.4, 3.5, 7.2.2).

929. Otoo, E.J.: A Multidimensional Digital Hashing Scheme for Files with Composite Keys; Proceedings ACM SIGMOD, Austin, TX, 14:214-231, (May 1986). (3.5.4).

930. Otoo, E.J.: Balanced Multidimensional Extendible Hash Tree; Proceedings ACM PODS, Cambridge, Mass., 5:100-113, (Mar 1986). (3.5.4).

931. Ottmann, T., Parker, D.S., Rosenberg, A.L., Six, H. and Wood, D.: Minimal-Cost Brother trees; SIAM J on Computing, 13(1):197-217, (Feb 1984). (3.4.2.3).

932. Ottmann, T., Rosenberg, A.L., Six, H. and Wood, D.: Binary Search Trees with Binary Comparison Cost; Int. J of Comp and Inf Sciences, 13(2):77-101, (Apr 1984). (3.4.1).

933. Ottmann, T., Schrapp, M. and Wood, D.: Purely Top-Down Updating Algorithms for Stratified Search Trees; Acta Informatica, 22(1):85-100, (1985). (3.4.1).

934. Ottmann, T., Six, H. and Wood, D.: On the Correspondence Between AVL Trees and Brother Trees; Computing, 23(1):43-54, (1979). (3.4.2.3, 3.4.1.3).

935. Ottmann, T., Six, H. and Wood, D.: One-Sided k-Height-Balanced Trees; Computing, 22(4):283-290, (1979). (3.4.1.3).

936. Ottmann, T., Six, H. and Wood, D.: Right Brother Trees; C.ACM, 21(9):769-776, (Sep 1978). (3.4.2.3).

937. Ottmann, T., Six, H. and Wood, D.: The Implementation of Insertion and Deletion Algorithms for 1-2 Brother Trees; Computing, 26:369-378, (1981). (3.4.2.3).

938. Ottmann, T. and Stucky, W.: Higher Order Analysis of Random 1-2 Brother Trees; BIT, 20(3):302-314, (1980). (3.4.2.3).

939. Ottmann, T. and Wood, D.: 1-2 Brother Trees or AVL Trees Revisited; Computer Journal, 23(3):248-255, (Aug 1980). (3.4.1.3, 3.4.2.3).

940. Ottmann, T. and Wood, D.: A Comparison of Iterative and Defined Classes of Search Trees; Int. J of Comp and Inf Sciences, 11(3):155-178, (June 1982). (3.4.1, 3.4.2).

941. Ottmann, T. and Wood, D.: Deletion in One-Sided Height-Balanced Search Trees; Int. J Computer Math, 6(4):265-271, (1978). (3.4.1.3).

942. Ottmann, T. and Wood, D.: How to update a balanced binary tree with a constant number of rotations; Proceedings Scandinavian Workshop in Algorithmic Theory, SWAT'90, Lecture Notes in Computer Science 447, Springer-Verlag, Bergen, Norway, 2:122-131, (July 1990). (3.4.1, 3.4.1.8).

943. Ouksel, M. and Scheuermann, P.: Implicit Data Structures for linear Hashing; Inf. Proc. Letters, 29(5):187-189, (Nov 1988). (3.3.14).

944. Ouksel, M. and Scheuermann, P.: Multidimensional B-Trees: Analysis of Dynamic Behavior; BIT, 21(4):401-418, (1981). (3.4.2, 3.5).

945. Ouksel, M. and Scheuermann, P.: Storage Mappings for Multidimensional Linear Dynamic Hashing; Proceedings ACM PODS, Atlanta GA, 2:90-105, (Mar 1983). (3.3.14).

946. Ouksel, M.: The interpolation-based grid file; Proceedings ACM PODS, Portland OR, 4:20-27, (Mar 1985). (3.3.13).

947. Overholt, K.J.: Efficiency of the Fibonacci Search Method; BIT, 13(1):92-96, (1973). (3.2).

948. Overholt, K.J.: Optimal Binary Search Methods; BIT, 13(1):84-91, (1973). (3.2.1).

949. Overmars, M.H., Smid, M., de Berg, M. and van Kreveld, M.: Maintaining Range Trees in Secondary Memory. Part I: Partitions; Acta Informatica, 27:423-452, (1990). (3.6).

950. Overmars, M.H. and van Leeuwen, J.: Dynamic Multidimensional Data Structures Based on Quad- and K-D Trees; Acta Informatica, 17(3):267-285, (1982). (2.2, 3.5.1, 3.5.2).

951. Overmars, M.H. and van Leeuwen, J.: Dynamizations of Decomposable Searching Problems Yielding Good Worst-Case Bounds; Lecture Notes in Computer Science 104, Springer-Verlag, :224-233, (1981). (2.2).

952. Overmars, M.H. and van Leeuwen, J.: Some Principles for Dynamizing Decomposable Searching Problems; Inf. Proc. Letters, 12(1):49-53, (Feb 1981). (2.2).

953. Overmars, M.H. and van Leeuwen, J.: Two General Methods for Dynamizing Decomposable Searching Problems; Computing, 26(2):155-166, (1981). (2.2).

954. Overmars, M.H. and van Leeuwen, J.: Worst-Case Optimal Insertion and Deletion Methods for Decomposable Searching Problems; Inf. Proc. Letters, 12(4):168-173, (Aug 1981). (2.2).

955. Overmars, M.H.: Dynamization of Order Decomposable Set Problems; J of Algorithms, 2(3):245-260, (Sep 1981). (2.2).

956. Overmars, M.H.: Efficient Data Structures for range searching on a grid; J of Algorithms, 9(2):254-275, (June 1988). (3.6.2).

957. Pagli, L.: Height-balanced Multiway Trees; Inform. Systems, 4:227-234, (1979). (3.4.1.3, 3.4.1.10).

958. Pagli, L.: Self Adjusting Hash Tables; Inf. Proc. Letters, 21(1):23-25, (July 1985). (3.3.8.5).

959. Palmer, E.M., Rahimi, M.A. and Robinson, R.W.: Efficiency of a Binary Comparison Storage Technique; J.ACM, 21(3):376-384, (July 1974). (3.4.1.1).

960. Pan, V.Y.: A Unified Approach to the Analysis of Bilinear Algorithms; J of Algorithms, 2(3):301-310, (Sep 1981). (6.3).

961. Pan, V.Y.: Computational Complexity of Computing Polynomials Over the Field of Real and Complex Numbers; Proceedings STOC-SIGACT, San Diego CA, 10:162-172, (May 1978). (6.4).

962. Pan, V.Y.: New Combinations of Methods for the Acceleration of Matrix Multiplication; Comput Math with Applic, 7:73-125, (1981). (6.3).

963. Pan, V.Y.: New Fast Algorithms for Matrix Operations; SIAM J on Computing, 9(2):321-342, (May 1980). (6.3).

964. Pan, V.Y.: New Methods for the Acceleration of Matrix Multiplication; Proceedings FOCS, San Juan PR, 20:28-38, (Oct 1979). (6.3).

965. Pan, V.Y.: Strassen's Algorithm is not Optimal: Trilinear Technique of Aggregating, Uniting and Canceling for Constructing Fast Algorithms for Matrix Operations; Proceedings FOCS, Ann Arbor MI, 19:166-176, (Oct 1978). (6.3).

966. Pan, V.Y.: The Additive and Logical Complexities of Linear and Bilinear Arithmetic Algorithms; J of Algorithms, 4(1):1-34, (Mar 1983). (6.3).

967. Pan, V.Y.: The Bit-Complexity of Arithmetic Algorithms; J of Algorithms, 2(2):144-163, (June 1981). (6.4).

968. Pan, V.Y.: The Techniques of Trilinear Aggregating and the Recent Progress in the Asymptotic Acceleration of Matrix Operations; Theoretical Computer Science, 33(1):117-138, (1984). (6.3).

969. Panny, W.: A Note on the higher moments of the expected behavior of straight insertion sort; Inf. Proc. Letters, 22(4):175-177, (Apr 1986). (4.1.2).

970. Papadakis, T., Munro, J.I. and Poblete, P.V.: Analysis of the expected search cost in skip lists; Proceedings Scandinavian Workshop in Algorithmic Theory, SWAT'90, Lecture Notes in Computer Science 447, Springer-Verlag, Bergen, Norway, 2:160-172, (July 1990). (3.1, 3.4.1).

971. Papadimitriou, C.H. and Bernstein, P.A.: On the Performance of Balanced Hashing Functions When Keys are Not Equiprobable; ACM TOPLAS, 2(1):77-89, (Jan 1980). (3.3.1).

972. Patt, Y.N.: Variable Length Tree Structures Having Minimum Average Search Time; C.ACM, 12(2):72-76, (Feb 1969). (3.4.4).

973. Payne, H.J. and Meisel, W.S.: An Algorithm for Constructing Optimal Binary Decision Trees; IEEE Trans. on Computers, 26(9):905-916, (1977). (3.4.1).

974. Pearson, P.K.: Fast Hashing of Variable-Length Text Strings; C.ACM, 33(6):677-680, (June 1990). (3.3.16, 3.3.1).

975. Peltola, E. and Erkio, H.: Insertion Merge Sorting; Inf. Proc. Letters, 7(2):92-99, (Feb 1978). (4.2.1, 4.2.5).

976. Perl, Y., Itai, A. and Avni, H.: Interpolation Search - A Log Log N Search; C.ACM, 21(7):550-553, (July 1978). (3.2.2).

977. Perl, Y. and Reingold, E.M.: Understanding the Complexity of Interpolation Search; Inf. Proc. Letters, 6(6):219-221, (Dec 1977). (3.2.2).

978. Perl, Y.: Optimum split trees; J of Algorithms, 5(3):367-374, (Sep 1984). (3.4.1.6).

979. Peters, J.G. and Kritzinger, P.S.: Implementation of Samplesort: A Minimal Storage Tree Sort; BIT, 15(1):85-93, (1975). (4.1.3).

980. Peterson, W.W.: Addressing for Random-Access Storage; IBM J Res. Development, 1(4):130-146, (Apr 1957). (3.2, 3.3).

981. Pflug, G.C. and Kessler, H.W.: Linear Probing with a Nonuniform Address Distribution; J.ACM, 34(2):397-410, (Apr 1987). (3.3.4).

982. Pinter, R.: Efficient String Matching with Don't-Care Patterns; Combinatorial Algorithms on Words, NATO ASI Series, Springer-Verlag, F12:11-29, (1985). (7.1).

983. Pippenger, N.: Computational Complexity in Algebraic Functions Fields; Proceedings FOCS, San Juan PR, 20:61-65, (Oct 1979). (6.2).

984. Pippenger, N.: On the Application of Coding Theory to Hashing; IBM J Res. Development, 23(2):225-226, (Mar 1979). (3.3).

985. Pippenger, N.: On the Evaluation of Powers and Monomials; SIAM J on Computing, 9(2):230-250, (May 1980). (6.2).

986. Pittel, B.: Asymptotical Growth of a Class of Random Trees; Annals of Probability, 13(2):414-427, (1985). (3.4.1).

987. Pittel, B.: Linear Probing: the probable largest search time grows logarithmically with the number of records; J of Algorithms, 8(2):236-249, (June 1987). (3.3.4).

988. Pittel, B.: On Growing Random Binary Trees; J of Mathematical Analysis and Appl, 103(2):461-480, (Oct 1984). (3.4.1.1).

989. Pittel, B.: Paths in a Random Digital Tree: Limiting Distributions; Advances Appl Probability, 18:139-155, (1986). (3.4.4).

990. Poblete, P.V. and Munro, J.I.: Last-Come-First-Served Hashing; J of Algorithms, 10(2):228-248, (June 1989). (3.3.3, 3.3.8.3, 3.3.9).

991. Poblete, P.V. and Munro, J.I.: The analysis of a fringe heuristic for binary search trees; J of Algorithms, 6(3):336-350, (Sep 1985). (3.4.1.6).

992. Poblete, P.V.: Approximating functions by their Poisson Transform; Inf. Proc. Letters, 23(3):127-130, (July 1987). (3.3.4, 4.1.7).

993. Poblete, P.V.: Fringe Techniques for Binary Search Trees; PhD Dissertation, Department of Computer Science, University of Waterloo, (1982). (3.4.1.6).

994. Pohl, I.: Minimean Optimality in Sorting Algorithms; Proceedings FOCS, Berkeley CA, 16:71-74, (Oct 1975). (4.1, 5.1).

995. Pooch, U.W. and Nieder, A.: A Survey of Indexing Techniques for Sparse Matrices; ACM C. Surveys, 5(2):109-133, (June 1973). (2.1).

996. Pope, A.: Bradford's Law and the Periodical Literature of Information Sciences; J American Society of Information Science, 26(4):207-213, (1975). (I.2).

997. Porter, T. and Simon, I.: Random Insertion into a Priority Queue Structure; IEEE Trans. Software Engineering, 1(3):292-298, (Sep 1975). (5.1.3).

998. Postmus, J.T., Rinnooy Kan, A.H.G. and Timmer, G.T.: An Efficient Dynamic Selection Method; C.ACM, 26(11):878-881, (Nov 1983). (5.2).

999. Power, L.R.: Internal Sorting Using a Mimimal Tree Merge Strategy; ACM TOMS, 6(1):68-79, (Mar 1980). (4.2).

1000. Pramanik, S. and Kin, M.H.: HCB-tree : a height Compressed B-tree for parallel processing; Inf. Proc. Letters, 29(5):213-220, (Nov 1988). (3.4.2).

1001. Pratt, V. and Yao, F.F.: On Lower Bounds for Computing the ith Largest Element; Proceedings SWAT (FOCS), Iowa City IO, 14:70-81, (Oct 1973). (5.2).

1002. Pratt, V.: The Power of Negative Thinking in Multiplying Boolean Matrices; Proceedings STOC-SIGACT, Seattle WA, 6:80-83, (Apr 1974). (6.3).

1003. Preparata, F.P.: A fast stable-sorting algorithm with absolutely minimum storage; Theoretical Computer Science, 1(2):185-190, (1975). (4.1).

1004. Price, C.E.: Table Lookup Techniques; ACM C. Surveys, 3(2):49-65, (1971). (3.2, 3.3, 3.4.1).

1005. Probert, R.L.: An Extension of Computational Duality to Sequences of Bilinear Computations; SIAM J on Computing, 7(1):91-98, (Feb 1978). (6.3).

1006. Probert, R.L.: Commutativity, Non-Commutativity and Bilinearity; Inf. Proc. Letters, 5(2):46-49, (June 1976). (6.3).

1007. Probert, R.L.: On the Additive Complexity of Matrix Multiplication; SIAM J on Computing, 5(2):187-203, (June 1976). (6.3).

1008. Probert, R.L.: On the Composition of Matrix Multiplication Algorithms; Proceedings Manitoba Conference on Num Math, Winnipeg, 6:357-366, (Sep 1976). (6.3).

1009. Proskurowski, A.: On the Generation of Binary Trees; J.ACM, 27(1):1-2, (Jan 1980). (3.4.1).

1010. Pugh, W.: Skip Lists: A probabilistic alternative to balanced trees; C.ACM, 33(6):668-676, (1990). (3.1, 3.4.1).

1011. Pugh, W.: Slow Optimally Balanced Search Strategies vs. Cached fast Uniformly Balanced Search Strategies; Inf. Proc. Letters, 34:251-254, (1990). (3.2).

1012. Quittner, P., Csoka, S., Halasz, S., Kotsis, D. and Varnai, K.: Comparison of Synonym Handling and Bucket Organization Methods; C.ACM, 24(9):579-583, (Sep 1981). (3.3.4, 3.3.11).

1013. Quitzow, K.H. and Klopprogge, M.R.: Space Utilization and Access Path Length in B-Trees; Inform. Systems, 5:7-16, (1980). (3.4.2).

1014. Radhakrishnan, T. and Kernizan, R.: Lotka's Law and Computer Science Literature; J American Society of Information Science, 30(1):51-54, (Jan 1979). (I.3).

1015. Radke, C.E.: The Use of Quadratic Residue Research; C.ACM, 13(2):103-105, (Feb 1970). (3.3.6).

1016. Raghavan, V.V. and Yu, C.T.: A Note on a Multidimensional Searching Problem; Inf. Proc. Letters, 6(4):133-135, (Aug 1977). (3.5).

1017. Raiha, K.J. and Zweben, S.H.: An Optimal Insertion Algorithm for One-Sided Height-Balanced Binary Search Trees; C.ACM, 22(9):508-512, (Sep 1979). (3.4.1.3).

1018. Ramakrishna, M.V. and Larson, P.: File Organization using Composite Perfect Hashing; ACM TODS, 14(2):231-263, (June 1989). (3.3.16).

1019. Ramakrishna, M.V. and Mukhopadhyay, P.: Analysis of Bounded Disorder File Organization; Proceedings ACM PODS, San Francisco, 8:117-125, (1988). (3.4.2, 3.4.3, 3.3).

1020. Ramakrishna, M.V.: An Exact Probability Model for Finite Hash Tables; Proceedings IEEE Conf. on Data Eng., Los Angeles, 4:362-368, (1988). (3.3).

1021. Ramakrishna, M.V.: Analysis of Random probing hashing; Inf. Proc. Letters, 31(2):83-90, (Apr 1989). (3.3.3).

1022. Ramakrishna, M.V.: Computing the probability of hash table/urn overflow; Comm. in Statistics - Theory and Methods, 16:3343-3353, (1987). (3.3).

1023. Ramakrishna, M.V.: Hashing in Practice, Analysis of Hashing and Universal Hashing; Proceedings ACM SIGMOD, Chicago IL, 17:191-199, (June 1988). (3.3.2, 3.3.11, 3.3.1).

1024. Ramamohanarao, K. and Sacks-Davis, R.: Partial match retrieval using recursive linear hashing; BIT, 25(3):477-484, (1985). (3.3.14, 3.6).

1025. Ramamohanarao, K., Lloyd, J.W. and Thom, J.A.: Partial-Match Retrieval Using Hashing and Descriptors; ACM TODS, 8(4):522-576, (1983). (3.5.4, 7.2.6).

1026. Ramamohanarao, K. and Lloyd, J.W.: Dynamic Hashing Schemes; Computer Journal, 25(4):478-485, (Nov 1982). (3.3.14).

1027. Ramamohanarao, K. and Sacks-Davis, R.: Recursive Linear Hashing; ACM TODS, 9(3):369-391, (1984). (3.3.14).

1028. Ramanan, P.V. and Hyafil, L.: New algorithms for selection; J of Algorithms, 5(4):557-578, (Dec 1984). (5.2).

1029. Rao, V.N.S., Iyengar, S.S. and Kashyap, R.L.: An average case analysis of MAT and inverted file; Theoretical Computer Science, 62(3):251-266, (Dec 1988). (3.4.3, 7.2.1).

1030. Rao, V.N.S., Vaishnavi, V.K. and Iyengar, S.S.: On the dynamization of data structures; BIT, 28(1):37-53, (1988). (2.2).

1031. Regener, E.: Multiprecision Integer Division Examples using Arbitrary Radix; ACM TOMS, 10(3):325-328, (1984). (6.1).

1032. Regnier, M.: Analysis of grid file algorithms; BIT, 25(2):335-357, (1985). (3.5.4).

1033. Regnier, M.: On the Average Height of Trees in Digital Search and Dynamic Hashing; Inf. Proc. Letters, 13(2):64-66, (Nov 1981). (3.4.4, 3.3.13).

1034. Reingold, E.M.: A Note on 3-2 Trees; Fibonacci Quarterly, 17(2):151-157, (Apr 1979). (3.4.2.1).

1035. Reiser, A.: A Linear Selection Algorithm for Sets of Elements with Weights; Inf. Proc. Letters, 7(3):159-162, (Apr 1978). (5.2).

1036. Remy, J.L.: Construction Evaluation et Amelioration Systematiques de Structures de Donnees; RAIRO Informatique Theorique, 14(1):83-118, (1980). (2.2).

1037. Revah, L.: On the Number of Multiplications/Divisions Evaluating a Polynomial with Auxiliary Functions; SIAM J on Computing, 4(3):381-392, (Sep 1975). (6.4).

1038. Richards, D. and Vaidya, P.: On the distribution of comparisons in sorting algorithms; BIT, 28(4):764-774, (1988). (4.1).

1039. Richards, D.: On the worst possible analysis of weighted comparison-based algorithms; Computer Journal, 31(3):276-278, (June 1988). (4.1).

1040. Richards, R.C.: Shape distribution of height-balanced trees; Inf. Proc. Letters, 17(1):17-20, (July 1983). (3.4.1.3).

1041. Rivest, R.L. and van de Wiele, J.P.: An $\Omega((n/\lg n)^{1/2})$ Lower Bound on the Number of Additions Necessary to Compute 0-1 Polynomials Over the Ring of Integer Polynomials; Inf. Proc. Letters, 8(4):178-180, (Apr 1979). (6.4).

1042. Rivest, R.L.: On Hash-Coding Algorithms for Partial-Match Retrieval; Proceedings FOCS, New Orleans LA, 15:95-103, (Oct 1974). (3.5.4, 7.2.6).

1043. Rivest, R.L.: On Self-Organizing Sequential Search Heuristics; C.ACM, 19(2):63-67, (Feb 1976). (3.1.2, 3.1.3).

1044. Rivest, R.L.: On the Worst-Case Behavior of String-Searching Algorithms; SIAM J on Computing, 6:669-674, (1977). (7.1).

1045. Rivest, R.L.: Optimal Arrangement of Keys in a Hash Table; J.ACM, 25(2):200-209, (Apr 1978). (3.3.8.2).

1046. Rivest, R.L.: Partial-Match Retrieval Algorithms; SIAM J on Computing, 5(1):19-50, (Mar 1976). (3.5.4, 7.2.6).

1047. Robertazzi, T.G. and Schwartz, S.C.: Best Ordering for Floating-Point Addition; ACM TOMS, 14(1):101-110, (Mar 1988). (6.1).

1048. Robinson, J.T.: Order Preserving Linear Hashing Using Dynamic Key Statistics; Proceedings ACM PODS, Cambridge, Mass., 5:91-99, (Mar 1986). (3.3.14).

1049. Robinson, J.T.: The k-d-B-tree: A Search Structure for Large Multidimensional Dynamic Indexes; Proceedings ACM SIGMOD, Ann Arbor MI, 10:10-18, (Apr 1981). (3.4.2, 3.5.2).

1050. Robson, J.M.: An Improved Algorithm for Traversing Binary Trees Without Auxiliary Stack; Inf. Proc. Letters, 2(1):12-14, (Mar 1973). (3.4.1).

1051. Robson, J.M.: Baer's Weight Balanced Trees do not Have Bounded Balance; Australian Computer Science Communications, 2(1):195-204, (1980). (3.4.1.5).

1052. Robson, J.M.: The Asymptotic Behaviour of the Height of Binary Search Trees; Australian Computer Science Communications, 4(1):88-98, (1982). (3.4.1.1).

1053. Robson, J.M.: The Height of Binary Search Trees; Australian Computer J, 11(4):151-153, (Nov 1979). (3.4.1.1).

1054. Rohrich, J.: A Hybrid of Quicksort with O(n log n) Worst Case Complexity; Inf. Proc. Letters, 14(3):119-123, (May 1982). (4.1.3).

1055. Romani, F. and Santoro, N.: On Hashing Techniques in a Paged Environment; Calcolo, 16(3), (1979). (3.3).

1056. Romani, F.: Some Properties of Disjoint Sums of Tensors Related to Matrix Multiplication; SIAM J on Computing, 11(2):263-267, (May 1982). (6.3).

1057. Rosenberg, A.L. and Snyder, L.: Minimal comparison 2-3 trees; SIAM J on Computing, 7(4):465-480, (Nov 1978). (3.4.2.1).

1058. Rosenberg, A.L. and Snyder, L.: Time- and Space-Optimality in B-Trees; ACM TODS, 6(1):174-193, (Mar 1981). (3.4.2).

1059. Rosenberg, A.L., Stockmeyer, L.J. and Snyder, L.: Uniform Data Encodings; Theoretical Computer Science, 11(2):145-165, (1980). (2.1).

1060. Rosenberg, A.L. and Stockmeyer, L.J.: Hashing Schemes for Extendible Arrays; J.ACM, 24(2):199-221, (Apr 1977). (3.3).

1061. Rosenberg, A.L. and Stockmeyer, L.J.: Storage Schemes for Boundedly Extendible Arrays; Acta Informatica, 7:289-303, (1977). (2.1).

1062. Rosenberg, A.L., Wood, D. and Galil, Z.: Storage Representations for Tree-Like Data Structures; Mathematical Systems Theory, 13(2):105-130, (1979). (2.1).

1063. Rosenberg, A.L.: Allocating Storage for Extendible Arrays; J.ACM, 21(4):652-670, (Oct 1974). (2.1).

1064. Rosenberg, A.L.: Data Encodings and their Costs; Acta Informatica, 9(3):273-292, (1978). (2.1).

1065. Rosenberg, A.L.: Encoding Data Structures in Trees; J.ACM, 26(4):668-689, (Oct 1979). (3.4.1).

1066. Rosenberg, A.L.: Managing Storage for Extendible Arrays; SIAM J on Computing, 4(3):287-306, (Sep 1975). (2.1).

1067. Rosenberg, A.L.: On Uniformly Inserting One Data Structure into Another; C.ACM, 24(2):88-90, (Feb 1981). (2.1, 2.2).

1068. Rotem, D.: Clustered Multiattribute Hash Files; Proceedings ACM PODS, Philadelfia PA, 8, (Mar 1989). (3.5.4).

1069. Rotem, D. and Varol, Y.L.: Generation of Binary Trees from Ballot Sequences; J.ACM, 25(3):396-404, (July 1978). (3.4.1).

1070. Rothnie, J.B. and Lozano, T.: Attribute Based File Organization in a Paged Memory Environment; C.ACM, 17(2):63-69, (Feb 1974). (3.3, 3.5).

1071. Ruskey, F. and Hu, T.C.: Generating Binary Trees Lexicographically; SIAM J on Computing, 6(4):745-758, (Dec 1977). (3.4.1).

1072. Ruskey, F.: Generating t-Ary Trees Lexicographically; SIAM J on Computing, 7(4):424-439, (Nov 1978). (3.4.1.10).

1073. Rytter, W.: A Correct Preprocessing Algorithm for Boyer-Moore String-Searching; SIAM J on Computing, 9:509-512, (1980). (7.1.3).

1074. Sack, J.R. and Strothotte, T.: A Characterization of Heaps and Its Applications; Information and Computation, 86(1):69-86, (May 1990). (5.1.3).

1075. Sack, J.R. and Strothotte, T.: An algorithm for merging heaps; Acta Informatica, 22(2):171-186, (1985). (5.1.3).

1076. Sacks-Davis, R., Ramamohanarao, K. and Kent, A.: Multikey access methods based on superimposed coding techniques; ACM TODS, 12(4):655-696, (1987). (3.5, 7.2.6).

1077. Sacks-Davis, R. and Ramamohanarao, K.: A Two Level Superimposed Coding Scheme for Partial Match Retrieval; Inform. Systems, 8:273-280, (1983). (3.5.4).

1078. Sacks-Davis, R. and Ramamohanarao, K.: A Two-Level Superimposed Coding Scheme for Partial Match Retrieval; Inform. Systems, 8(4):273-280, (1983). (3.5.4, 7.2.6).

1079. Sager, T.J.: A Polynomial Time Generator for Minimal Perfect Hash Functions; C.ACM, 28(5):523-532, (May 1985). (3.3.16).

1080. Salowe, J.S. and Steiger, W.L.: Simplified stable merging tasks; J of Algorithms, 8(4):557-571, (Dec 1987). (4.3.2).

1081. Salowe, J.S. and Steiger, W.L.: Stable unmerging in linear time and Constant space; Inf. Proc. Letters, 25(5):285-294, (July 1987). (4.3).

1082. Salzberg, B.: Merging sorted runs using large main memory; Acta Informatica, 27(3):195-216, (1989). (4.4).

1083. Samadi, B.: B-trees in a system with multiple views; Inf. Proc. Letters, 5(4):107-112, (Oct 1976). (3.4.2).

1084. Samet, H.: A Quadtree Medial Axis Transform; C.ACM, 26(9):680-693, (Sep 1983). (3.5.1.1).

1085. Samet, H.: Data Structures for Quadtree Approximation and Compression; C.ACM, 28(9):973-993, (Sep 1985). (3.5.1.1).

1086. Samet, H.: Deletion in Two-Dimensional Quad Trees; C.ACM, 23(12):703-710, (Dec 1980). (3.5.1.1).

1087. Samet, H.: The Quadtree and Related Hierarchical Data Structures; ACM C. Surveys, 16(2):187-260, (June 1984). (3.5.1.1).

1088. Samson, W.B. and Davis, R.H.: Search Times Using Hash Tables for Records with Non-Unique Keys; Computer Journal, 21(3):210-214, (Aug 1978). (3.3.6).

1089. Samson, W.B.: Hash Table Collision Handling on Storage Devices with Latency; Computer Journal, 24(2):130-131, (May 1981). (3.3.4, 3.3.5).

1090. Santoro, N. and Sidney, J.B.: Interpolation Binary Search; Inf. Proc. Letters, 20(4):179-182, (May 1985). (3.2.1, 3.2.2).

1091. Santoro, N.: Chain Multiplication of Matrices Approximately or Exactly the Same Size; C.ACM, 27(2):152-156, (Feb 1984). (6.3).

1092. Santoro, N.: Extending the Four Russians' Bound to General Matrix Multiplication; Inf. Proc. Letters, 10(2):87-88, (Mar 1980). (6.3).

1093. Santoro, N.: Full Table Search by Polynomial Functions; Inf. Proc. Letters, 5(3):72-74, (Aug 1976). (3.3.6).

1094. Sarwate, D.V.: A Note on Universal Classes of Hash Functions; Inf. Proc. Letters, 10(1):41-45, (Feb 1980). (3.3.1).

1095. Sassa, M. and Goto, E.: A Hashing Method for Fast Set Operations; Inf. Proc. Letters, 5(2):31-34, (1976). (3.3).

1096. Savage, J.E.: An Algorithm for the Computation of Linear Forms; SIAM J on Computing, 3(2):150-158, (June 1974). (6.3, 6.4).

1097. Saxe, J.B. and Bentley, J.L.: Transforming Static Data Structures to Dynamic Data Structures; Proceedings FOCS, San Juan PR, 20:148-168, (Oct 1979). (2.2).

1098. Saxe, J.B.: On the Number of Range Queries in k-Space; Discr App Math, 1(3):217-225, (1979). (3.6.2).

1099. Schaback, R.: On the Expected Sublinearity of the Boyer-Moore Algorithm; SIAM J on Computing, 17(4):648-658, (1988). (7.1.3).

1100. Schachtel, G.: A Noncommutative Algorithm for Multiplying 5 x 5 Matrices Using 103 Multiplications; Inf. Proc. Letters, 7(4):180-182, (June 1978). (6.3).

1101. Schay, G. and Raver, N.: A Method for Key-to-Address Transformation; IBM J Res. Development, 7:121-126, (1963). (3.3).

1102. Schay, G. and Spruth, W.G.: Analysis of a File Addressing Method; C.ACM, 5(8):459-462, (Aug 1962). (3.3.4).

1103. Scheurmann, P. and Ouksel, M.: Multidimensional B-trees for Associative Searching in Database Systems; Inform. Systems, 7:123-137, (1982). (3.4.2.5, 3.5).

1104. Scheurmann, P.: Overflow Handling in Hashing Tables: A Hybrid Approach; Inform. Systems, 4:183-194, (1979). (3.3).

1105. Schkolnick, M.: Secondary Index Optimization; Proceedings ACM SIGMOD, San Francisco CA, 4:186-192, (May 1975). (3.4.3).

1106. Schkolnick, M.: A Clustering Algorithm for Hierarchical Structures; ACM TODS, 2(1):27-44, (Mar 1977). (3.4.3).

1107. Schkolnick, M.: The Optimal Selection of Secondary Indices for Files; Inform. Systems, 1:141-146, (1975). (3.4.3).

1108. Schlumberger, M. and Vuillemin, J.: Optimal Disk Merge Patterns; Acta Informatica, 3(1):25-35, (1973). (4.3, 4.4).

1109. Schmidt, J.P. and Siegel, A.: On Aspects of Universality and Performance for Closed Hashing; Proceedings STOC-SIGACT, Seattle, Washington, 21:355-366, (1989). (3.3.16, 3.3.1).

1110. Schmidt, J.P. and Siegel, A.: The Analysis of Closed Hashing under Limited Randomness; Proceedings STOC-SIGACT, Baltimore MD, 22:224-234, (May 1990). (3.3.2, 3.3.4, 3.3.5, 3.3.1).

1111. Schnorr, C.P. and van de Wiele, J.P.: On the Additive Complexity of Polynomials; Theoretical Computer Science, 10(1):1-18, (1980). (6.4).

1112. Schnorr, C.P.: How Many Polynomials Can be Approximated Faster than they can be Evaluated?; Inf. Proc. Letters, 12(2):76-78, (Apr 1981). (6.4).

1113. Scholl, M.: New File Organizations Based on Dynamic Hashing; ACM TODS, 6(1):194-211, (Mar 1981). (3.3.13, 3.3.14).

1114. Schonhage, A., Paterson, M.S. and Pippenger, N.: Finding the Median; JCSS, 13(2):184-199, (Oct 1976). (5.2).

1115. Schonhage, A.: Fast Multiplication of Polynomials Over Fields of Characteristic 2; Acta Informatica, 7:395-398, (1977). (6.4).

1116. Schonhage, A.: Partial and Total Matrix Multiplication; SIAM J on Computing, 10(3):434-455, (Aug 1981). (6.3).

1117. Schoor, A.: Fast Algorithm for Sparse Matrix Multiplication; Inf. Proc. Letters, 15(2):87-89, (Sep 1982). (6.3).

1118. Schulte Monting, J.: Merging of 4 or 5 Elements with n Elements; Theoretical Computer Science, 14(1):19-37, (1981). (4.3.3).

1119. Scowen, R.S.: Algorithm 271, Quickersort; C.ACM, 8(11):669-670, (Nov 1965). (4.1.3).

1120. Sedgewick, R.: A new upper bound for shellsort; J of Algorithms, 7(2):159-173, (June 1986). (4.1.4).

1121. Sedgewick, R.: Data Movement in Odd-Even Merging; SIAM J on Computing, 7(3):239-272, (Aug 1978). (4.2, 4.3).

1122. Sedgewick, R.: Implementing Quicksort Programs; C.ACM, 21(10):847-856, (Oct 1978). (4.1.3).

1123. Sedgewick, R.: Quicksort With Equal Keys; SIAM J on Computing, 6(2):240-267, (June 1977). (4.1.3).

1124. Sedgewick, R.: Quicksort; PhD Dissertation, Computer Science Department, Stanford University, (May 1975). (4.1.3).

1125. Sedgewick, R.: The Analysis of Quicksort Programs; Acta Informatica, 7:327-355, (1977). (4.1.3).

1126. Seeger, B. and Kriegel, H.P.: Techniques for design and implementation of efficient spatial data structures; Proceedings VLDB, Los Angeles CA, 14:360-371, (1988). (3.5).

1127. Seiferas, J. and Galil, Z.: Real-time recognition of substring repetition and reversal; Mathematical Systems Theory, 11:111-146, (1977). (7.1).

1128. Sellers, P.: An Algorithm for the Distance Between Two Finite Sequences; J of Combinatorial Theory (A), 16:253-258, (1974). (7.1.8).

1129. Sellers, P.: On the theory and computation of evolutionary distances; SIAM J Appl Math, 26:787-793, (1974). (7.1).

1130. Sellers, P.: The Theory and Computation of Evolutionary Distances: Pattern Recognition; J of Algorithms, 1:359-373, (1980). (7.1).

1131. Sellis, T., Roussopoulos, N. and Faloutsos, C.: The R^+-tree: A dynamic index for multidimensional objects; Proceedings VLDB, Brighton, England, 13:507-518, (1987). (3.5).

1132. Selmer, E.S.: On shellsort and the Frobenius problem; BIT, 29(1):37-40, (1989). (4.1.4).

1133. Senko, M.E., Lum, V.Y. and Owens, P.J.: A File Organization Model (FOREM); Proceedings Information Processing 68, Edinburgh, :514-519, (1969). (3.4.3).

1134. Senko, M.E.: Data Structures and Data Accessing in Data Base Systems: Past, Present and Future; IBM Systems J, 16(3):208-257, (1977). (3.4.3).

1135. Severance, D.G. and Carlis, J.V.: A Practical Approach to Selecting Record Access Paths; ACM C. Surveys, 9(4):259-272, (1977). (3.4.3).

1136. Severance, D.G. and Duhne, R.: A Practitioner's Guide to Addressing Algorithms; C.ACM, 19(6):314-326, (June 1976). (3.3).

1137. Shaw, M. and Traub, J.F.: On the Number of Multiplications for the Evaluation of a Polynomial and Some of its Derivatives; J.ACM, 21(1):161-167, (Jan 1974). (6.4).

1138. Shaw, M. and Traub, J.F.: Selection of Good Algorithms from a Family of Algorithms for Polynomial Derivative Evaluation; Inf. Proc. Letters, 6(5):141-145, (Oct 1977). (6.4).

1139. Sheil, B.A.: Median Split Trees: A Fast Lookup Technique for Frequently Occurring Keys; C.ACM, 21(11):947-958, (Nov 1978). (3.4.1.6).

1140. Shell, D.L.: A High-Speed Sorting Procedure; C.ACM, 2(7):30-32, (July 1959). (4.1.4).

1141. Shell, D.L.: Optimizing the Polyphase Sort; C.ACM, 14(11):713-719, (Nov 1971). (4.4.4).

1142. Sherk, M.: Self-adjusting k-ary search trees; Proceedings Workshop in Algorithms and Data Structures, Lecture Notes in Computer Science 382, Springer-Verlag, Ottawa, Canada, 1:75-96, (Aug 1989). (3.4.1.6, 3.4.1.10).

1143. Shirg, M.: Optimum ordered Bi-weighted binary trees; Inf. Proc. Letters, 17(2):67-70, (Aug 1983). (3.4.1.7).

1144. Shneiderman, B. and Goodman, V.: Batched Searching of Sequential and Tree Structured Files; ACM TODS, 1(3):268-275, (1976). (3.4.2, 3.1, 3.4.3).

1145. Shneiderman, B.: A Model for Optimizing Indexed File Structures; Int. J of Comp and Inf Sciences, 3(1):93-103, (Mar 1974). (3.4.3).

1146. Shneiderman, B.: Jump Searching: A Fast Sequential Search Technique; C.ACM, 21(10):831-834, (Oct 1978). (3.1.5).

1147. Shneiderman, B.: Polynomial Search; Software - Practice and Experience, 3(2):5-8, (1973). (3.1).

1148. Siegel, A.: On Universal Classes of Fast High Performance Hash Functions, Their Time-Space Tradeoff, and their Applications; Proceedings FOCS, Research Triangle Park, NC, 30:20-27, (1989). (3.3.1).

1149. Silva-Filho, Y.V.: Average Case Analysis of Region Search in Balanced k-d Trees; Inf. Proc. Letters, 8(5):219-223, (June 1979). (3.5.2).

1150. Silva-Filho, Y.V.: Optimal Choice of Discriminators in a Balanced k-d Binary Search Tree; Inf. Proc. Letters, 13(2):67-70, (Nov 1981). (3.5.2).

1151. Singleton, R.C.: An Efficient Algorithm for Sorting with Minimal Storage; C.ACM, 12(3):185-187, (Mar 1969). (4.1.3).

1152. Six, H. and Wegner, L.M.: Sorting a random access file in situ; Computer Journal, 27(3):270-275, (Aug 1984). (4.4).

1153. Six, H.: Improvement of the m-way Search Procedure; Angewandte Informatik, 15(1):79-83, (Feb 1973). (3.1.5).

1154. Skiena, S.S.: Encroaching lists as a measure of presortedness; BIT, 28(2):775-784, (1988). (4.1.8).

1155. Sleator, D.D., Tarjan, R.E. and Thurston, P.W.: Rotation distance, Triangulations, and Hyperbolic Geometry; Proceedings STOC-SIGACT, Berkeley CA, 18:122-135, (May 1986). (3.4.1.8).

1156. Sleator, D.D. and Tarjan, R.E.: A Data Structure for Dynamic Trees; JCSS, 26(3):362-391, (June 1983). (3.4.1).

1157. Sleator, D.D. and Tarjan, R.E.: Self-Adjusting Binary Search Trees; J.ACM, 32(3):652-686, (July 1985). (3.4.1.6, 5.1.6).

1158. Sleator, D.D. and Tarjan, R.E.: Self-Adjusting Heaps; SIAM J on Computing, 15(1):52-69, (Feb 1986). (5.1.3).

1159. Slisenko, A.: Determination in real time of all the periodicities in a word; Soviet Math Dokl, 21:392-395, (1980). (7.1).

1160. Slisenko, A.: Recognition of palindromes by multihead Turing machines; Dokl. Steklov Math. Inst., Akad Nauk SSSR, 129:30-202, (1973). (7.1).

1161. Slough, W. and Efe, K.: Efficient algorithms for tree reconstruction; BIT, 29(2):361-363, (1989). (3.4.1).

1162. Smid, M. and Overmars, M.H.: Maintaining Range Trees in Secondary memory. Part II: Lower bounds; Acta Informatica, 27:423-452, (1990). (3.6).

1163. Smith, J. and Weiss, S.: Formatting Texts Accessed Randomly; Software - Practice and Experience, 17(1):5-16, (Jan 1987). (7.2.7).

1164. Snir, M.: Exact balancing is not always good; Inf. Proc. Letters, 22(2):97-102, (Jan 1986). (2.2.2.1).

1165. Snyder, L.: On Uniquely Represented Data Structures; Proceedings FOCS, Providence RI, 18:142-146, (Oct 1977). (3.4.1).

1166. Snyder, L.: On B-Trees Re-Examined; C.ACM, 21(7):594, (July 1978). (3.4.2).

1167. Sobel, S.: Oscillating Sort - a New Sort Merging Technique; J.ACM, 9:372-374, (1962). (4.4.5).

1168. Solomon, M. and Finkel, R.A.: A Note on Enumerating Binary Trees; J.ACM, 27(1):3-5, (Jan 1980). (3.4.1).

1169. Sorenson, P.G., Tremblay, J.P. and Deutscher, R.F.: Key-to-Address Transformation Techniques; Infor, 16(1):1-34, (1978). (3.3.1).

1170. Soule, S.: A Note on the Nonrecursive Traversal of Binary Trees; Computer Journal, 20(4):350-352, (1977). (3.4.1).

1171. Sprugnoli, R.: On the Allocation of Binary Trees to Secondary Storage; BIT, 21(3):305-316, (1981). (3.4.1.1).

1172. Sprugnoli, R.: Perfect Hashing Functions: A Single Probe Retrieving Method for Static Sets; C.ACM, 20(11):841-850, (Nov 1977). (3.3.16).

1173. Sprugnoli, R.: The analysis of a simple in-place merging algorithm; J of Algorithms, 10(3):366-380, (Sep 1989). (4.3.2).

1174. Stanfel, L.: Tree Structures for Optimal Searching; J.ACM, 17(3):508-517, (1970). (3.4.1).

1175. Stanfel, L.: Optimal Tree Lists for Information Storage and Retrieval; Inform. Systems, 2:65-70, (1976). (3.4.4, 7.2.1).

1176. Stasko, J.T. and Vitter, J.S.: Paring Heaps: Experiments and Analysis; C.ACM, 30(3):234-249, (Mar 1987). (5.1.3).

1177. Stephenson, C.J.: A Method for Constructing Binary Search Trees by Making Insertions at the Root; Int. J of Comp and Inf Sciences, 9(1):15-29, (Feb 1980). (3.4.1).

1178. Stockmeyer, L.J.: The Complexity of Approximate Counting; Proceedings STOC-SIGACT, Boston Mass, 15:118-126, (Apr 1983). (6.1).

1179. Stockmeyer, P.K. and Yao, F.F.: On the Optimality of Linear Merge; SIAM J on Computing, 9(1):85-90, (Feb 1980). (4.3.3).

1180. Stout, Q.F. and Warren, B.L.: Tree Rebalancing in Optimal Time and Space; C.ACM, 29(9):902-908, (Sep 1986). (3.4.1, 3.4.1.8).

1181. Strassen, V.: The Asymptotic Spectrum of Tensors and the Exponent of Matrix Multiplication; Proceedings FOCS, Toronto, Canada, 27:49-54, (Oct 1986). (6.3).

1182. Strassen, V.: Gaussian Elimination is not Optimal; Numer Math, 13:354-356, (1969). (6.3).

1183. Strassen, V.: Polynomials with Rational Coefficients Which are Hard to Compute; SIAM J on Computing, 3(2):128-149, (June 1974). (6.4).

1184. Strong, H.R., Markowsky, G. and Chandra, A.K.: Search Within a Page; J.ACM, 26(3):457-482, (July 1979). (3.4.1, 3.4.2, 3.4.3).

1185. Strothotte, T., Eriksson, P. and Vallner, S.: A note on constructing min-max heaps; BIT, 29(2):251-256, (1989). (5.1.3).

1186. Sundar, R.: Worst-Case data structures for the priority queue with Attrition; Inf. Proc. Letters, 31(2):69-75, (Apr 1989). (5.1).

1187. Suraweera, F. and Al-anzy, J.M.: Analysis of a modified Address calculations sorting algorithm; Computer Journal, 31(6):561-563, (Dec 1988). (4.2.3).

1188. Sussenguth, E.H.: Use of Tree Structures for Processing Files; C.ACM, 6(5):272-279, (1963). (3.4.4).

1189. Szpankowski, W.: Average Complexity of Additive Properties for Multiway Tries: A Unified Approach; Proceedings CAAP, Lecture Notes in Computer Science 249, Pisa, Italy, 14:13-25, (1987). (3.4.4).

1190. Szpankowski, W.: Digital data structures and order statistics; Proceedings Workshop in Algorithms and Data Structures, Lecture Notes in Computer Science 382, Springer-Verlag, Ottawa, Canada, 1:206-217, (Aug 1989). (3.4.4).

1191. Szpankowski, W.: How much on the average is the Patricia trie better?; Proceedings Allerton Conference, Monticello, IL, 24:314-323, (1986). (3.4.4.5).

1192. Szpankowski, W.: On an Alternative Sum Useful in the Analysis of Some Data Structures; Proceedings SWAT 88, Halmstad, Sweden, 1:120-128, (1988). (3.4.4).

1193. Szpankowski, W.: Some results on V-ary asymmetric tries; J of Algorithms, 9(2):224-244, (June 1988). (3.4.4).

1194. Szwarcfiter, J.L. and Wilson, L.B.: Some Properties of Ternary Trees; Computer Journal, 21(1):66-72, (Feb 1978). (3.4.1.10, 4.2.6).

1195. Szwarcfiter, J.L.: Optimal multiway search trees for variable size keys; Acta Informatica, 21(1):47-60, (1984). (3.4.1.10).

1196. Szymanski, T.G.: Hash table reorganization; J of Algorithms, 6(3):322-355, (Sep 1985). (3.3).

1197. Tai, K.C. and Tharp, A.L.: Computed Chaining A Hybrid of Direct and Open Addressing; Proceedings AFIPS, Anaheim CA, 49:275-282, (1980). (3.3, 3.3.10).

1198. Tainiter, M.: Addressing for Random-Access Storage with Multiple Bucket Capacities; J.ACM, 10:307-315, (1963). (3.3.4).

1199. Takaoka, T.: An On-line Pattern Matching Algorithm; Inf. Proc. Letters, 22:329-330, (1986). (7.1.2).

1200. Tamminen, M.: Analysis of N-Trees; Inf. Proc. Letters, 16(3):131-137, (Apr 1983). (3.4.2).

1201. Tamminen, M.: Comment on Quad- and Octtrees; C.ACM, 27(3):248-249, (Mar 1984). (3.5.1.1).

1202. Tamminen, M.: Extendible Hashing with Overflow; Inf. Proc. Letters, 15(5):227-232, (Dec 1982). (3.3.13).

1203. Tamminen, M.: Order Preserving Extendible Hashing and Bucket Tries; BIT, 21(4):419-435, (1981). (3.3.13, 3.4.4).

1204. Tamminen, M.: On search by address computation; BIT, 25(1):135-147, (1985). (3.3.13, 3.3.14).

1205. Tamminen, M.: Two levels are as good as any; J of Algorithms, 6(1):138-144, (Mar 1985). (4.2.5).

1206. Tan, K.C. and Hsu, L.S.: Block Sorting of a Large File in External Storage by a 2-Component Key; Computer Journal, 25(3):327-330, (Aug 1982). (4.4).

1207. Tan, K.C.: On Foster's Information Storage and Retrieval Using AVL Trees; C.ACM, 15(9):843, (Sep 1972). (3.4.1.3).

1208. Tang, P.T.P.: Table-Driven Implementation of the Exponential Function in IEEE Floating Point Arithmetic; ACM TOMS, 15(2):144-157, (1989). (6.2).

1209. Tanner, R.M.: Minimean Merging and Sorting: An Algorithm; SIAM J on Computing, 7(1):18-38, (Feb 1978). (4.3, 4.2).

1210. Tarhio, J. and Ukkonen, E.: Boyer-Moore approach to approximate string matching; Proceedings Scandinavian Workshop in Algorithmic Theory, SWAT'90, Lecture Notes in Computer Science 447, Springer-Verlag, Bergen, Norway, 2:348-359, (July 1990). (7.1.8).

1211. Tarjan, R.E. and Yao, A.C-C.: Storing a Sparse Table; C.ACM, 22(11):606-611, (Nov 1979). (3.3.16, 3.4.4).

1212. Tarjan, R.E.: Algorithm Design; C.ACM, 30(3):204-213, (Mar 1987). (2.2).

1213. Tarjan, R.E.: Sorting Using Networks of Queues and Stacks; J.ACM, 18(2):341-346, (Apr 1972). (4.2).

1214. Tarjan, R.E.: Updating a Balanced Search Tree in O(1) Rotations; Inf. Proc. Letters, 16(5):253-257, (June 1983). (3.4.2.2, 3.4.1.8).

1215. Tarter, M.E. and Kronmal, R.A.: Non-Uniform Key Distribution and Address Calculation Sorting; Proceedings ACM-NCC, Washington DC, 21:331-337, (Aug 1966). (4.1.6, 4.2.3).

1216. Tenenbaum, A.M. and Nemes, R.M.: Two Spectra of Self-Organizing Sequential Algorithms; SIAM J on Computing, 11(3):557-566, (Aug 1982). (3.1.2).

1217. Tenenbaum, A.M.: Simulations of Dynamic Sequential Search Algorithms; C.ACM, 21(9):790-791, (Sep 1978). (3.1.3).

1218. Thanh, M., Alagar, V.S. and Bui, T.D.: Optimal Expected-Time algorithms for merging; J of Algorithms, 7(3):341-357, (Sep 1986). (4.3.2).

1219. Thanh, M. and Bui, T.D.: An Improvement of the Binary Merge Algorithm; BIT, 22(4):454-462, (1982). (4.3.3).

1220. Tharp, A.L. and Tai, K.C.: The Practicality of Text Signatures for Accelerating String Searching Software; Software - Practice and Experience, 12:35-44, (1982). (7.2.6).

1221. Tharp, A.L.: Further Refinement of the Linear Quotient Hashing Method; Inform. Systems, 4:55-56, (1979). (3.3.8.1).

1222. Thompson, K.: Regular Expression Search Algorithm; C.ACM, 11:419-422, (1968). (7.1.6).

1223. Ting, T.C. and Wang, Y.W.: Multiway Replacement Selection Sort with Dynamic Reservoir; Computer Journal, 20(4):298-301, (Nov 1977). (4.4.1).

1224. Todd, S.: Algorithm and Hardware for a Merge Sort Using Multiple Processors; IBM J Res. Development, 22(5):509-517, (Sep 1978). (4.2.1).

1225. Torn, A.A.: Hashing with overflow index; BIT, 24(3):317-332, (1984). (3.3).

1226. Trabb Pardo, L.: Stable Sorting and Merging with Optimal Space and Time Bounds; SIAM J on Computing, 6(2):351-372, (June 1977). (4.3.2, 4.1).

1227. Tropf, H. and Herzog, H.: Multidimensional Range Search in Dynamically Balanced Trees; Angewandte Informatik, 2:71-77, (1981). (3.6.2).

1228. Tsakalidis, A.K.: AVL-trees for localized search; Information and Control, 67(1-3):173-194, (Oct 1985). (3.4.1.3).

1229. Tsi, K.T. and Tharp, A.L.: Computed chaining: A hybrid of Direct Chaining and Open Addressing; Inform. Systems, 6:111-116, (1981). (3.3).

1230. Tzoreff, T. and Vishkin, U.: Matching Patterns in Strings Subject to Multilinear Transformations; Theoretical Computer Science, 60:231-254, (1988). (7.3).

1231. Ukkonen, E. and Wood, D.: A simple on-line algorithm to approximate string matching; (Report A-1990-4)Helsinki, Finland, (1990). (7.1.8).

1232. Ukkonen, E.: Algorithms for Approximate String Matching; Information and Control, 64:100-118, (1985). (7.1.8).

1233. Ukkonen, E.: Finding Approximate Patterns in Strings; J of Algorithms, 6:132-137, (1985). (7.1.8).

1234. Ukkonen, E.: On Approximate String Matching; Proceedings Int. Conf. on Foundations of Computation Theory, Lecture Notes in Computer Science 158, Springer-Verlag, Borgholm, Sweden, :487-495, (1983). (7.1.8).

1235. Ullman, J.D.: A Note on the Efficiency of Hashing Functions; J.ACM, 19(3):569-575, (July 1972). (3.3.1).

1236. Unterauer, K.: Dynamic Weighted Binary Search Trees; Acta Informatica, 11(4):341-362, (1979). (3.4.1.4).

1237. Vaishnavi, V.K., Kriegel, H.P. and Wood, D.: Height Balanced 2-3 Trees; Computing, 21:195-211, (1979). (3.4.2.1).

1238. Vaishnavi, V.K., Kriegel, H.P. and Wood, D.: Optimum Multiway Search Trees; Acta Informatica, 14(2):119-133, (1980). (3.4.1.10).

1239. van de Wiele, J.P.: An Optimal Lower Bound on the Number of Total Operations to Compute 0-1 Polynomials Over the Field of Complex Numbers; Proceedings FOCS, Ann Arbor MI, 19:159-165, (Oct 1978). (6.4).

1240. van der Nat, M.: A Fast Sorting Algorithm, a Hybrid of Distributive and Merge Sorting; Inf. Proc. Letters, 10(3):163-167, (Apr 1980). (4.2.5).

1241. van der Nat, M.: Binary Merging by Partitioning; Inf. Proc. Letters, 8(2):72-75, (Feb 1979). (4.3).

1242. van der Nat, M.: Can Integers be Sorted in Linear Worst Case Time?; Angewandte Informatik, 25(11):499-501, (Nov 1983). (4.2.4).

1243. van der Nat, M.: On Interpolation Search; C.ACM, 22(12):681, (Dec 1979). (3.2.2).

1244. van der Pool, J.A.: Optimum Storage Allocation for a File in Steady State; IBM Systems J, 17(1):27-38, (1973). (3.3.11).

1245. van der Pool, J.A.: Optimum Storage Allocation for a File with Open Addressing; IBM Systems J, 17(2):106-114, (1973). (3.3.4).

1246. van der Pool, J.A.: Optimum Storage Allocation for Initial Loading of a File; IBM Systems J, 16(6):579-586, (1972). (3.3.11).

1247. van Emde-Boas, P., Kaas, R. and Zijlstra, E.: Design and Implementation of an Efficient Priority Queue; Mathematical Systems Theory, 10:99-127, (1977). (5.1.4).

1248. van Emde-Boas, P.: Preserving Order in a Forest in Less than Logarithmic Time and Linear Space; Inf. Proc. Letters, 6(3):80-82, (June 1977). (5.1.4).

1249. van Emden, M.H.: Algorithm 402, qsort; C.ACM, 13(11):693-694, (Nov 1970). (4.1.3).

1250. van Emden, M.H.: Increasing the Efficiency of Quicksort; C.ACM, 13(9):563-567, (Sep 1970). (4.1.3).

1251. van Leeuwen, J. and Overmars, M.H.: Stratified Balanced Search Trees; Acta Informatica, 18(4):345-359, (1983). (3.4.1, 3.4.2).

1252. van Leeuwen, J. and Wood, D.: Dynamization of Decomposable Searching Problems; Inf. Proc. Letters, 10(2):51-56, (Mar 1980). (2.2.2).

1253. van Wyk, C.J. and Vitter, J.S.: The Complexity of Hashing with Lazy Deletion; Algorithmica, 1(1):17-29, (1986). (3.3).

1254. Veklerov, E.: Analysis of Dynamic Hashing with Deferred Splitting; ACM TODS, 10(1):90-96, (Mar 1985). (3.3.13, 3.3.14).

1255. Verkamo, A.I.: Performance of Quicksort Adapted for virtual Memory use; Computer Journal, 30(4):362-371, (Aug 1987). (4.1.3).

1256. Veroy, B.S.: Average Complexity of Divide-and-Conquer algorithms; Inf. Proc. Letters, 29(6):319-326, (Dec 1988). (3.4.2).

1257. Veroy, B.S.: Expected Combinatorial Complexity of Divide-and-Conquer Algorithms; Proceedings SCCC Int. Conf. in Computer Science, Santiago, Chile, 8:305-314, (July 1988). (2.2.2.1).

1258. Vishkin, U.: Deterministic Sampling: A New Technique for Fast Pattern Matching; Proceedings STOC-SIGACT, Baltimore MD, 22:170-180, (May 1990). (7.1).

1259. Vitter, J.S. and Chen, W-C.: Optimal algorithms for a model of direct chaining; SIAM J on Computing, 14(2):490-499, (May 1985). (3.3.10).

1260. Vitter, J.S.: A Shared-Memory Scheme for Coalesced Hashing; Inf. Proc. Letters, 13(2):77-79, (Nov 1981). (3.3.12).

1261. Vitter, J.S.: Analysis of Coalesced Hashing; PhD Dissertation, Stanford University, (Aug 1980). (3.3.12).

1262. Vitter, J.S.: Analysis of the Search Performance of Coalesced Hashing; J.ACM, 30(2):231-258, (Apr 1983). (3.3.12).

1263. Vitter, J.S.: Deletion Algorithms for Hashing that Preserve Randomness; J of Algorithms, 3(3):261-275, (Sep 1982). (3.3.12).

1264. Vitter, J.S.: Implementations for Coalesced Hashing; C.ACM, 25(12):911-926, (Dec 1982). (3.3.12).

1265. Vitter, J.S.: Tuning the Coalesced Hashing Method to Obtain Optimum Performance; Proceedings FOCS, Syracuse NY, 21:238-247, (Oct 1980). (3.3.12).

1266. Vuillemin, J.: A Data Structure for Manipulating Priority Queues; C.ACM, 21(4):309-314, (Apr 1978). (5.1.7).

1267. Vuillemin, J.: A Unifying Look at Data Structures; C.ACM, 23(4):229-239, (Apr 1980). (2.1).

1268. Wagner, R.E. and Fischer, M.J.: The string-to-string correction problem; J.ACM, 21:168-178, (1974). (7.1.8).

1269. Wagner, R.E.: Indexing Design Considerations; IBM Systems J, 17(4):351-367, (1973). (3.4.2, 3.4.3).

1270. Wagner, R.E.: On the complexity of the extended string-to-string correction problem; Proceedings STOC-SIGACT, New York, 7:218-223, (1975). (7.1.8).

1271. Wainwright, R.L.: A Class of Sorting Algorithms Based on Quicksort; C.ACM, 28(4):396-403, (Apr 1985). (4.1.3).

1272. Walah, T.R.: How Evenly Should one divide to conquer quickly?; Inf. Proc. Letters, 19(4):203-208, (Nov 1984). (2.2.2.1).

1273. Walker, W.A. and Wood, D.: Locally Balanced Binary Trees; Computer Journal, 19(4):322-325, (Nov 1976). (3.4.1.6).

1274. Warren, H.S.: Minimal Comparison Sorting by Choosing Most Efficient Comparisons; Inf. Proc. Letters, 2(5):129-130, (Dec 1973). (4.1.8).

1275. Waterman, M.S.: General Methods of Sequence Comparison; Bulletin of Mathematical Biology, 46:473-500, (1984). (7.2.1).

1276. Waters, S.J.: Analysis of Self-Indexing, Disc Files; Computer Journal, 18(3):200-205, (Aug 1975). (3.2.2).

1277. Webb, D.A.: The Development and Application of an Evaluation Model for Hash Coding Systems; PhD Dissertation, Syracuse University, (1972). (3.3).

1278. Weems, B.P.: A Study of page Arrangements for Extendible Hashing; Inf. Proc. Letters, 27(5):245-248, (Apr 1988). (3.3.13).

1279. Wegman, M.N. and Carter, J.L.: New Classes and Applications of Hash Functions; Proceedings FOCS, San Juan PR, 20:175-182, (Oct 1979). (3.3.1).

1280. Wegner, L.M.: A generalized, one-way-stackless quicksort; BIT, 27(1):44-48, (1987). (4.1.3).

1281. Wegner, L.M.: Sorting a Linked List with Equal Keys; Inf. Proc. Letters, 15(5):205-208, (Dec 1982). (4.2.2).

1282. Weiner, P.: Linear Pattern Matching Algorithm; Proceedings FOCS, Iowa City IA, 14:1-11, (Oct 1973). (7.2.2).

1283. Weiss, M.A. and Navlakha, J.K.: Distribution of keys in a binary heap; Proceedings Workshop in Algorithms and Data Structures, Lecture Notes in Computer Science 382, Springer-Verlag, Ottawa, Canada, 1:510-516, (Aug 1989). (5.1.3).

1284. Weiss, M.A. and Sedgewick, R.: Bad Cases for Shaker-sort; Inf. Proc. Letters, 28(3):133-136, (July 1988). (4.1.1).

1285. Weiss, M.A. and Sedgewick, R.: More on Shellsort Increment Sequences; Inf. Proc. Letters, 34:267-270, (1990). (4.1.4).

1286. Weiss, M.A. and Sedgewick, R.: Tight Lower Bounds for Shellsort; Proceedings SWAT 88, Halmstad, Sweden, 1:255-262, (1988). (4.1.4).

1287. Wessner, R.L.: Optimal Alphabetic Search Trees with Restricted Maximal Height; Inf. Proc. Letters, 4(4):90-94, (Jan 1976). (3.4.1.7).

1288. Whitt, J.D. and Sullenberger, A.G.: The Algorithm Sequential Access Method: an Alternative to Index Sequential; C.ACM, 18(3):174-176, (Mar 1975). (3.2.2, 3.4.3).

1289. Wikstrom, A.: Optimal Search Trees and Length Restricted Codes; BIT, 19(4):518-524, (1979). (3.4.1.7).

1290. Wilber, R.: Lower Bounds for Accessing Binary Search Trees with Rotations; Proceedings FOCS, Toronto, Canada, 27:61-70, (Oct 1986). (3.4.1.8).

1291. Willard, D.E. and Lueker, G.S.: Adding Range Restriction Capability to Dynamic Data Structures; J.ACM, 32(3):597-617, (July 1985). (3.6).

1292. Willard, D.E.: Good Worst-Case Algorithms for Inserting and Deleting Records in Dense Sequential Files; Proceedings ACM SIGMOD, Washington DC, 15:251-260, (May 1986). (3.4.3).

1293. Willard, D.E.: Log-logarithmic worst-case range queries are possible in space $\Theta(N)$; Inf. Proc. Letters, 17(2):81-84, (Aug 1983). (3.6.2).

1294. Willard, D.E.: Maintaining Dense Sequential Files in a Dynamic Environment; Proceedings STOC-SIGACT, San Francisco CA, 14:114-121, (May 1982). (3.1.1, 3.4.3).

1295. Willard, D.E.: Multidimensional Search Trees that Provide New Types of Memory Reductions; J.ACM, 34(4):846-858, (Oct 1987). (3.5).

1296. Willard, D.E.: New Data Structures for Orthogonal Range Queries; SIAM J on Computing, 14(1):232-253, (Feb 1985). (3.5.3).

1297. Willard, D.E.: New Trie Data Structures Which Support Very fast Search operations; JCSS, 28(3):379-394, (June 1984). (3.5.3).

1298. Willard, D.E.: Polygon Retrieval; SIAM J on Computing, 11(1):149-165, (Feb 1982). (3.5).

1299. Williams, F.A.: Handling Identifiers as Internal Symbols in Language Processors; C.ACM, 2(6):21-24, (June 1959). (3.3.12).

1300. Williams, J.G.: Storage Utilization in a Memory Hierarchy when Storage Assignment is Performed by a Hashing Algorithm; C.ACM, 14(3):172-175, (Mar 1971). (3.3).

1301. Williams, J.W.J.: Algorithm 232; C.ACM, 7(6):347-348, (June 1964). (4.1.5, 5.1.3).

1302. Williams, R.: The Goblin Quadtree; Computer Journal, 31(4):358-363, (Aug 1988). (3.5.1.1).

1303. Wilson, L.B.: Sequence Search Trees: Their Analysis Using Recurrence Relations; BIT, 16(3):332-337, (1976). (3.4.1.1, 3.4.1).

1304. Winograd, S.: A New Algorithm for Inner Product; IEEE Trans. on Computers, C17(7):693-694, (July 1968). (6.3).

1305. Winograd, S.: The Effect of the Field of Constants on the Number of Multiplications; Proceedings FOCS, Berkeley CA, 16:1-2, (Oct 1975). (6.2).

1306. Winters, V.G.: Minimal perfect hashing in polynomial time; BIT, 30(2):235-244, (1990). (3.3.16).

1307. Wise, D.S.: Referencing Lists by an Edge; C.ACM, 19(6):338-342, (June 1976). (3.1.1).

1308. Wogulis, J.: Self-Adjusting and split sequence Hash Tables; Inf. Proc. Letters, 30(4):185-188, (Feb 1989). (3.3.6, 3.3.8.5).

1309. Wong, C.K. and Chandra, A.K.: Bounds for the string editing problem; J.ACM, 23(1):13-16, (Jan 1976). (7.1.8).

1310. Wong, C.K. and Yue, P.C.: Free Space Utilization of a Disc File Organization Method; Proceedings Princeton Conf. on Information Sciences, Princeton, 7:5-9, (1973). (3.4.2).

1311. Wong, J.K.: Some Simple In-Place Merging Algorithms; BIT, 21(2):157-166, (1981). (4.3.2).

1312. Wong, K.F. and Strauss, J.C.: An Analysis of ISAM Performance Improvement Options; Manag. Datamatics, 4(3):95-107, (1975). (3.4.3).

1313. Wood, D.: Extremal Cost Tree Data Structures; Proceedings SWAT 88, Halmstad, Sweden, 1:51-63, (1988). (3.4.1.3, 3.4.2.1, 3.4.2.3).

1314. Woodall, A.D.: A Recursive Tree Sort; Computer Journal, 14(1):103-104, (1971). (4.2.6).

1315. Wright, W.E.: Average Performance of the B-Tree; Proceedings Allerton Conference, Monticello, IL, 18:233-241, (1980). (3.4.2).

1316. Wright, W.E.: Binary Search Trees in Secondary Memory; Acta Informatica, 15(1):3-17, (1981). (3.4.1.1, 3.4.1.3).

1317. Wright, W.E.: Some Average Performance Measures for the B-tree; Acta Informatica, 21(6):541-558, (1985). (3.4.2).

1318. Xunuang, G. and Yuzhang, Z.: A New Heapsort Algorithm and the Analysis of its Complexity; Computer Journal, 33(3):281, (June 1990). (4.1.5).

1319. Yang, W.P. and Du, M.W.: A backtracking method for constructing perfect hash functions from a set of mapping functions; BIT, 25(1):148-164, (1985). (3.3.16).

1320. Yang, W.P. and Du, M.W.: A Dynamic Perfect Hash Function defined by an Extended Hash Indicator Table; Proceedings VLDB, Singapore, 10:245-254, (1984). (3.3.16).

1321. Yao, A.C-C. and Yao, F.F.: Lower Bounds on Merging Networks; J.ACM, 23(3):566-571, (July 1976). (4.3).

1322. Yao, A.C-C. and Yao, F.F.: On the Average-Case Complexity of Selecting k-th Best; SIAM J on Computing, 11(3):428-447, (Aug 1982). (5.2).

1323. Yao, A.C-C. and Yao, F.F.: The Complexity of Searching an Ordered Random Table; Proceedings FOCS, Houston TX, 17:173-177, (Oct 1976). (3.2.2).

1324. Yao, A.C-C.: A Note on the Analysis of Extendible Hashing; Inf. Proc. Letters, 11(2):84-86, (Oct 1980). (3.3.13).

1325. Yao, A.C-C.: An Analysis of (h,k,1)-Shellsort; J of Algorithms, 1(1):14-50, (1980). (4.1.4).

1326. Yao, A.C-C.: On optimal arrangements of keys with double hashing; J of Algorithms, 6(2):253-264, (June 1985). (3.3.5, 3.3.9).

1327. Yao, A.C-C.: On Random 2-3 Trees; Acta Informatica, 9(2):159-170, (1978). (3.4.2.1).

1328. Yao, A.C-C.: On Selecting the K largest with Median tests; Algorithmica, 4(2):293-300, (1989). (5.2).

1329. Yao, A.C-C.: On the Evaluation of Powers; SIAM J on Computing, 5(1):100-103, (Mar 1976). (6.2).

1330. Yao, A.C-C.: Probabilistic Computations - Toward a Unified Measure of Complexity; Proceedings FOCS, Providence RI, 18:222-226, (Oct 1977). (2.2.2.1).

1331. Yao, A.C-C.: Should Tables Be Sorted?; J.ACM, 28(3):615-628, (July 1981). (3.2.1, 3.3.16).

1332. Yao, A.C-C.: Space-Time Tradeoff for Answering Range Queries; Proceedings STOC-SIGACT, San Francisco CA, 14:128-136, (May 1982). (3.6.2).

1333. Yao, A.C-C.: The Complexity of Pattern Matching for A Random String; SIAM J on Computing, 8:368-387, (1979). (7.1).

1334. Yao, A.C-C.: Uniform Hashing is Optimal; J.ACM, 32(3):687-693, (July 1985). (3.3.2).

1335. Yap, C.K.: New Upper Bounds for Selection; C.ACM, 19(9):501-508, (Sep 1976). (5.2).

1336. Yongjin, Z. and Jianfang, W.: On Alphabetic-Extended Binary Trees with Restricted Path Length; Scientia Sinica, 22(12):1362-1371, (Dec 1979). (3.4.1).

1337. Yuba, T. and Hoshi, M.: Binary Search networks: A new method for key searching; Inf. Proc. Letters, 24(1):59-66, (Apr 1987). (3.2).

1338. Yue, P.C. and Wong, C.K.: Storage Cost Considerations in Secondary Index Selection; Int. J of Comp and Inf Sciences, 4(4):307-327, (1975). (3.4.3).

1339. Yuen, T-S. and Du, D.H-C.: Dynamic File Structure for Partial Match Retrieval Based on Overflow Bucket Sharing; IEEE Trans. Software Engineering, SE-12(8):801-810, (Aug 1986). (3.5.4).

1340. Yuval, G.: A Simple Proof of Strassen's Result; Inf. Proc. Letters, 7(6):285-286, (Oct 1978). (6.3).

1341. Zaki, A.S.: A Comparative Study of 2-3 Trees and AVL Trees; Int. J of Comp and Inf Sciences, 12(1):13-33, (1983). (3.4.1.3, 3.4.2.1).

1342. Zaki, A.S.: A space saving insertion algorithm for 2-3 trees; Computer Journal, 27(4):368-372, (Nov 1984). (3.4.2.1).

1343. Zave, D.A.: Optimal Polyphase Sorting; SIAM J on Computing, 6(1):1-39, (Mar 1977). (4.4.4).

1344. Zerling, D.: Generating Binary Trees Using Rotations; J.ACM, 32(3):694-701, (July 1985). (3.4.1, 3.4.1.8).

1345. Zhu, R.F. and Takaoka, T.: A Technique for Two-Dimensional Pattern Matching; C.ACM, 32(9):1110-1120, (Sep 1989). (7.3.2).

1346. Ziviani, N., Olivie, H.J. and Gonnet, G.H.: The analysis of an improved symmetric Binary B-Tree algorithm; Computer Journal, 28(4):417-425, (Aug 1985). (3.4.2.2).

1347. Ziviani, N. and Tompa, F.W.: A Look at Symmetric Binary B-trees; Infor, 20(2):65-81, (May 1982). (3.4.1.3, 3.4.2.2).

1348. Ziviani, N.: The Fringe Analysis of Search Trees; PhD Dissertation, Department of Computer Science, University of Waterloo, (1982). (3.4.1.1, 3.4.1.3, 3.4.2, 3.4.2.1, 3.4.2.2).

1349. Zvegintzov, N.: Partial-Match Retrieval in an Index Sequential Directory; Computer Journal, 23(1):37-40, (Feb 1980). (3.4.3, 3.6.2).

1350. Zweben, S.H. and McDonald, M.A.: An Optimal Method for Deletions in One-Sided Height-Balanced Trees; C.ACM, 21(6):441-445, (June 1978). (3.4.1.3).

APPENDIX IV

Algorithms Coded in Pascal and C

The following entries are selected algorithms which are coded in a language different from that used in the main entries.

IV.1 Searching algorithms

3.1.1: Insertion for arrays (C)

```
void insert(key, r)
typekey key;  dataarray r;

{ extern int n;
  if (n>=m) Error /*** Table is full ***/;
  else r[n++].k = key;
}
```

3.1.2: Insertion for lists (C)

```
datarecord *insert(new, list)
typekey new; datarecord *list;

{ extern int n;
n++;
return(NewNode(new, list));
}
```

3.1.2: Self-organizing (Transpose) sequential search (C)

```
int search(key, r)
typekey key; dataarray r;

{ extern int n;
  int i;
  datarecord tempr;

for (i=0; i<n-1 && r[i].k != key; i++);
if (key == r[i].k) {
    if (i>0) {
        /*** Transpose with predecessor ***/
        tempr = r[i];
        r[i] = r[i-1];
        r[--i] = tempr;
        };
    return(i);    /*** found(r[i]) ***/
    }
else return(-1);   /*** notfound(key) ***/
}
```

3.2.1: Binary search for arrays (C)

```
int search(key, r)
typekey key; dataarray r;

{ int high, i, low;
```

```
for (low=(-1), high=n;  high-low > 1; )
     {
     i = (high+low) / 2;
     if (key <= r[i].k)  high = i;
          else           low  = i;
     }
if (key==r[high].k)  return(high);
     else            return(-1);
}
```

3.2.1: Insertion in a sorted array (C)

```
void insert(new, r)
typekey new;  dataarray r;

{ extern int n;
  int i;
  if (n>=m) Error     /*** table is full ***/;
  else { for (i=n++;  i>=0 && r[i].k>new;  i--)   r[i+1] = r[i];
         r[i+1].k = new;
       }
}
```

3.3.4: Linear probing hashing: search (C)

```
int search(key, r)
typekey key;  dataarray r;

{ int i, last;

i = hashfunction(key) ;
last = (i+n-1) % m;
while (i!=last && !empty(r[i]) && r[i].k!=key)
     i = (i+1) % m;
if (r[i].k==key)  return(i);
     else        return(-1);
}
```

3.3.4: Linear probing hashing: insertion (C)

```
void insert(key, r)
typekey key;  dataarray r;

{ extern int n;
  int i, last;

i = hashfunction(key) ;
last = (i+m-1) % m;
while (i!=last && !empty(r[i]) && !deleted(r[i]) && r[i].k!=key)
     i = (i+1) % m;
if (empty(r[i]) || deleted(r[i]))
          {
          /*** insert here ***/
          r[i].k = key;
          n++;
          }
else Error      /*** table full, or key already in table ***/;
}
```

3.3.5: Double hashing: search (C)

```
int search(key, r)
typekey key;  dataarray r;

{ int i, inc, last;

i = hashfunction(key) ;
inc = increment(key);
last = (i+(n-1)*inc) % m;
while (i!=last && !empty(r[i]) && r[i].k!=key)
     i = (i+inc) % m;
if (r[i].k==key) return(i);
     else          return(-1);
}
```

3.3.5: Double hashing: insertion (C)

```
void insert(key, r)
typekey key; dataarray r;

{ extern int n;
  int i, inc, last;

i = hashfunction(key) ;
inc = increment(key);
last = (i+(m-1)*inc) % m;
while (i!=last && !empty(r[i]) && !deleted(r[i]) && r[i].k!=key)
     i = (i+inc) % m;
if (empty(r[i]) || deleted(r[i]))
          {
          /*** insert here ***/
          r[i].k = key;
          n++;
          }
else Error      /*** table full, or key already in table ***/;
}
```

3.3.8.1: Brent's reorganization scheme: insertion (C)

```
void insert(key, r)
typekey key; dataarray r;

{ extern int n;
  int i, inc, ii, init, j, jj;

init = hashfunction(key);
inc  = increment(key);
for (i=0; i<=n; i++)
     for (j=i; j>=0; j--)
          {
          jj = (init + j*inc) % m;
          ii = (jj + (i-j)*increment(r[jj].k)) % m;
          if (empty(r[ii]) || deleted(r[ii]))
               {
               /*** move record forward ***/
               r[ii] = r[jj];
```

```
                /*** insert new in r[jj] ***/
                r[jj].k = key;
                n++;
                return;
                }
            };
    Error    /*** table is full ***/;
    }
```

3.4.1: Data structure definition for binary trees (C)

```
typedef struct btnode {          /*** binary tree definition ***/
    typekey k;                   /*** key ***/
    struct btnode *left,*right; /*** pointers to subtrees ***/
    } node, *tree;
```

3.4.1.1: Binary tree search (C)

```
search(key, t)
typekey key;
tree t;
{
while(t != NULL)
    if (t->k == key)
        { found(t); return; }
    else if (t->k < key) t = t->right;
            else         t = t->left;
notfound(key);
}
```

3.4.1.1: Binary tree insertion (C)

```
tree insert(key, t)
typekey key;
tree t;
{
```

```
if(t==NULL) t = NewNode(key, NULL, NULL);
else if(t->k == key)
    Error; /*** key already in table ***/
else if(t->k < key) t->right = insert(key, t->right);
    else            t->left  = insert(key, t->left);
return(t);
}
```

Note that the insertion algorithm returns the new tree, as 'C' does not have **var** variables.

3.4.1.3: Height balanced tree left rotation (C)

```
tree lrot(t)
tree t;

{ tree temp;
  int a;

  temp = t;
  t = t->right;
  temp->right = t->left;
  t->left = temp;
  /*** adjust balance ***/
  a = temp->bal;
  temp->bal = a - 1 - max(t->bal, 0);
  t->bal = min(a-2, min(a+t->bal-2, t->bal-1));
  return(t);
}
```

3.4.1.4: Weight balanced tree insertion (C)

```
tree insert(key, t)
typekey key;
tree t;

{ if(t == NULL) {
        t = NewNode(key, NULL, NULL);
        t->weight = 2;
  }
```

```
    else if(t->k == key)
         Error; /*** Key already in table ***/
    else { if(t->k < key) t->right = insert(key, t->right);
                    else t->left  = insert(key, t->left);
         t->weight = wt(t->left) + wt(t->right);
         t = checkrots(t);
         }
    return(t);
  }
```

3.4.1.4: Weight balanced tree deletion (C)

```
tree delete(key, t)
typekey key;
tree t;

{ if(t == NULL) Error; /*** key not found ***/
  else {
      /*** search for key to be deleted ***/
      if(     t->k < key) t->right = delete(key, t->right);
      else if(t->k > key) t->left  = delete(key, t->left);

      /*** key found, delete if a descendant is NULL ***/
      else if(t->left  == NULL) t = t->right;
      else if(t->right == NULL) t = t->left;

      /*** no descendant is null, rotate on heavier side ***/
      else if(wt(t->left) > wt(t->right))
           { t = rrot(t);  t->right = delete(key, t->right); }
      else { t = lrot(t);  t->left  = delete(key, t->left); }

      /*** reconstruct weight information ***/
      if(t != NULL) {
          t->weight = wt(t->left) + wt(t->right);
          t = checkrots(t);
          }
      }
  return(t);
}
```

3.4.1.4: Weight balanced tree left rotation (C)

```
tree lrot(t)
tree t;

{ tree temp;

  temp = t;
  t = t->right;
  temp->right = t->left;
  t->left = temp;
  /*** adjust weight ***/
  t->weight = temp->weight;
  temp->weight = wt(temp->left) + wt(temp->right);
  return(t);
}
```

The Pascal data structure used to define B-trees is

3.4.2: B-tree data structure (Pascal)

```
btree = ↑node;
node  = record
    d : 0..2*M;
    k : array [1..2*M] of typekey;
    p : array [0..2*M] of btree
    end;
```

Note that the lexicographical order is given by the fact that all the keys in the subtree pointed by $p[i]$ are greater than $k[i]$ and less than $k[i+1]$.

3.4.2: B-tree search (Pascal)

```
procedure search(key : typekey; t : btree);

var i : integer;
begin
if t=nil then {*** Not Found ***}
    notfound(key)
else with t↑ do begin
```

```
        i := 1;
        while (i<d) and (key>k[i]) do i := i+1;
        if key = k[i] then   {*** Found ***}
            found(t↑, i)
        else if key < k[i] then search(key, p[i-1])
            else                search(key, p[i])
        end
    end;
```

3.4.2: B-tree insertion (Pascal)

```
function NewNode(k1 : typekey; p0, p1 : btree) : btree;

var t : btree;
begin
    new(t);
    t↑.p[0] := p0;
    t↑.p[1] := p1;
    t↑.k[1] := k1;
    t↑.d := 1;
    NewNode := t
end;

procedure insert(key : typekey; var t : btree);

var ins : typekey;
    NewTree : btree;

  function InternalInsert(t : btree) : typekey;
  var i, j : integer;
      ins : typekey;
      tempr : btree;
  begin
  if t=nil then begin {*** The bottom of the tree has been reached:
                           indicate insertion to be done ***}
      InternalInsert := key;
      NewTree := nil
  end
  else with t↑ do begin
      InternalInsert := NoKey;
      i := 1;
      while (i<d) and (key>k[i]) do     i := i+1;
```

```
    if key = k[i] then
         Error {*** Key already in table ***}
    else begin
         if key > k[i] then i := i+1;
         ins := InternalInsert(p[i-1]);
         if ins <> NoKey then
         {*** the key in "ins" has to be inserted in present node ***}
              if d<2*M then      InsInNode(t, ins, NewTree)
              else      {*** Present node has to be split ***}
              begin
                {*** Create new node ***}
                if i<=M+1 then begin
                     tempr := NewNode(k[2*M], nil, p[2*M]);
                     d := d-1;
                     InsInNode(t, ins, NewTree)
                end
                else tempr := NewNode(ins, nil, NewTree);
                {*** move keys and pointers ***}
                for j:=M+2 to 2*M do
                      InsInNode(tempr, k[j], p[j]);
                d := M;
                tempr↑.p[0] := p[M+1];
                InternalInsert := k[M+1];
                NewTree := tempr
              end
         end
    end
  end;

begin
   ins := InternalInsert(t);
   {*** check for growth at the root ***}
   if ins <> NoKey then t := NewNode(ins, t, NewTree)
end;
```

The insertion code uses the function *InsertInNode*, described below.

3.4.2: Insert an entry in a B-tree node (Pascal)

```
procedure InsInNode(t : btree; key : typekey; ptr : btree);

label   999;
var     j : integer;
```

```
begin
with t↑ do begin
    j := d;
    while j >= 1 do
        if key < k[j] then begin
            k[j+1] := k[j];
            p[j+1] := p[j];
            j := j-1
            end
        else goto 999;  {*** break ***}
    999:
    k[j+1] := key;
    p[j+1] := ptr;
    d := d+1
    end
end;
```

3.4.2: Auxiliary functions for B-tree insertion (C)

```
btree NewNode(k1, p0, p1)
typekey k1;
btree p0, p1;

{btree tempr;
    tempr = (btree)malloc(sizeof(node));
    tempr->p[0] = p0;
    tempr->p[1] = p1;
    tempr->k[0] = k1;
    tempr->d = 1;
    return(tempr);
}

InsInNode(t, key, ptr)
btree t, ptr;
typekey key;

{int j;
    for(j=t->d; j>0 && key<t->k[j-1]; j--)    {
        t->k[j] = t->k[j-1];
        t->p[j+1] = t->p[j];
        }
    t->d++;
```

```
        t->k[j] = key;
        t->p[j+1] = ptr;
    }
```

IV.2 Sorting algorithms

4.1.2: Linear insertion sort (Pascal)

```
procedure sort(var r : ArrayToSort; lo, up : integer);

var i, j : integer;
    tempr : ArrayEntry;
    flag : boolean;
begin
for i:=up-1 downto lo do begin
    tempr := r[i];
    j := i+1;
    flag := true;
    while (j<=up) and flag do
        if tempr.k > r[j].k then begin
            r[j-1] := r[j];
            j := j+1
            end
        else flag := false;
    r[j-1] := tempr
    end
end;
```

The above algorithm is slightly more complicated than the C version, as the internal loop cannot test for the double condition in a single statement. This forces the use of the boolean variable *flag*.

4.1.2: Linear insertion sort with sentinel (Pascal)

```
procedure sort(var r : ArrayToSort; lo, up : integer);

var i, j : integer;
```

```
    tempr : ArrayEntry;
begin
r[up+1].k := MaximumKey;
for i:=up-1 downto lo do begin
    tempr := r[i];
    j := i+1;
    while tempr.k > r[j].k do begin
        r[j-1] := r[j];
        j := j+1
        end;
    r[j-1] := tempr
    end
end;
```

4.1.3: Quicksort (with bounded stack usage) (Pascal)

```
procedure sort(var r : ArrayToSort; lo, up : integer);

var  i, j : integer;
     tempr : ArrayEntry;
begin
while up>lo do begin
     i := lo;
     j := up;
     tempr := r[lo];
     {*** Split file in two ***}
     while i<j do begin
          while r[j].k > tempr.k do
               j := j-1;
          r[i] := r[j];
          while (i<j) and (r[i].k<=tempr.k) do
               i := i+1;
          r[j] := r[i]
          end;
     r[i] := tempr;
     {*** Sort recursively, the smallest first ***}
     if i-lo < up-i then begin
               sort(r,lo,i-1);
               lo := i+1
               end
          else begin
               sort(r,i+1,up);
```

```
                up := i-1
                end
        end
    end;
```

4.1.3: Quicksort (with bounded stack usage) (C)

```
sort(r, lo, up)
ArrayToSort r;
int lo, up;

{int i, j;
ArrayEntry tempr;
while (up>lo) {
    i = lo;
    j = up;
    tempr = r[lo];
    /*** Split file in two ***/
    while (i<j) {
        for (; r[j].k > tempr.k; j--);
        for (r[i]=r[j]; i<j && r[i].k<=tempr.k; i++);
        r[j] = r[i];
        }
    r[i] = tempr;
    /*** Sort recursively, the smallest first ***/
    if (i-lo < up-i) { sort(r,lo,i-1); lo = i+1; }
        else    { sort(r,i+1,up); up = i-1; }
    }
}
```

The above version of Quicksort is designed to prevent the growth of the recursion stack in the worst case (which could be $O(n)$). This is achieved by changing the second recursive call into a while loop, and selecting the smallest array to be sorted recursively.

4.1.4: Shellsort (Pascal)

```
procedure sort(var r : ArrayToSort; lo, up : integer);

label   999;
```

```
var d, i, j : integer;
    tempr : ArrayEntry;
begin
d := up-lo+1;
while d>1 do begin
    if d<5 then    d := 1
        else d := trunc(0.45454*d);
    {*** Do linear insertion sort in steps size d ***}
    for i:=up-d downto lo do begin
        tempr := r[i];
        j := i+d;
        while j <= up do
            if tempr.k > r[j].k then begin
                r[j-d] := r[j];
                j := j+d
                end
            else goto 999;  {*** break ***}
        999:
        r[j-d] := tempr
        end
    end
end;
```

As this algorithm is a composition using linear insertion sort (see Section 4.1.2), the same comments can be applied.

For a predetermined, not computable, sequence of increments, the Shellsort algorithm becomes:

4.1.4: Shellsort for fixed increments (C)

```
int Increments[ ] = {34807,15823,7193,3271,1489,
                     677,307,137,61,29,13,5,2,1,0};
    sort(r, lo, up)
    ArrayToSort r;
    int lo, up;

    {int d, i, id, j;
    ArrayEntry tempr;
    for (id=0; (d=Increments[id]) > 0; id++) {
        /*** Do linear insertion sort in steps size d ***/
        for (i=up-d; i>=lo; i--) {
            tempr = r[i];
            for (j=i+d; j<=up && (tempr.k>r[j].k); j+=d)
```

```
                r[j-d] = r[j];
            r[j-d] = tempr;
            }
        }
    }
```

4.1.5: Heapsort (C)

```
sort(r, lo, up)
ArrayToSort r;
int lo, up;

{int i;
/*** construct heap ***/
for (i=up/2; i>1; i--) siftup(r,i,up);
/*** repeatedly extract maximum ***/
for (i=up; i>1; i--)  {
    siftup(r,1,i);
    exchange(r, 1, i);
    }
};
```

4.1.6: Interpolation sort (Pascal)

```
procedure sort(var r : ArrayToSort; lo, up : integer);

var  iwk : ArrayIndices;
     out : ArrayToSort;
     tempr : ArrayEntry;
     i, j : integer;
     flag : boolean;
begin

iwk[lo] := lo-1;
for i:=lo+1 to up do iwk[i] := 0;
for i:=lo to up do begin
     j := phi(r[i].k, lo, up);
     iwk[j] := iwk[j]+1
     end;
```

```
for i:=lo to up-1 do iwk[i+1] := iwk[i+1] + iwk[i];
for i:=up downto lo do begin
    j := phi(r[i].k, lo, up);
    out[iwk[j]] := r[i];
    iwk[j] := iwk[j]-1
    end;
for i:=lo to up do r[i] := out[i];
{*** Linear-insertion sort phase ***}
for i:=up-1 downto lo do begin
    tempr := r[i];
    j := i+1;
    flag := true;
    while (j<=up) and flag do
        if tempr.k > r[j].k then begin
            r[j-1] := r[j];
            j := j+1
            end
        else flag := false;
    r[j-1] := tempr
    end;
end;
```

4.1.6: Interpolation function (Pascal)

```
function phi(key : typekey; lo, up : integer) : integer;
var  i : integer;
begin
i := trunc((key-MinKey) * (up-lo+1.0) / (MaxKey-MinKey)) + lo;
phi := i;
if i>up then phi := up
    else if i<lo then phi := lo
end;
```

4.1.6: Interpolation (in-place) sort (C)

```
sort(r, lo, up)
ArrayToSort r;
int lo, up;
```

```
{ArrayIndices iwk;
ArrayEntry tempr;
int i, j;

for (i=lo; i<=up; i++)  {iwk[i] = 0;  r[i].k = -r[i].k;}
iwk[lo] = lo-1;
for (i=lo; i<=up; i++)  iwk[phi(-r[i].k,lo,up)]++;
for (i=lo; i<up;  i++)  iwk[i+1] += iwk[i];
for (i=up; i>=lo; i--)  if (r[i].k<0)
    do  {
        r[i].k = -r[i].k;
        j = iwk[phi(r[i].k, lo, up)]--;
        tempr = r[i];
        r[i] = r[j];
        r[j] = tempr;
        } while (i != j);
for (i=up-1; i>=lo; i--) {
    tempr = r[i];
    for (j=i+1; j<=up && (tempr.k>r[j].k); j++)
        r[j-1] = r[j];
    r[j-1] = tempr;
    }
};
```

The above algorithm only works for positive keys.

4.1.7: Linear probing sort (C)

```
sort(r, lo, up)
ArrayToSort r;
int lo, up;

{ArrayToSort r1;
int i, j, uppr;
uppr = up + (UppBoundr-up)*3/4;
for (j=lo; j<=up; j++)  r1[j] = r[j];
for (j=lo; j<=UppBoundr; j++) r[j].k = NoKey;
for (j=lo; j<=up; j++) {
    for (i=phi(r1[j].k,lo,uppr); r[i].k != NoKey; i++) {
        if (r1[j].k < r[i].k) {
            r1[j-1] = r[i];
            r[i] = r1[j];
            r1[j] = r1[j-1];
```

```
                };
            if (i > UppBoundr) Error;
            }
        r[i] = r1[j];
        };
    for (j=i=lo; j<=UppBoundr; j++)
        if (r[j].k != NoKey)
            r[i++] = r[j];
    while (i <= UppBoundr)
            r[i++].k = NoKey;
    };
```

4.2.1: Merge sort (Pascal)

```
function sort(var r : list; n : integer) : list;
label 999;
var     fi, la, temp : list;

begin
if r = nil then sort := nil
else    if n>2 then
    sort := merge(sort(r, n div 2), sort(r, (n+1) div 2))
else begin
    fi := r;    la := r;
    r := r↑.next;
    {*** Build list as long as possible ***}
    while r <> nil do
        if r↑.k >= la↑.k then begin
            la↑.next := r;
            la := r;
            r := r↑.next;
            end
        else if r↑.k <= fi↑.k then begin
            temp := r;
            r := r↑.next;
            temp↑.next := fi;
            fi := temp
            end
        else    goto 999;
999:
    la↑.next := nil;
    sort := fi
```

```
    end
end;
```

The above algorithm is similar to the one in the main entry, except that at the bottom level of recursion, it tries to construct the longest possible list of ordered elements. To achieve this, it compares the next element in the list against the head and the tail of the list being constructed. Consequently, this algorithm will improve significantly when used to sort partially ordered (or reverse-ordered) files.

4.2.1: Merge sort (C)

```
list sort(n)
int n;

{
list fi, la, temp;
extern list r;
if (r == NULL) return(NULL);
else if (n>1)
    return(merge(sort(n/2), sort((n+1)/2)));
else    {
    fi = r; la = r;
    /*** Build list as long as possible ***/
    for (r=r->next; r!=NULL;)
        if (r->k >= la->k) {
            la->next = r;
            la = r;
            r = r->next;
            }
        else if (r->k <= fi->k) {
            temp = r;
            r = r->next;
            temp->next = fi;
            fi = temp;
            }
        else break;
    la->next = NULL;
    return(fi);
    }
};
```

Owing to the absence of var variables in C, the list to be sorted is stored in a global variable named *r*.

4.2.4: Radix sort (C)

```
list sort(r)
list r;

{
list head[M], tail[M];
int i, j, h;
for (i=D; i>0; i--) {
    for (j=0; j<M; j++) head[j] = NULL;
    while (r != NULL) {
        h = charac(i, r->k);
        if (head[h]==NULL) head[h] = r;
        else    tail[h] ->next = r;
        tail[h] = r;
        r = r->next;
        };
    /*** Concatenate lists ***/
    r = NULL;
    for (j=M-1; j>=0; j--)
        if (head[j] != NULL) {
            tail[j] ->next = r;
            r = head[j];
            }
    };
return(r);
};
```

The above algorithm uses the function *charac* which returns the *i*th character of the given key. The global constant *M* gives the range of the alphabet (or characters). The constant or variable *D* gives the number of characters used by the key.

4.2.4: Top-down radix sort (C)

```
list sort(s, j)
list s;
int j;
```

```
{
int i;
list head[M], t;
struct rec aux;
extern list Last;
if (s==NULL) return(s);
if (s->next == NULL) {Last = s; return(s);}
if (j>D)    {
    for (Last=s; Last->next!=NULL; Last = Last->next);
    return(s);
    }
for (i=0; i<M; i++) head[i] = NULL;
/*** place records in buckets ***/
while (s != NULL) {
    i = charac(j, s->k);
    t = s;
    s = s->next;
    t->next = head[i];
    head[i] = t;
    }
/*** sort recursively ***/
t = &aux;
for (i=0; i<M; i++)
    if (head[i]!=NULL) {
        t->next = sort(head[i], j+1);
        t = Last;
    }
return(aux.next);
}
```

4.3.1: List merging (C)

```
list merge(a, b)
list a, b;

{
list temp;
struct rec aux;
temp = &aux;
while (b != NULL)
    if (a == NULL) { a = b; break; }
    else if (b->k > a->k)
```

```
                { temp = temp ->next = a; a = a ->next; }
          else { temp = temp ->next = b; b = b ->next; };
    temp ->next = a;
    return(aux.next);
    };
```

4.3.2: Array merging into same or third array (Pascal)

```
procedure merge (a,b : RecordArray; var c : RecordArray; na,nb : integer);
{*** Merges the arrays a and b into c (increasing order assumed)
        a or b may coincide with c ***}
begin
while (na>=1) or (nb>=1) do
    if na<1 then
        while nb>0 do begin
            c[nb] := b[nb];
            nb := nb-1
            end {while}
    else if nb<1 then
        while na>0 do begin
            c[na] := a[na];
            na := na-1
            end {while}
    else if a[na].k < b[nb].k then begin
        c[na+nb] := b[nb];
        nb := nb-1
        end {if...then}
    else begin
        c[na+nb] := a[na];
        na := na-1
        end; {else}
end;
```

IV.3 Selection algorithms

5.1.1: Sorted list extraction (Pascal)

```
function extract(var pq : list) : typekey;
begin
if pq=nil then Error {*** Extracting from empty queue ***}
else begin
     extract := pq↑.k;
     pq := pq↑.next
     end
end;
```

5.1.1: Sorted list insertion (Pascal)

```
procedure insert(new : list; var pq : list);
label 9999;
var  p : list;

begin
if pq=nil then pq := new
else if pq↑.k < new↑.k then begin
     new↑.next := pq;
     pq := new
     end
else begin
     p := pq;
     while p↑.next <> nil do begin
          if p↑.next↑.k < new↑.k then begin
               new↑.next := p↑.next;
               p↑.next := new;
               goto 9999
               end;
          p := p↑.next
          end;
     p↑.next := new
     end;
9999:
end;
```

5.1.1: Unsorted list extraction (Pascal)

```
function extract(var pq : list) : typekey;
var  max, p : list;

begin
if pq=nil then Error {*** Extraction from an empty list ***}
else if pq↑.next = nil then begin
      extract := pq↑.k; pq := nil  end
else begin
      max := pq; p := pq;
      while p↑.next <> nil do begin
            if max↑.next↑.k < p↑.next↑.k then max := p;
            p := p↑.next
            end;
      if max↑.next↑.k < pq↑.k then begin
            extract := pq↑.k;   pq := pq↑.next   end
      else begin
            extract := max↑.next↑.k;
            max↑.next := max↑.next↑.next
            end
      end
end;
```

5.1.1: Unsorted list insertion (Pascal)

```
procedure insert(new : list; var pq : list);
begin
new↑.next := pq;
pq := new
end;
```

5.1.2: P-trees deletion (Pascal)

```
procedure delete (var pq : tree);
begin
if pq = nil then Error {*** deletion on an empty queue ***}
else if pq↑.left = nil then pq := nil
```

```
else if pq↑.left↑.left = nil then begin
        pq↑.left := pq↑.right;
        pq↑.right := nil
        end
else  delete(pq↑.left)
end;
```

5.1.2: P-trees insertion (Pascal)

```
procedure insert (new : tree; var pq : tree);
label 9999;
var  p : tree;

begin
if pq = nil then  pq := new
else if pq↑.k >= new↑.k then begin
                {*** Insert above subtree ***}
                new↑.left := pq;
                pq := new
                end
else  begin
        p := pq;
        while p↑.left <> nil do
                if p↑.left↑.k >= new↑.k then begin
                        {*** Insert in right subtree ***}
                        insert(new, p↑.right);
                        goto 9999
                        end
                else  p := p↑.left;
        {*** Insert at bottom left ***}
        p↑.left := new
        end;
9999:
end;
```

5.1.2: P-trees, inspection of top of queue (Pascal)

```
function inspect (pq : tree) : typekey;
begin
```

```
if pq = nil then Error {*** Inspecting an empty queue ***};
while pq↑.left <> nil do pq := pq↑.left;
inspect := pq↑.k
end;
```

5.1.3: Heap insertion (C)

```
insert(new, r)
RecordArray r;
ArrayEntry new;

{int i, j;
extern int n;
n++;
for (j=n; j>1; j=i) {
        i = j/2;
        if (r[i].k >= new.k) break;
        r[j] = r[i];
        }
r[j] = new;
};
siftup(r, i, n)
RecordArray r;
int i, n;

{ArrayEntry tempr;
int j;
        while ((j=2*i) <= n) {
                if (j<n && r[j].k < r[j+1].k) j++;
                if (r[i].k < r[j].k) {
                        tempr = r[j];
                        r[j] = r[i];
                        r[i] = tempr;
                        i = j;
                        }
                else break;
                }
};
```

5.1.3: Heap deletion (C)

```
delete(r)
RecordArray r;

{
extern int n;
if (n<1) Error /*** extracting from an empty Heap ***/;
else {
      r[1] = r[n];
      siftup(r, 1, --n);
      }
};
```

5.1.5: Pagodas merging (C)

```
tree merge(a, b)
tree a, b;

{
tree bota, botb, r, temp;
if (a==NULL) return(b);
else if (b==NULL) return(a);
else {
      /*** Find bottom of a's rightmost edge ***/
      bota = a->right; a->right = NULL;
      /*** bottom of b's leftmost edge ***/
      botb = b->left; b->left = NULL;
      r = NULL;
      /*** Merging loop ***/
      while (bota!=NULL && botb!=NULL)
            if (bota->k < botb->k) {
                  temp = bota->right;
                  if (r==NULL) bota->right = bota;
                        else {bota->right = r->right;
                              r->right = bota;
                              };
                  r = bota;
                  bota = temp;
                  }
            else {temp = botb->left;
```

```
                        if (r==NULL) botb->left = botb;
                             else {botb->left = r->left;
                                   r->left = botb;
                                   };
                        r = botb;
                        botb = temp;
                        };
                /*** one edge is exhausted, finish merge ***/
                if (botb==NULL) {
                     a->right = r->right;
                     r->right = bota;
                     return(a);
                     }
                else {b->left = r->left;
                     r->left = botb;
                     return(b);
                     }
                }
        };
```

5.1.5: Pagodas insertion (C)

```
tree insert(new, pq)
tree new, pq;
{
new->left = new; new->right = new;
return(merge(pq, new));
};
```

5.1.5: Pagodas deletion (C)

```
tree delete(pq)
tree pq;

{
tree le, ri;
if (pq==NULL) Error /*** Deletion on empty queue ***/;
else {
      /*** Find left descendant of root ***/
      if (pq->left == pq) le = NULL;
```

```
        else {
                le = pq->left;
                while (le->left != pq) le = le->left;
                le->left = pq->left;
                };
    /*** Find right descendant of root ***/
    if (pq->right == pq) ri = NULL;
        else {
                ri = pq->right;
                while (ri->right != pq) ri = ri->right;
                ri->right = pq->right;
                };
    /*** merge them ***/
    return(merge(le, ri));
    }
};
```

5.1.6.1: Leftist trees deletion (C)

```
tree merge(a, b)
tree a, b;
{
if (a == NULL) return(b);
else if (b == NULL) return(a);
else if (a->k > b->k) {
    a->right = merge(a->right, b);
    fixdist(a);
    return(a);
    }
else {
    b->right = merge(a, b->right);
    fixdist(b);
    return(b);
    }
};

tree delete(pq)
tree pq;
{
if (pq == NULL) Error /*** delete on an empty queue ***/;
else return(merge(pq->left, pq->right));
};
```

5.1.6.1: Leftist trees insertion (C)

```
tree insert(new, pq)
tree new, pq;

{
if (pq==NULL) return(new);
else if (pq->k > new->k) {
            pq->right = insert(new, pq->right);
            fixdist(pq);
            return(pq);
            }
else {
      new->left = pq;
      return(new);
      }
};
```

5.1.6.1: Leftist trees distance (C)

```
int distance(pq)
tree pq;
{ return(pq==NULL ? 0 : pq->dist); };

fixdist(pq)
tree pq;
{
tree temp;
if (distance(pq->left) < distance(pq->right)) {
      temp = pq->right;
      pq->right = pq->left;
      pq->left = temp;
      };
pq->dist = distance(pq->right) + 1;
};
```

5.1.6.2: Binary priority queues deletion (C)

```
tree delete(pq)
tree pq;

{tree temp;
if (pq == NULL) Error /*** deletion on an empty queue ***/;
else if (pq->right == NULL)
     return(pq->left);
else {
     /*** promote left descendant up ***/
     pq->k = pq->left->k;
     pq->left = delete(pq->left);
     /*** rearrange according to constraints ***/
     if (pq->left == NULL) {
          pq->left = pq->right; pq->right = NULL; };
     if (pq->right != NULL)
          if (pq->left->k < pq->right->k) {
               /*** descendants in wrong order ***/
               temp = pq->right;
               pq->right = pq->left;
               pq->left = temp;
               }
     return(pq);
     }
};
```

5.1.6.2: Binary priority queues insertion (C)

```
tree insert(new, pq)
tree new, pq;

{
if (pq == NULL) return(new);
else if (pq->k <= new->k) {
          new->left = pq;
          return(new);
          }
else if (pq->left == NULL)
          pq->left = new;
else if (pq->left->k <= new->k)
```

```
                pq->left = insert(new, pq->left);
    else        pq->right = insert(new, pq->right);
    return(pq);
    };
```

5.1.6.2: Merging of binary priority queues (C)

```
function merge (a, b : tree) : tree;
var     temp : tree;
begin
if a=nil then merge := b
else if b=nil then merge := a
else begin
    if a↑.k < b↑.k then begin
        temp := a;   a := b;   b := temp  end;
    a↑.right := merge(a↑.right, b);
    if a↑.left <> nil then
        if a↑.left↑.k < a↑.right↑.k then begin
            temp := a↑.right;
            a↑.right := a↑.left;
            a↑.left := temp
            end
    end
end;
```

IV.4 Text algorithms

7.1: Composition to search external files (Pascal)

```
function extsearch(pat: PATTERN): integer;
var offs, i, m, nb, nr: integer;
    buff: TEXT;
    found: boolean;

    function fillbuff: integer;
    var j: integer;
    begin
      j := nb+1;
      while (j <= BUFSIZ-nb) and not eof(input) do begin
```

```
            read(buff[j]);
            j := j+1;
        end;
        fillbuff := j-nb-1;
        for i:=j to BUFSIZ do buff[i] := chr(0);
    end;

begin
    found := FALSE;
    m := length(pat);
    if m = 0 then begin
        extsearch := 1;
        found := TRUE;
        end;
    if m >= BUFSIZ then begin {*** Buffer is too small ***}
        extsearch := -1;
        found := TRUE;
        end;
    {*** Assume that the file is open and positioned ***}
    offs := 0;      {*** number of characters already read ***}
    nb := 0;        {*** number of characters in buffer ***}
    while not found do begin
        if nb >= m then begin
            {*** try to match ***}
            i := search(pat,buff);
            if i <> 0 then begin
                extsearch := i+offs; {*** found ***}
                found := TRUE;
                end;
            for i:=1 to m-1 do buff[i] := buff[i+nb-m+2];
            offs := offs + nb-m+1;
            nb := m-1;
            end;
        {*** read more text ***}
        if not found then begin
            nr := fillbuff;
            if nr <= 0 then begin
                extsearch := 0; {*** not found ***}
                found := TRUE;
                end;
            nb := nb + nr;
            end;
        end;
end;
```

7.1.1: Brute force string searching (C)

```
char *search(pat, text)
char *pat, *text;

{ int m;

  if(*pat == EOS) return(text);
  m = strlen(pat);
  for(; *text != EOS; text++)
     if(strncmp(pat, text, m) == 0) return(text);
  return(NULL);
}
```

7.1.2: Knuth–Morris–Pratt string searching (Pascal)

```
function search(pat: PATTERN; text: TEXT): integer;

var next: array [1..MAXPATLEN] of integer;
    i, j, m, n: integer;
    found: boolean;

    procedure preprocpat;

    var k, l: integer;
    begin
      m := length(pat);
      l := 1;
      k := 0; next[1] := 0;
      repeat begin
          if (k=0) or (pat[l]=pat[k]) then begin
                l := l+1; k := k+1;
                if pat[k]=pat[l] then next[l] := next[k]
                else                  next[l] := k;
                end
           else k := next[k];
           end
      until (l > m);
    end;
begin
  found := FALSE; search := 0;
```

```
    m := length(pat);
    if m=0 then begin
         search := 1;  found := TRUE; end;
    preprocpat;

    n := length(text);
    j := 1; i := 1;
    while not found and (i <= n) do begin
        if (j=0) or (pat[j] = text[i]) then begin
            i := i+1; j := j+1;
            if j > m then begin
                search := i-j+1;
                found := TRUE;
                end;
            end
        else j := next[j];
        end;
end;
```

7.1.3: Boyer–Moore–Horspool string searching (Pascal)

```
function search(pat: PATTERN; text: TEXT): integer;

var i, j, k, m, n: integer;
    skip: array [0..MAXCHAR] of integer;
    found: boolean;
begin
  found := FALSE; search := 0;
  m := length(pat);
  if m=0 then begin
       search := 1;  found := TRUE; end;
  for k:=0 to MAXCHAR do skip[k] := m;   {*** Preprocessing ***}
  for k:=1 to m-1 do     skip[ord(pat[k])] := m-k;

  k := m;  n := length(text);              {*** Search ***}
  while not found and (k <= n) do begin
      i := k; j := m;
      while (j >= 1) do
          if text[i] <> pat[j] then j := -1
          else begin
               j := j-1;  i := i-1; end;
      if j = 0 then begin
```

```
            search := i+1; found := TRUE; end;
        k := k + skip[ord(text[k])];
        end;
    end;
```

7.1.5: Karp–Rabin string searching (C)

```
#define B 131

    char *search(pat, text)
    char *pat, *text;

    { int hpat, htext, Bm, j, m;

      if(pat[0]==EOS) return(text);
      Bm = 1;
      hpat = htext = 0;

      for(m=0; text[m] != EOS && pat[m] != EOS; m++) {
          Bm *= B;
          hpat = hpat*B + pat[m];
          htext = htext*B + text[m];
          }

      if(text[m]==EOS && pat[m]!=EOS) return(NULL);

      for(j=m; TRUE; j++) {
          if(hpat==htext && strncmp(text+j-m,pat,m)==0)
              return(text+j-m);
          if(text[j]==EOS) return(NULL);
          htext = htext*B - text[j-m]*Bm + text[j];
          }
    }
```

7.1.8: Brute force string searching with mismatches (Pascal)

```
    function search(k: integer; pat: PATTERN; text: TEXT): integer;

    var i, j, m, n, count: integer;
```

```
    found: boolean;
  begin
    found := FALSE; search := 0;
    m := length(pat);
    if m=0 then begin
        search := 1;  found := TRUE; end;
    n := length(text);
    j := 1; i := 1;
    while (i<=n-m+1) and not found do begin
        count := 0;  j := 1;
        while (j <= m) and (count <= k) do  begin
            if text[i+j-1] <> pat[j] then count := count + 1;
            j := j + 1;
            end;
        if count <= k then begin
            search :=  i;  found := TRUE; end;
        i := i + 1;
        end
  end;
```

Index